Feb. 10	No Class— Lincoln's Birthday (Maybe the CBOT)
Feb. 17	Exam I
Feb. 24	Chapter 5/ Technical Price Analysis and Review of Options
Feb. 24	Chapter 6/ Pricing Strategy
Mch. 3	Chapter 7/ Basis: The Economics of When and Where To Sell
Mch. 10	Exam II and Chapter 9/ Livestock Marketing
Mch. 17	No Classes— Spring Break
Mch. 24	Chapter 10/ 730- Day Marketing
Mch. 31	Review
April 7	Final

**** All tests are the property of the instructor ****

(15) USDA, "Revising U.S. Exports: Why Is It Taking So Long?" *Farmline,* Vol. VII, No. 9 (Washington, D.C.: Economic Research Service, Sept., 1986).

(16) Phil Paarlberg and Otto Doering, "U.S. Agricultural Exports: Why So Low?" *Purdue Agricultural Economics Report* (West Lafayette, Ind.: Cooperative Extension Service, Purdue University, Nov., 1986).

(17) Geoffrey S. Shepherd, Gene A. Futrell, and J. Robert Strain, *Marketing Farm Products,* 6th ed. (Ames: Iowa State University Press, 1976) 16.

(18) Richard L. Kohls and Joseph N. Uhl, *Marketing of Agricultural Products,* 6th ed. (New York: Macmillan Publishing Co., Inc., 1985) 23.

(19) Shepherd, Futrell, and Strain, *Marketing Farm Products* 18.

(20) Shepherd, Futrell, and Strain, *Marketing Farm Products* 7-13.

(21) Wayne Purcell, *Agricultural Marketing: Systems Coordination, Cash and Futures Prices* (Reston, Va.: Reston Publishing Co., 1979) 12.

Chapter 2

Marketing Tools

MARKET PRICE LEVEL AND BASIS PRICE LEVEL	26
KEY BUSINESS EVENTS	27
MARKETING TOOLS	27
Cash Sale	27
On-Farm Storage	29
Hired Storage	29
Delayed Pricing	31
Deferred Payment Contracts	34
Basis Contracts	34
Cash Sale and "Buying the Board"	37
Cash Contracts, Forward Sale, Booking	37
Futures Contracts	40
Delivery Against Contracts: Cash Settlement	42
Offsetting Trades	44
Parallelism of Cash and Futures Prices	45
Buying Hedge	46
Margins: Settling the Accounts	48
Hedged-to-Arrive Contracts	49
Agricultural Options	50
SUMMARY	51

Marketing Tools

It takes 90 to 200 days to grow a grain crop or finish a pen of cattle. It can take less than two minutes to sell the product. But, in order to maximize profit, farmers should take up to 24 months to market their products. This book builds toward a *730-Day Marketing Plan.*

Successful marketing is more than just selling. The key to profitable marketing requires recognition of profitable opportunities and use of the correct tool to capture those opportunities.

In large measure, farmers are price takers because of the competitive nature of their business. Any single producer acting alone cannot affect market price, either up or down. The producer takes what the market offers. Watching the panorama of changing prices, producers must decide WHEN and HOW to respond to market offers. WHEN they react to pricing opportunities forms the structure of their marketing strategy (discussed later). HOW they react depends upon the marketing tools they have at their disposal.

Different tools accomplish different objectives. Successful marketing requires knowing what marketing tools are available and knowing what they can accomplish. To better understand the tools, we begin our discussion by defining two components of pricing and the business events that occur in executing a sale. Later in this chapter, we will dissect these business events and describe various marketing tools for making profitable sales.

MARKET PRICE LEVEL AND BASIS PRICE LEVEL

The pricing transaction has two separate but related components. One concerns the market price that is discovered on the commodity exchanges and is referred to here as price level. The other concerns the difference between a cash market price and a selected futures contract price. This price spread is called "basis" and is referred to here as basis level.

Each price component has its own price-making forces. The market price level on futures exchanges is set by anticipated worldwide future demand and future supply conditions as interpreted by traders active on the exchanges. The price at a local market (spot market) is influenced by the general market price level, but it responds to local conditions, here and now, and therefore may be higher or lower than the futures price level. Most of the differences in prices between a central market and those available to a producer located in the country can be attributed to transportation and storage costs. However, market imperfections and price distortions account for a part of the difference also. It is these distortions which provide profitable opportunities for astute marketers. Awareness that local price imperfections can and do exist is knowledge that farmers can use profitably.

To help distinguish between the two prices, let us examine the needs of a continuous-flow corn-oil processor in Dayton, Ohio. This processor draws most of its corn as raw material for processing from a 200-mile radius of the plant. If farmers in that region are reluctant sellers, the processor must increase the cash bid, or "spot" price, in order to convince sellers to part with their grain. There may be willing sellers at a lower price in Iowa, but frozen rivers and rail car shortages could prevent Iowa corn from satisfying the Ohio processor's current needs. Since it is costly to shut down and start up this operation, the processor is forced (within limits) to bid a premium relative to the price level that exists in other parts of the country. In market terms, local basis would be strong. Later when rivers thaw and transportation problems are solved, there will be no reason for the processor to continue bidding a premium. The processor would lower its "spot" price relative to the market price level. The spread between local cash and central market cash price would deteriorate, and basis would weaken.

Farmers are concerned with capturing advantageous prices in both price level and basis level markets. They seek both a high price level and a strong local basis when they make pricing decisions. When the local cash price is strong relative to the futures price level, the basis portion of the pricing decision is good and should be captured with one of the marketing tools available. When price level is high, but basis is weak, another tool should be used. The discussion will explain how marketing tools fix either the price level or basis level component of price while leaving the other open to fluctuation. Some tools fix both price level and basis level, thereby eliminating all price fluctuation.

KEY BUSINESS EVENTS

The key business events of commodity transactions are:

DELIVERY — physical transfer of the commodity from buyer to seller.
PRICING — agreement on an acceptable price by buyer and seller.
TITLE TRANSFER — determining who owns the commodity at what instant in time.
SETTLEMENT — payment from buyer to seller.

All of these business events can occur together at a single instant, or these events can be separated by many months. Pricing need not occur in conjunction with actual delivery of the commodity; it can be done before planting or delayed until after harvest. Transferring title of ownership may occur when grain is physically exchanged, but it need not be simultaneous. Payment can occur at a time far removed from the other events. By separating the business events, the farmer has greater opportunity to execute each one at the most advantageous time.

MARKETING TOOLS

Cash Sale

With cash sale, the seller delivers the commodity to a buyer and takes an immediate cash payment in exchange. All business events — *delivery, pricing, title transfer, and settlement — occur in one transaction*. The commodity must be physically in existence at the time of the transfer. Cash sales may be made directly from the field, immediately after harvest, or later from storage.

A cash sale removes all price fluctuation. Both price level and basis level are fixed. There is no chance to profit from additional price improvement; likewise, there is no chance to lose from future price decline. Picking the day to make a cash sale can be an excruciating experience because it finalizes a year of plowing, planting, and praying. Because cash sales so completely remove pricing options, many producers split their sales into several decisions, hoping that the average price achieved will be higher than a one-time sales decision.

Compared to the other market alternatives, this method is almost business risk – free for the seller, assuming the buyer's check does not bounce. The seller retains title, and often physical ownership, until the sale is made. To minimize the risk of non-payment, producers should accept checks drawn on a local bank and cash them immediately. If the buyer declares bankruptcy in the interim between issuing and clearing the check, the seller will be only one of many unsecured creditors who claim title to the assets of the buyer.

The cash price quoted for a bushel of grain is based upon grading standards as defined by the United States Grain Standards Act. If the grain as delivered does not meet those standards, discounts will be calculated and the price received will be less than the quoted price. Producers should understand the discounts and reasons for them. They will thus be in a position to manage production and harvest practices to minimize or, better yet, optimize discounts.

Grading standards for corn are outlined in Table 2-1.[1] (Soybeans and wheat have similar standards, as shown in tables at the end of the chapter.) Grade differences depend upon test weight, moisture, broken grain and foreign material, and damaged kernels. In addition, mustiness, souring, heating, or insects lower quality. Elevators usually publish discount tables and reduce the price paid for any conditions that are substandard.

Table 2-1: Grades and Grade Requirements for Corn

Grade	Minimum Test Weight per Bushel	Maximum Limits of—			
		Moisture	Broken Corn and Foreign Material	Damaged Kernels	
				Total	Heat-Damaged Kernels
	(lb.)	.... (%)			
U.S. No. 1	56.0	14.0	2.0	3.0	0.1
U.S. No. 2	54.0	15.5	3.0	5.0	0.2
U.S. No. 3	52.0	17.5	4.0	7.0	0.5
U.S. No. 4	49.0	20.0	5.0	10.0	1.0
U.S. No. 5	46.0	23.0	7.0	15.0	3.0
U.S. Sample grade	U.S. Sample grade shall be corn which does not meet the requirements for any of the grades from U.S. No. 1 to U.S. No. 5, inclusive; or which contains stones; or which is musty, or sour, or heating; or which has any commercially objectionable foreign odor; or which is otherwise of distinctly low quality.				

Source: USDA, *The Official United States Standards for Grain* (Washington: GPO, 1978) 2.3.

Moisture discounts are usually higher than other discounts and cause the most misunderstanding between farmers and elevator managers. Furthermore, there is little standardization of procedure among elevators in their methods of dealing with wet grain, drying, and shrinkage calculations. Though they may be quoting the same bid price, the net price received by the producer after deductions can vary considerably from elevator to elevator. The producer must determine which offer represents the best selling opportunity.[2]

On-Farm Storage

On-farm storage facilities permit delay of delivery, pricing, title transfer, and settlement transactions. Rather than forcing the farmer to deliver grain at harvest, storage provides greater flexibility in fulfilling business events. Storage bins are typically used to speculate on the possibility of rising prices. Certainly no one would store at harvest if he or she expected prices to be lower in the future.

Unpriced grain held in storage is subject to both price level and basis level fluctuations. Price level may rise or fall; basis level may strengthen or weaken over time while grain is held in storage.

Using grain-drying, -handling, and -holding facilities as extensions of the production process can help facilitate timely harvest. Harvest losses increase substantially after corn dries down to 22 percent moisture in the field. At a price level of $2.85, the loss during the time corn drops only 2 percent in moisture (from 22% to 20%) could cost 7 cents per bushel on an average yield. This drop represents a large portion of the annual costs of owning grain-drying and -handling facilities.

However, the production process ends the moment dry grain is deposited into a storage bin. At that point a producer with on-farm storage is performing a marketing function and should view this activity as that of a storage space merchandiser. The farmer-storer incurs the investment and ownership costs of the facility. Farmers also incur quality risk through the chance of physical deterioration, and price risk due to price fluctuations. However, they minimize business risk by retaining title and physical ownership. Full storage bins produce a sense of satisfaction and financial security when winter snows begin to blow, and they provide a commonly accepted method for speculating on price change. Unfortunately, the results of this speculative activity are often not properly evaluated. When 10,000 bushels of soybeans worth $9.00 at harvest are sold six months later for $7.00, a loss of $20,000 (plus holding costs) has been incurred. Some would argue that the farmer never had the $20,000 to begin with. In reality the bypassed opportunity represents a real loss when it comes time to pay bills.

Hired Storage

Hired storage is *delivery today;* title transfer, pricing, and payment are to be finalized at some time in the future. The commodity changes hands physically, but business transactions are delayed. This method allows sellers to speculate on the cash market without the capital investment required by on-farm storage facilities; instead they pay a space rental fee to the elevator operator. Neither price level nor basis level is fixed by storing unpriced grain.

The producer usually deposits grain with the commercial elevator and draws a *warehouse receipt,* as evidence that the holder of the receipt is entitled to hold, receive, or dispose of the goods represented by the receipt (Figure 2-1). Negotiable warehouse receipts are documents of title; the transfer of the warehouse receipts will transfer title to the grain they represent. Loss or theft of warehouse receipts represents loss of title to the goods. Warehouse receipts are generally accepted by lenders as gilt-edged collateral against which they will lend a high percentage of market value of the grain represented. Thus, these receipts can be used to generate a cash flow while retaining title (subject to lien) of the commodity.

Most all states have grain warehouse licensing requirements for commercial grain warehousers. They make it unlawful to accept grain for storage and to issue warehouse receipts without being licensed by either the state or federal government. Federally licensed warehouses are inspected twice annually to verify the

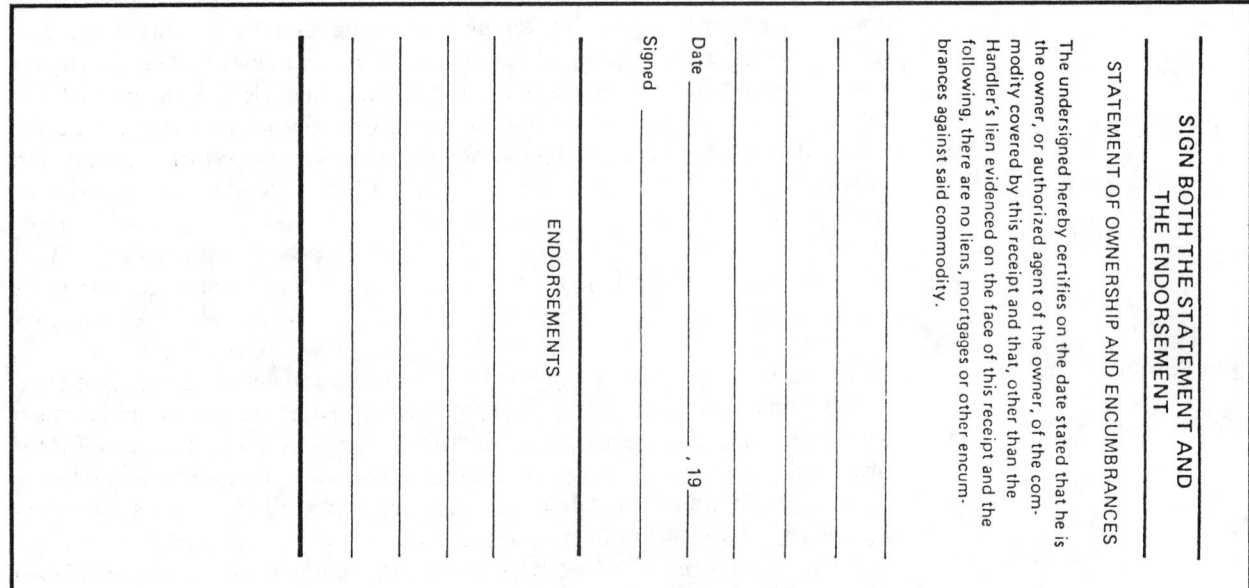

Figure 2-1: Grain Warehouse Receipt. (Top — Front; Bottom — Back.)

presence of grain in the warehouses. Inspections are a measure of protection for the producer. To qualify for licensing, the grain warehouse must be bonded to guarantee financial integrity. However, producers should not be misled by an elevator's poster which says "Licensed and Bonded," until they know the amount of the bond. In many states, the bond is only a small fraction of the elevator's capacity in terms of value. In the event of a shortage, claims against the bond can quickly exceed its value. At that point, the warehouse receipt is no more valuable than the financial integrity of the warehouser himself or herself.

As evidence of title, the warehouse receipt clearly distinguishes the storage transaction from a "delayed pricing" sale. The receipt indicates that the grain still belongs to the farmer or holder of the warehouse receipt.

Delayed Pricing

Delayed pricing, or priced later sales, involves *delivery and title transfer today*, while pricing and payment are delayed until a time in the future. When farmers deliver grain to the elevator and choose the delayed pricing option, they are given a delayed price contract which specifies the quantity and quality of grain delivered and outlines the charges assessed to the seller. The business events of delivery and title transfer occur simultaneously; pricing and payment are delayed until later.

Many producers deposit grain in the elevator, accept a scale ticket marked D.P. (delaying pricing), and assume that they have stored the grain to be priced later. In fact, they have *not* stored the grain; they *have sold it*. Title to the grain has been transferred to the elevator. Except for pricing and payment, the sale is complete. An example of a D.P. agreement, shown in Figure 2-2, clearly states that title is being transferred.

The delayed price agreement gives farmers the privilege of choosing a price that is being paid by the elevator any day in the future, generally up to the end of the marketing year for the crop. If the local cash price is higher later in the marketing season than when farmers delivered the grain, they gain from the higher level. If it is lower, they receive less. Thus, farmers can use D.P. to speculate on local cash price. For this privilege the elevator assesses a service charge, a delayed pricing fee. The price farmers receive is equal to local cash price less the service fees. Farmers receive payment for the grain at the time they establish the price.

Delayed pricing permits producers to speculate on the cash market without investing in on-farm facilities or paying elevator storage charges. Neither price level nor basis is fixed. However, delayed pricing fees often equal the basis gain and thereby limit the farmers' opportunity to capture basis improvement.

For the elevator operator, delayed pricing can mean larger operating volume. Because it is a sale and not a storage transaction, the elevator operator need not keep the grain on hand. It can be shipped out immediately to allow elevator operators to move more grain through their facility during the season and therefore not limit their business volume to their storage capacity. Some elevators ration storage space by accepting only part of the producer's deliveries for storage, thus forcing the rest of the delivered crop into the market as a cash sale or delayed price transaction.

Since title to the grain passes to the elevator at the time of delivery, elevator operators are free to dispose of the grain in any manner and at any time that it is most profitable to do so. There are three avenues they may follow. First, elevators may keep the delayed price grain on hand and treat it the same as stored grain. When the farmer decides to fix a price, the elevator sells the grain and passes the proceeds to the farmer. Second, since title has been transferred, the elevator operator can sell the D.P. grain and have interest-free use of the money until the farmer prices the grain. Finally, elevator operators can store the grain in either their own facilities or with terminal elevators, cause warehouse receipts to be issued against the grain, and pledge them as collateral for loans to the elevator.

Delayed pricing can be a business quagmire for farmers. Because title has been transferred to the elevator operator, the sale is often not covered by the grain warehouse licensing and handling requirements of state and federal agencies. Having relinquished title to the grain, the farmer is an unsecured creditor of the

---

DELAYED PRICE AGREEMENT

Date: _____   Agreement No. _____

Commodities under agreement: _____

It is hereby agreed that I, the undersigned seller, may from time to time by my own choice, sell and deliver to the undersigned buyer agricultural commodities as listed above, on which the price is to be established at a later date. I pledge that the commodities delivered pursuant to this agreement shall be free of any lien or encumbrance.

In selling commodities under this agreement, I, the seller, fully understand that *I am transferring title* to the undersigned buyer upon delivery, and that after delivery I am a creditor of the buyer for the market value of commodities so delivered until the price is established and settlement is completed. If the buyer defaults in his/her obligation for settlement I may be a common (unsecured) creditor of the buyer for the value of the commodities not settled for.

Upon demand by the seller, the buyer is obligated to pay his/her regular bid price on the date of demand for the commodities being priced by the seller which have been delivered under this agreement, less any service charge which is due and payable to the buyer.

A separate contract may be entered into between buyer and seller to set forth the terms and conditions of establishing the price. This contract shall also list the charges to be assessed against the seller for services rendered by the buyer from the time of delivery of the commodity until the price is established by the seller. Said contract may be printed on and be a part of the inbound scale ticket, or may be contained on a separate form which shall be supplied to the seller either before or at the time of delivery of the commodity under contract.

If a separate contract is not used, the condition and charges for services under this agreement for which the seller shall be liable to the buyer are as follows:

_____ ¢ per bushel from the date of delivery through _____ , 19 _____ , then
_____ ¢ per bushel per (day) (month) after the above date until the date of pricing.

If, after the signing of this agreement and the delivery of commodities thereon, the buyer causes a change to be made in the method of pricing or the service charge as listed hereon which would apply to the delivery of commodities after said change, a new agreement shall be signed between the buyer and the seller showing the date of the new agreement and the new service charges which are then in effect.

Section 926.29(C) of the Ohio Revised Code states that the seller under a delayed price agreement *may demand* as security for payment an amount that, at the time of delivery, is equal to the government loan rate for the commodity or 75% of the current bid price, whichever is less.

I am ☐ I am not ☐ demanding a payment guarantee from the buyer.

_____   _____
Seller   Buyer

_____   _____
Address   Address

_____   _____
Authorized Signature   Authorized Signature

UNDER THE REQUIREMENTS OF ODA RULE 901:7-2-17(D): Agricultural commodities may be delivered on a delayed price agreement for a period of one year after its effective date. Commodities delivered thereafter shall be covered by a new agreement which must be executed not later than the first delivery under the new agreement.

Figure 2-2: Delayed Pricing Agreement.

elevator until paid. The scale ticket and delayed price agreement are no more than a promise to pay when a price is established. If the elevator becomes insolvent, the D.P. promise is secondary to those holding warehouse receipts and mortgage interests in the available grain and other assets. In bankruptcy, after those other liens have been satisfied, the holders of delayed price agreements will share whatever remains with other general creditors. Bankrupt elevators, by definition, do not have sufficient assets to cover the claims of all creditors; D.P. obligations are seldom paid.

Delayed pricing service fees charged by elevators often are more flexible than their published storage rates. Their fees are highest during harvest and drop in late winter or early spring; they may be zero at certain times of the marketing year. D.P. service charges generally reflect the market's need for cash grain, as determined by the basis market. Free delayed pricing is most commonly offered late in the marketing year, when small harvests are in prospect. Farmers need to clear out storage bins for the new crop and typically feel there is price strength in the market. Under these conditions, basis is usually strong. Elevators entice grain movement from farmers to themselves with low or zero D.P. fees. Later on, basis typically weakens with the harvest movement of grain, irrespective of whether the general price level has increased or not. If elevator operators protect themselves with a hedge against price level change, delayed pricing is a method for them to profit by speculating the basis. If they receive grain when the basis is strong, but the farmer prices the crop when basis is weak, the elevator, rather than the farmer, profits from basis change. On the other hand, if basis strengthens between delivery and pricing, elevator operators stand to lose from their basis speculation.

In order to calculate whether delayed pricing is a profitable marketing alternative, a farmer needs to look at potential costs and returns. Soybeans might be selling at $5.00 per bushel at a local elevator during October. The price of a July futures contract might be trading at $5.80 that same day. The farmer could capture the 80-cent difference by delivering D.P. grain and simultaneously selling a July futures to hedge away price level change.* The farmer's costs are the D.P. service charge, plus the interest which would have been received on money in the bank instead of keeping it invested in grain, plus the brokerage fee on the futures contract, plus interest on funds required as margin on the futures transaction. A typical D.P. service charge on soybeans for central Ohio is 24 cents per bushel for 120 days and 3 cents per month thereafter. The charge would total 36 cents by July. The farmer's additional carrying charge is the $5.00 per bushel tied up for 8 months — money that could be used to pay off loans or be placed in interest-bearing investments. Using an annual interest rate of 12 percent, the interest cost of storage for 8 months would be 40 cents. To hedge in the return, broker fees on futures transactions cost about 1 cent per bushel, and interest on the margin deposit costs 1.6 cents. Total carrying costs would, therefore, be 78.6 cents, leaving a 1.4-cent profit if the farmer were to use D.P. and hedge in a return. With such a low locked-in profit margin, most producers do not place the hedge in the futures. Instead, they deliver D.P. grain in order to SPECULATE that local price level will rise and thereby afford them higher pricing opportunities.

Farmers accept the dual risks of lower prices and the financial integrity of the buyer in using this marketing tool as a method to speculate. There are other ways that farmers can speculate on price without speculating on the financial strength

---

*Assuming basis converges to zero by July.

of the elevator operator as well. These other methods permit farmers to capture advantageous basis markets when they exist, basis appreciation that is not limited by D.P. fees.

As a marketing tool, delayed pricing can be a very risky and expensive way to speculate.

Deferred Payment Contracts

For sales with deferred payment provisions, *delivery, title exchange, and pricing* transactions are completed at the same instant. Only settlement is deferred until a later date, usually in the next tax year. As with a cash sale, price level and basis level are both fixed at the time delayed payment contracts are executed.

Deferred payments can provide a tax advantage. With proper stipulations in the agreement, the Internal Revenue Service has accepted the fact that the producer did not have right of payment in the year of the sale. As a result, the IRS has concluded that a producer could be taxed only on income from the sale in the year when payment was actually received. Tax rulings have made it clear that to achieve the tax deferral objective, there must be a valid contract that calls for payment during the following year. Otherwise, the IRS will consider the income "constructively received" in the year of the sale. The contract cannot be supported by a note or security that the seller could redeem for cash. Or if the farmer has the right to collect payment at any time after delivery of the commodity and all that is lacking is demand for payment, the farmer would have to include the sale in the tax year of delivery, regardless of when actual payment is received.[3]

Deferred payments can also be risky. Since delivery and transfer of title precede payment, deferred payment contractual arrangements carry the same business risk as delayed pricing sales. Sellers are unsecured creditors until cash is received. They should have confidence in the financial position of the buyers before agreeing to defer payment.

An example of the Deferred Payment contract and its provisions are detailed in Figure 2-3.

Basis Contracts

Executing a sale with a basis contract allows *delivery, title transfer, partial settlement of payment,* and a *basis price* fixed relative to a futures contract price. This sale method allows the farmer to establish a price level at a later date, hopefully when the price level has risen. It thus allows speculation on price level, but not basis level.

The basis contract in Figure 2-4 specifies delivery date, quality, amount, and delivery point. The difference between it and a cash contract is in the price notation. Price on a basis contract is stated relative to a specified futures contract, rather than as a fixed price. For example, the price may be "30 cents under March Soybeans," rather than a $7.52 flat price, as could be the case in a cash contract. Consequently, if the March futures contract price rises, the seller will get a higher price, equal to 30 cents less than the March contract, however high it rises. On the other hand, if the March futures contract price declines, the seller will receive a lower price, still 30 cents below the March futures price.

Farmers have the option of selecting the day they wish to fix the price level for their grain. They may choose any day following execution of the contract until about the first day of the month of maturity of the futures contract specified. Any time before maturity, they simply notify the buyer to complete the transaction and forward remaining payment.

DEFERRED PAYMENT CONTRACT

John Doe Grain Co., Anytown, Ohio, a corporation, hereinafter known as Buyer, and _____ , hereinafter known as Seller, covenant and agree as follows.
(a partnership / an association)

Buyer agrees to purchase from Seller approximately _____ bushels of _____ at the agreed price of $ _____ per bushel under the following terms and conditions to wit:

Seller shall deliver the above described grain within _____ days to Buyer's elevator in Anytown, Ohio.

Seller hereby warrants that title to the above grain is free from encumbrance of any nature whatsoever and that Seller has good and merchantable title, with the right to dispose of the same. Seller hereby authorizes the Buyer to contract for the sale or disposition of said grain for its own account either before or after delivery to the elevator above designated without further authorization from Seller except this contract, and buyer shall not be required to account for or pay the balance of said purchase price until the _____ day of _____ , 19 _____ .

Seller hereby acknowledges receipt of the sum of $ _____ to apply on this contract, and the balance shall not be due until the _____ day of _____ , 19 _____ , it being the intent of all parties hereto that said sale shall not be completed until final payment by the Buyer to the Seller, and it is specifically agreed that Seller shall, under no circumstances, be entitled to the balance due herein until the _____ day of _____ , 19 _____ .

Buyer agrees that it will pay the balance due without interest on the due date as fixed herein to the Seller.

It is further agreed by and between the Buyer and Seller that this contract shall be binding upon the heirs, administrators, and executors of the respective parties and that this contract cannot be assigned.

Dated this _____ day of _____ , 19 _____ .

JOHN DOE GRAIN COMPANY

_____   By _____
Witness   Buyer

_____   _____
Witness   Seller

Figure 2-3: Deferred Payment Contract.

Or farmers can continue to delay the pricing decision. They can roll forward the basis contract to another futures contract month to continue to hold the open speculative position. For example, they would roll forward to the July contract, if the price level increase they were expecting had not occurred by March, and if they still believed it would do so by July. The elevator typically assesses the seller the brokerage costs of the rollover.

The most important advantage of the basis contract method of sale is that it permits the producer to split pricing decisions into a basis level sale and a price level sale. When the basis is strong, but price is low and the producer feels it will rise, execution of a basis contract captures the advantageous basis level and allows continued speculation on price level. After the price level has risen, a producer can make the price level commitment. The producer thus has the opportunity to capture high prices in both markets if the price level does, in fact, rise as expected.

With basis contracting, farmers can select a delivery date during periods when few competing jobs need to be done on the farm. The risks of quality deterioration and shrinkage are eliminated by making delivery to the buyer. This may be especially advantageous with low-quality grain. Farmers can move the grain without having to fix price level.

Risks of non-payment are also reduced with a basis contract, because farmers can draw a large percentage (70 to 80 percent) of the payment at the time of delivery of the grain. They are paid the remainder of the crop value when they price the grain by notifying the grain company that they want to complete the transaction. If the price level drops to the point that the advance payment exceeds

---

BASIS PURCHASE CONTRACT

DATE _____   CONTRACT NO. _____

UNDERSIGNED SELLER hereby sells _____ bushels of _____ to BUYER subject to the following terms and conditions:
grade and grain

1. SELLER has, as of the date of this contract, delivered to BUYER the quantity and type of grain indicated above. The delivery of grain is documented on scale tickets at BUYER'S scale.

Title and all other rights of ownership and control over the grain under this contract are transferred to BUYER.

2. BUYER hereby grants the SELLER option to establish final cash price in accordance with the following terms and conditions:

(a) Basis portion of final cash price is hereby established at _____ ¢ under the _____ Future Option on the Chicago Board of Trade.

(b) SELLER may price grain between 9:40 a.m. and 1:00 p.m. on any day the Chicago Board of Trade is open between today and the _____ day of _____ .

(c) SELLER may, at his/her option and expense, spread to a basis to another Chicago Board of Trade contract on or prior to date in section (b) above.

(d) If SELLER does not establish final cash price on or before above date, the BUYER will establish price based on the closing range of the first trading day following the above date on the Chicago Board of Trade.

(e) All bushels shall be priced or spread forward at one time.

3. Final price shall be subject to grade discounts at market scale of BUYER at time of delivery. Shrinkage, drying, and any other charges that may be applicable shall be deducted before determining the final quantity and price of grain sold under this contract.

4. The BUYER agrees to pay the SELLER _____ dollars per bushel at the time this contract is entered into between the BUYER and SELLER, and to pay the final remaining balance on the day this contract is priced as outlined in Item 2, section (b).

5. SELLER hereby warrants that title to the above grain is free from encumbrance of any nature whatsoever, and the SELLER has good and merchantable title, with the right to dispose of same. It is further agreed by and between the BUYER and SELLER that this contract shall be binding upon the heirs, administrators, and executors of the respective parties and that this contract cannot be assigned without given consent of the other party.

_____   By _____
Seller   Representing Buyer

_____   _____
Phone Number   Dated

Figure 2-4: Basis Purchase Contract.

the remaining value of the crop, the farmer can continue to speculate on price by returning part of the advance to the grain company. Otherwise the grain company can cancel the basis contract and settle the account.

Cash Sale and "Buying the Board"

Producers can achieve the same results as basis contracting by making a cash sale *and* substituting a paper position in the futures market for the physical grain. The cash sale portion has the same business and legal implications noted earlier; delivery, title transfer, pricing, and settlement are completed. Both price level and basis level are fixed by the cash sale. In order to maintain a speculative position, hoping to profit from a price level increase, the producer "Buys the Board" equal to the number of bushels in his or her cash sale.* When the futures price level rises to the price objective, the farmer liquidates the futures position and profits from the price level rise. Should the price level decline, the farmer can liquidate the futures contract and absorb the loss; or the farmer can continue holding the position by making deposits to cover the declining value. The futures contract has a finite life. At some point in time, the contract expires. The farmer must eventually close out the position and withdraw the profits or absorb the losses.

With this marketing tool, the farmer replaces the physical grain with a paper position in the futures market. The cash sale fixes the basis level and requires delivery; the futures contract retains an open price level position, subject to fluctuation.

Business risk is minimized, since full value payment is made on the cash sale. But the futures transaction requires deposits and brokerage fees to execute the futures position. It is also subject to margin calls to cover adverse price movements.

Cash Contracts, Forward Sale, Booking

Cash contracting, forward sale, and booking are synonymous terms for a marketing tool that completes the sale by fixing *price, title exchange,* and *delivery date(s)* on the day of execution; only physical delivery and settlement remain to be completed in the future. Since the latter two typically occur simultaneously, payment risks are minimized. Nevertheless, an insolvent buyer at time of delivery can make the contract worthless, and a favorable contracted price is no longer enforceable.

The advantage of this tool is that the producer can establish a favorable price without having the commodity on hand. Farmers can contract to sell intended production many months in advance of physical delivery. The delivery time need not even be during harvest; delivery can occur from storage anytime. By removing price uncertainty with a forward contract, the producer can more accurately project net returns from an enterprise before a crop is planted or livestock placed in a feedlot.

With cash contracting, both price level and basis level are fixed by the commitment. The price stated on the contract is what the producer will receive when delivery is made. All price speculation is eliminated with the cash forward contract. However, both the price level and the basis level are seldom favorable at the same time. That is the major disadvantage to this tool; it does not allow splitting the pricing decision between price level and basis level sales.

---

*"Buying the Board" is defined as executing long futures positions, explained later in this chapter.

One of the risks of executing a cash contract is the possibility that prices will rise after contracting. Farmers bemoan their "loss in profit" when prices rise and they cannot cancel their delivery obligation. They should not cash contract until they are satisfied with the price being offered. They must give up the opportunity to reap higher prices in return for protection against the risk of lower price.

Production risks are magnified by the fact that farmers are obligated to deliver a specified quantity and quality. Should drought, hail, or other disaster occur, farmers are legally bound to purchase the commodity from someone else to deliver against the contract, or make payment to compensate the contract buyer for any loss that might be incurred. For this reason, producers are often reluctant to sell what they do not have, or else they cash contract only a portion of their expected production.

Figure 2-5 shows production and price changes that can occur after forward pricing commitments are made. Prices may rise or decline; actual production may exceed or fall short of the contracted amount. The example assumes 100 bushels contracted at $3.40 per bushel, for $340 gross receipts. Higher or lower production or price changes the gross actually realized.

Outcome I has both higher price ($3.60) than contracted and higher than expected production (120 bu.). If a producer contracted 100 bushels but harvests 120, the extra 20 bushels can be sold at the higher price, but the 100-bushel contracted amount sells at the contract price. Producers bemoan this result because they could have received $24.00 more per acre had they not contracted. The result rates Good, however, because it is typically profitable. In the aggregate, bumper production seldom accompanies high price, so the probability of this combination of events occurring is low.

Outcome II presents the strongest argument for cash contracting; contract price is higher than harvest time market price ($2.80), and production is higher than expected (120 bu.). The extra 20 bushels must be sold at the lower market price and elicit the response, "I should have contracted more." The sale of the 100-bushel base amount produced the happy result of protecting income against lower price. In the aggregate, higher production usually accompanies lower price, as this combination depicts. The results rate Excellent because of the high probability of this combination of events and the income protection afforded by it.

Outcome III illustrates what many farmers fear the most — higher market price ($3.60) and low yields (80 bu.). Production shortfalls must be offset by purchases at higher price to fill the contract. There is a double loss from not having the commodity and having sold what little there is at the lower contracted price. These results must be rated as Bad.

One solution to guard against production shortfalls is to contract less than full production, such as 30 to 50 percent of expectations. Another is to buy crop insurance. It is available across the U.S. for practically all crops. For anyone who cannot bear risk, both production insurance and price protection can mean the difference between remaining in the business or not.

Outcome IV details the combination of lower price ($2.80) and lower production (80 bu.). The results are Good for contracting, because price protection compensates for production problems. Producers enter the open market to replace the shortfall in production at a lower price than they get by delivering against their contract. They buy low-cost grain from neighbors and sell it on contract at a higher price.

The extent of forward pricing anticipated production from a growing commodity depends upon the individual producer. An individual may be highly spec-

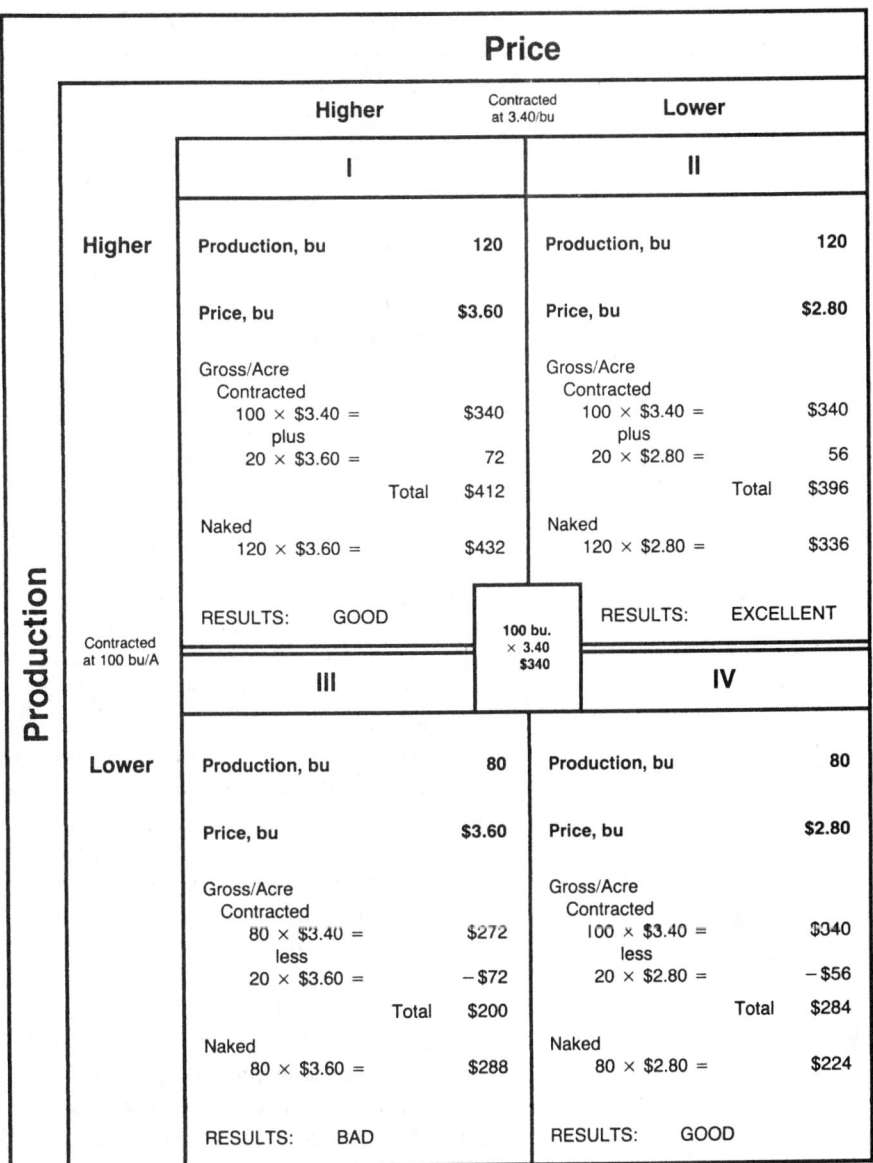

Figure 2-5: Effects on Income of Price Change and Production Differences from Contracted Amounts.

ulative or ultra-conservative, financially stretched or free of debt. The farmer may be located in a marginally productive area or in one that never has had a crop failure. All these factors come into play in determining how much of expected production one is willing to commit by forward pricing. In some years, the highest price for the next production season occurs shortly after harvesting the old crop. Instead of weighing the likelihood of successful versus disastrous production and the relative profits compared with possible losses, many farmers avoid "selling what they do not have" and thereby forego profitable opportunities. They demonstrate their faith in a successful crop by borrowing heavily to buy seed, fertilizers, pesticides, and fuel, but they will not forward contract production even though prices may be at historically high levels. Fear of Outcome III and the quest for Outcome I should not deter farmers from making profitable sales when the market offers profitable opportunities, even though the crop is not on hand. Forward selling of intended production is common practice in other business enterprises.

Farmers must accept the legal implications of contract pricing. Forward contracts exchange title and are legally enforceable. The Uniform Commercial Code (UCC), now adopted in every state except Louisiana, provides that contracts for the sale of goods such as grain and livestock, which involve $500 or more, are legally not enforceable unless they are in writing. To protect themselves, elevators use the standard form of contract, or confirmation, illustrated in Figure 2-6. The confirmation normally includes the amount, quality, agreed price, and terms of delivery and thereby helps to avoid misunderstandings. The UCC also specifies that if the transaction is between two "merchants," the seller need not sign and return the confirmation in order to be bound by its terms. Most courts have looked at the market sophistication of the farmers involved and concluded that they should be called "merchants." If there are second thoughts or if the confirmation does not sound like the trade made verbally, the seller should contact the buyer immediately in an attempt to resolve any discrepancies. If the difference cannot be resolved, the seller has 10 days after receiving the confirmation to reject the offer in writing.

The terms of the contract should be reviewed by an attorney. The omitted provisions may cause more problems than the stipulated ones. For example, what if harvest is delayed by weather? Generally, the contracts will allow the buyer to accept or refuse later delivery at the buyer's option. What if the elevator is full and the elevator cannot accept delivery as called for? Who pays storage charges and who pays for yield loss or deterioration in quality while the crop stands in the field until the buyer can take delivery? It is unlikely that legal risks of marketing can be totally eliminated, but they should be recognized and evaluated by the producer-seller *before* a problem occurs.

Futures Contracts

A futures contract is an agreement entered into today to complete a business transaction on some future date. It specifies the date and place of the transaction, and the quantity and quality of the goods to be delivered, along with the price at which the transaction is to be consummated. Neither title nor money changes hands until the date specified. A futures contract represents an obligation made on a particular day to enter into the cash transaction at a later date.

Contracting for future delivery is not an uncommon business practice, despite the fact that people often ask: "How can you sell something you do not have, or buy something that does not exist?" For example, a futures contract is established between a building contractor and a home buyer for the construction of a house. Contractors are selling something they do not have — the buyer is buying something that does not exist. Contractors enter into a contractual agreement with a customer that at some future date they will have erected a house on the customer's lot. The customer agrees to accept delivery on that date. Plans, specifications, and payment provisions are detailed in the contract. Upon entering the agreement, the contractor has neither a house nor the concrete, lumber, paint, or other materials to build the house. The builder knows, however, where to acquire the materials and hire the labor. The contractor knows how long past construc-

GRAIN SALES CONTRACT

CONTRACT NO. _____

THIS AGREEMENT, made and entered into this _____ day of _____ , 19 _____ , by and between _____ of (city and state) _____ (hereinafter called Seller), and _____ of (city and state) _____ (hereinafter called Buyer), wherein each of the parties agrees to the following terms and conditions of the sale of Seller's _____ (hereinafter called grain).

    1. Seller agrees to sell and deliver to Buyer at Buyer's elevator or to Buyer's designated recipient for Buyer's account on or before the _____ day of _____ , 19 _____ , _____ bushels of _____ , Grade No. _____ , said grain to be clean, sound, and merchantable according to the established grading standards of the USDA, and subject to Buyer's discounts applicable at the day of delivery, for the agreed-on price of _____ DOLLARS ($ _____) per bushel, less any pre-payment, discount, or dockage arising from the condition of the grain. All grain to meet Pure Food and Drug Act requirements.

    2. Seller agrees that in event Buyer or its designated recipient is incapable of receiving all or part of said grain on the date it is offered for delivery by Seller, Seller will hold all or so much of the grain as cannot be received and deliver it to Buyer or its designated recipient upon receipt of notice that the grain can be delivered.

    3. Seller and Buyer agree that the selling price quotation for the grain herein is binding upon both parties and that market fluctuations after the date of this contract shall have no effect on either party's obligation hereunder and will not excuse non-performance of this contract by either party.

    4. Seller and Buyer agree that this is a contract for delivery of Seller's grain and is not a futures contract which can be purchased back by Seller. Seller's failure or inability to deliver said grain in no way relieves him/her or his/her obligation of delivery of said grain, and Seller agrees Buyer may enforce this contract by an action for specific performance, except should delivery of grain be made impossible solely because of a natural disaster. Buyer agrees that upon proper notification from Seller, prior to the delivery date or within the extended delivery period, if Buyer exercises his/her option to extend the contract, to permit Seller to reimburse Buyer for any and all losses Buyer may suffer from Seller's partial performance of this contract in lieu of delivery of all or part of said grain.

    5. In the event of default by either party, all sums due shall be payable without relief from valuation and appraisement laws and with reasonable attorney fees, and with interest at the rate of eight per cent (8%) per year, calculated monthly from the date of default.

    6. Seller agrees that Buyer may extend the due date for delivery of the grain beyond the aforementioned date at Buyer's sole option on written notification of such extension to Seller mailed to Seller's address as shown on this contract.

    7. Seller agrees that Buyer or its designated recipient shall have the sole option of accepting or rejecting any grain in excess of the contract amount and if accepted, it is agreed Buyer will pay Seller Buyer's closing market price on the date of delivery if delivered to Buyer's elevator, and Buyer's closing market price on the date of receipt of notification of such excess delivery from either Seller or Buyer's designated recipient if delivered to designated recipient.

    8. Seller warrants that all grain delivered to Buyer will be free and clear of all liens and encumbrances and that Seller has and will pass merchantable title.

    9. The word "bushels" as used herein means the equivalent number of pounds of said grain as established by the Bureau of Weights and Measures.

    10. This Agreement constitutes the entire understanding of the parties and may be modified only by a signed writing of both parties, excepting Buyer's right to extend this contract, and is binding on the heirs, assigns, and successors of the parties.

SELLER: _____   BUYER: _____
                  NAME

_____   BY: _____
                ADDRESS   TITLE

Figure 2-6: Cash Contract.

tion projects have taken and anticipates the cost likely to be incurred in the building process. A contractor enters into a futures contract to build a house based upon expectations about events that will transpire in due time. The lot owner wishing to own a home enters this agreement in anticipation of taking delivery and making payment on some future date. Any futures market, whether in houses or commodities, is an intentions and expectations market.

A farmer or rancher should have no more trepidation in executing a commodity futures contract than in signing a contract with a house builder. There is an intention to produce the commodity by some future date. There is the expectation that the cash transaction will be completed, as specified, in due time.

Commodity futures contracting is the process of setting a price in the present for a business transaction that will be completed in the future. *Price, quantity, delivery date, and delivery point* are fixed as part of the contract. Actual delivery and payment remain to be completed.

This marketing tool provides price protection similar to cash contracting but differs in its mechanics and pricing effect. In cash contracting, the farmer deals with a local elevator. With futures contracts, a farmer deals with an account executive or a broker directly through a commodity exchange. Cash contracting specifies a flat price the producer is to receive; both price level and basis level are fixed. A futures contract fixes the price level but permits speculation on basis level. With cash contracting, the farmer knows exactly what price is to be received; with futures contracts, the farmer knows only approximately what will be received, since basis level remains to be established.

As a commodity futures contract approaches maturity, final settlement must be made. Settlement of a futures contract differs from settlement of a cash contract, however. With the latter, the buyer expects delivery of the commodity from the seller and is usually inflexible in that requirement. Futures contracts can be settled by one of two methods:

1. Delivery Against the Futures Contract by completing the transaction as specified in the contract.
2. Offsetting Trades by making an opposite futures transaction to cancel the original contract and thus eliminate the obligation for fulfilling it.

The flexibility of settlement provided by these two alternatives provides futures contracts a distinct advantage over cash contracts.

Delivery Against Contracts: Cash Settlement. Cash settlement of a futures contract requires delivery by the seller and payment by the buyer. The producer physically delivers the commodity to the delivery point on the date specified on the contract. The buyer accepts title and pays full value on the transaction.

Location of delivery points is critically important to the seller. Delivery points are usually located in commercial centers (Figure 2-7), far removed from the farm in both distance and mechanics of operation. Chicago and Toledo are the major grain delivery points. For cattle and hogs, delivery is at key livestock marketing points throughout middle America, since this area produces the largest proportion of the nation's livestock. The cotton futures market specifies delivery at the key warehousing and marketing points for the largest U.S. cotton regions. Kansas City, Missouri, is the delivery point for hard red winter wheat, and Minneapolis and Duluth for spring wheat. The frozen orange juice contract requires delivery at regulated warehouses at or near Florida processing plants.

Marketing Tools

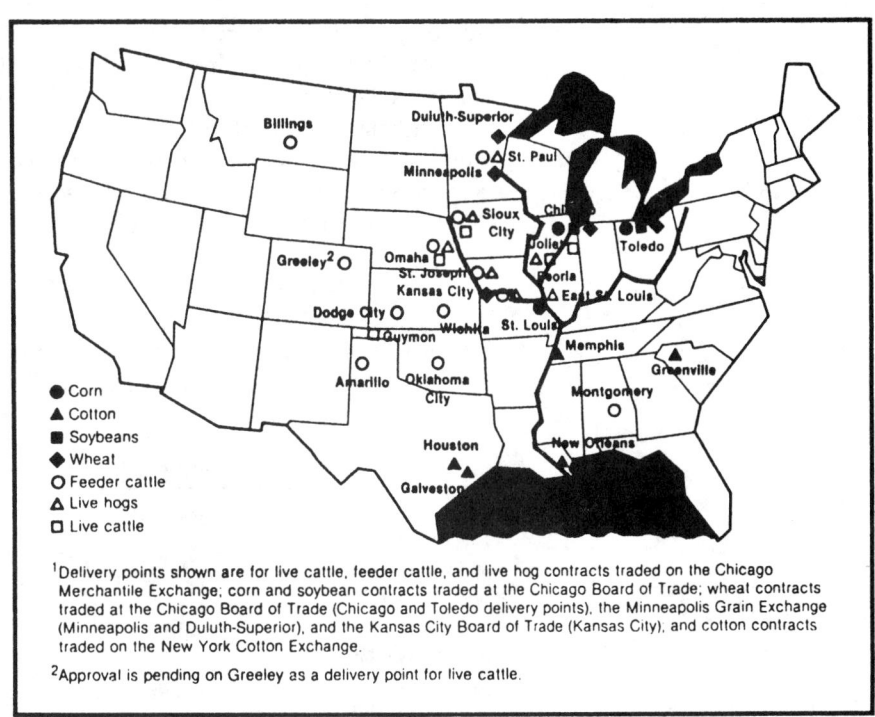

Figure 2-7: Farmer Delivery Points Located on Main Shipping Channels and at Major Cash Markets. Source: *Farmline* Reprint, "Futures and the Farmer" (Washington, D.C.: USDA, 1980) 10.

The seller must pay transportation charges from the farm or ranch to the delivery point. In addition, the seller must pay handling charges to the terminal operator and brokerage fees for services rendered. The net price realized at the farm gate is the contracted futures price less these charges. (This difference between futures price and realized price is the subject of basis discussion in Chapter 7.)

A trader who buys a futures contract and holds it for delivery does not expect trucks to back up to his or her home or office and unload the actual commodity. Even if the buyer wants to accept delivery against the futures contract, the buyer is not given the commodity; instead, a warehouse receipt of ownership is issued from the facility that accepted the seller's delivery at the delivery point.

Commodity buyers rarely take delivery. Since many types and grades are deliverable against the contract and since the option to deliver or not remains with the seller, the buyer does not know in advance exactly what would be delivered. Processors require a careful selection of specific grades that may not be matched by the goods being delivered.

Those who do accept delivery on futures contracts are typically merchants or dealers. These traders find it practical to accept delivery, because they deal in many grades and types of the actual commodity and feel confident that they can find a buyer for whatever grade the seller delivers.

In actual practice, less than 1 percent of most futures contracts are settled by delivery, as shown in Table 2-2. The fact that delivery of the actual commodity can be made to satisfy futures obligation links together a price relationship between cash markets and futures markets. If cash and futures prices get too far out of line with each other, there are traders (arbiters, practicing arbitrage) who earn their livelihood by buying one and selling the other to bring them back into proper economic relationship.

The purpose of futures trading is to shift price risk rather than provide a mechanism for cash sale. Actual delivery and acceptance of a physical commodity to satisfy a futures contract is not essential for futures to fulfill that purpose; the mere possibility of physical exchange insures the proper price relationship.

Table 2-2: Delivery of Selected Commodities
to Satisfy Futures Contracts, 1985-1986

Contract	Volume of Trading[1]	Total Contracts Settled by Delivery or Cash Settlement[2]	Percentage Settled by Delivery
Live Cattle[3] (40,000 lbs.)	4,985,000	3,075	.06%
Live Hogs[3] (30,000 lbs.)	1,886,000	281	.01%
Wheat[4] (5,000 bu.)	2,190,000	10,365	.47%
Corn[4] (5,000 bu.)	6,188,000	24,315	.39%
Soybeans[4] (50,000 bu.)	6,715,000	118,135	1.76%
Cotton[5] (50,000 lbs.)	876,000	1,135	.13%
Orange Juice[5] (15,000 lbs.)	236,000	2,210	.94%

Source: Annual Report, 1986 CFTC (Washington: USPO), 1987.
[1] Rounded to nearest thousand.
[2] Excludes commodities redelivered against other futures commitments.
[3] Chicago Mercantile Exchange.
[4] Chicago Board of Trade.
[5] New York Cotton Exchange.

Offsetting Trades. Producers generally do not find it advantageous to make delivery. Their products may not match the grade or quantity specified on futures contracts, and the costs of making delivery to a distant terminal may be prohibitive. It is usually more advantageous for them to simply offset the transaction in the futures market and deliver their product locally. Settlement, by the offsetting trades, eliminates the mechanical problems of delivery, without impinging on the effectiveness of transferring risk.

Producers who SELL a futures contract in essence OWE delivery of the commodity. At any time before maturity of the contract (even the next day), they can BUY back the futures contract and thereby offset the original commitment. This is called "covering" the position with an offsetting trade. Producers buy back the same number of contracts they sold. They are then free to sell the cash commodity to anyone, since its futures delivery commitment has been canceled.

The flexibility of canceling futures contract commitments with offsetting trades fosters hedging. Commodity producers or owners transfer the economic effects of price change from themselves to others by hedging. To understand the mechanics of price risk transfer afforded by hedging, we need to understand the parallel relationship between cash prices and futures prices.

Parallelism of Cash and Futures Prices. Futures trading creates neither a loss nor a profit for a producer who has a commodity to sell. It merely fixes a price level that will eventually be realized. This truism exists because of the parallel relationship between the cash price and the futures contract price of a commodity. The two prices tend to fluctuate together. When cash prices increase, futures prices increase; when cash prices decline, futures prices decline, as long as basis is constant.*

Cash price determines the inventory value of the growing or stored crops or livestock on hand. Rising cash price increases inventory value; falling price lowers inventory value. It is the erosion in value that a producer seeks to avoid by **hedging** to fix a price level in the futures market.**

Hedging involves **selling** futures contracts equal to the amount of grain or number of livestock that a producer owns or expects to produce. The farmer-hedger OWNS grain, "knee high on the Fourth of July," and he or she OWES delivery of it against futures contracts. When the price level declines, as shown in Figure 2-8, two values change. First, the inventory value of grain in the field decreases. A drop in the cash price from $3.10 per bushel to $2.35 results in a loss in value of 75 cents per bushel. With 100,000 expected bushels of corn in the field, its total value decreases $75,000. Over the same time, however, the hedge account gained in value. The futures contract was sold (to establish the hedge) when the price level was $3.50 per bushel. It can be offset (bought back) at $2.75 per bushel. Over the period, $75,000 was credited to the brokerage account. The 75-cent loss in inventory value was offset by the 75-cent gain on the futures contracts.

The producer covers the futures contract by buying it back and sells the grain on the local cash market for $2.35. The 75 cents in price protection realized from the futures market position is added to local cash price of $2.35 per bushel; the hedger pays bills with $3.10 corn. Neighbors who *speculated* in the cash market and lost when prices plunged try to pay their bills with $2.35 corn.

In a rising market, the gains and losses are reversed. Anticipating lower prices, the farmer sells a futures contract at $3.50 per bushel. But the futures market rallies to $4.00 per bushel and the cash price rises from $3.10 to $3.60. When the hedger covers the hedge position by buying back the contract sold, there is a 50-cent-per-bushel deficit in the futures transaction. This is, however, recovered by the 50-cent increase in the cash price. By deducting the futures account deficit from the cash price received, the realized net cash price is once again $3.10.

To be sure, there is some regret on the part of the hedger who pays the bills with $3.10 corn. Neighbors are paying their bills with $3.60 corn. In this case they speculated in the cash market and won.

In these two examples of a "perfect" hedge (no change in basis), the hedger was guaranteed a $3.10 price. Neighbors who speculated in both instances "enjoyed" $2.35 and $3.60 corn.

---

*Parallelism assumes a perfect hedge without gain or loss from basis change. The assumption of a perfect hedge will be relaxed and explained in Chapter 7.

**This is an example of a selling hedge; a buying hedge works by the same principle, but in reverse. It will be illustrated later.

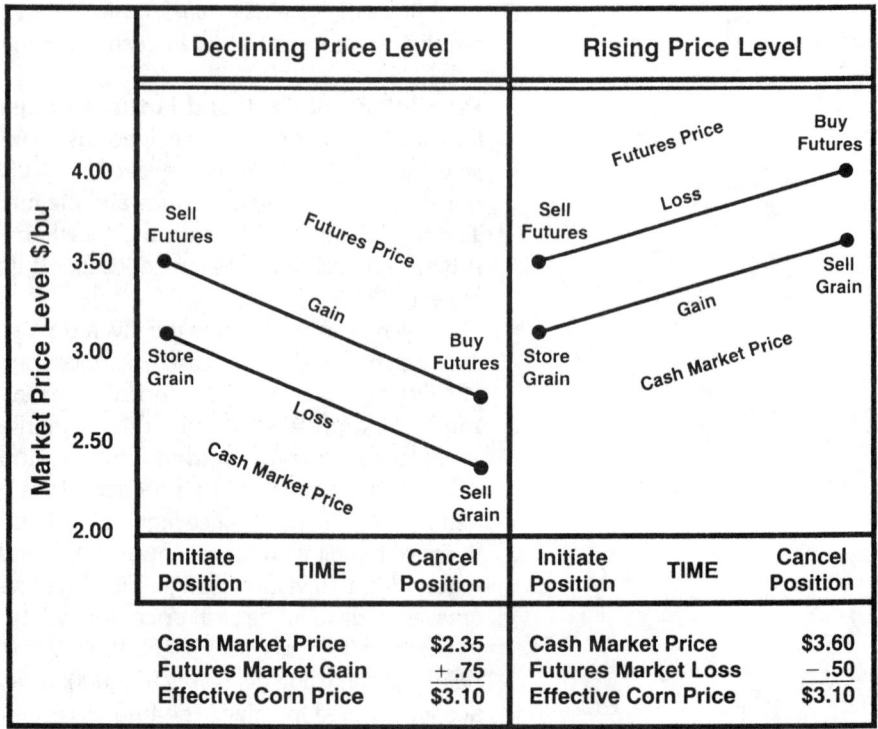

Figure 2-8: Parallel Movement of Futures and Cash Prices, Illustrating Offsetting Gains and Losses in a Selling Hedge.

In both the rising and falling market examples, the farmer received a net price of $3.10. Hedging guaranteed that price level and eliminated the possibility of higher or lower price. The hedger gave up the opportunity to enjoy a higher price in return for the protection from suffering from a lower price. The price was fixed at a level that the producer decided was satisfactory when he or she set the hedge.

Farmer reluctance to hedge planned or growing production arises from production risk. Drought during the growing season could cut yields, reducing total production. Even though price per bushel increases, as illustrated in Figure 2-8, the hedger might not have enough bushels (multiplied by the higher price) to offset the futures position deficit. Total inventory value gain would be less than total futures deficit. Farmers hedging a growing enterprise do not have the luxury of assurance that production goals will be met. This problem can be remedied, as we will explain later.

Buying Hedge. Another type of hedge is a Buying Hedge. Buyers of raw materials utilize futures contracts to fix the cost of the resources they plan to use. For example, a High Plains cattle feeder can hedge the costs of feeder animals, feed corn, and soybean meal that will be used to produce slaughter-weight cattle. A **buying hedge** involves "buying" futures contracts equal to the amount of feed ingredients that a feeder expects to use. Having determined feed corn requirements, the livestock feeder would establish an equal, but opposite, position in the corn futures market. The feeder is "short" cash corn needed to feed the cattle. In order to fix feed cost before placing cattle in the lot, the hedger "buys" corn futures contracts. The feeder "OWES" feed to the cattle and "OWNS" the right to receive corn via futures contracts. The parallel movement of prices in the futures and cash markets, as shown in Figure 2-9, illustrates how hedging fixes the cost of feed corn.

Suppose the corn market quoted a futures price of $3.50 per bushel and a cash price of $3.00 when the feeder sets the hedge. In a rising market, both prices would be higher when buying cash corn from the market. The feeder might have to pay $3.60 for feed corn, incurring a $0.60-per-bushel higher cost. However, the futures contract price would also have risen. The hedger offsets the $3.50 "buy" position with a "sell" transaction at $4.10, producing a $0.60 gain in the futures account. Subtracting the futures gain from cash purchase cost produces a $3.00 net cost of the feed ingredient.

In a declining market, the gains and losses would be reversed. The feeder buys cash corn at a price of $2.60, or $0.40 less than planned. Upon covering the futures contract at $3.10, however, there is a $0.40 loss in the position. Adding the futures loss to the lower purchase price of corn produces a net cost of $3.00 for the feed ingredient.

In both market situations, the feed has a net cost of $3.00 per bushel. That is the cost that the feeder decided would be satisfactory when establishing the hedge position. The feeder gave up the opportunity to buy corn at a lower cost in return for protection against having to pay more for it. Gains in one market were

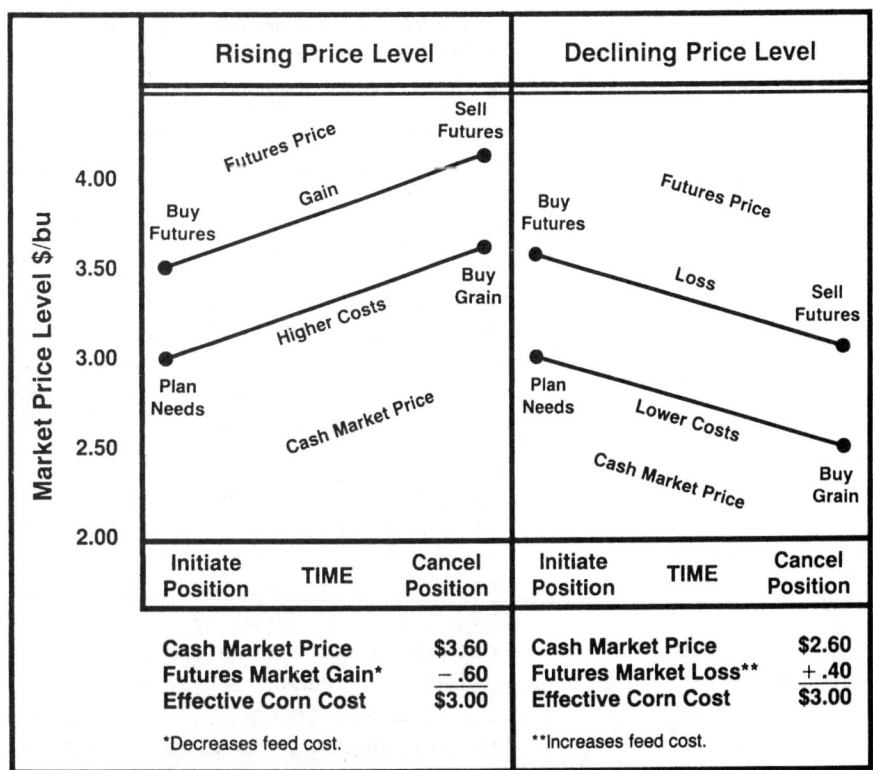

Figure 2-9: Parallel Movement of Futures and Cash Prices, Illustrating Offsetting Gains and Losses in a Buying Hedge.

offset by losses in the other; the price was fixed by the buying hedge just the same as in the example of the selling hedge.

The various positions and terminology of Selling and Buying Hedges are summarized in Figure 2-10.

Cash market price and futures market price **tend** to rise and fall together in a parallel manner. In reality, they do not move in perfect harmony over time. Later on we will remove the assumption of parallel movement of prices in futures and cash markets to show how one might profit from the convergence or divergence of the two prices.

	Selling Hedge		Buying Hedge	
Participants:	Farmers growing corn, wheat, cotton, soybeans. Livestock feeders selling cattle and hogs.		Cattle feeders buying feeder animals. Livestock producers buying corn and meal as feed ingredients.	
	CASH POSITION	FUTURES POSITION	CASH POSITION	FUTURES POSITION
Initiate position	Plant crop Buy feeders Breed sows	"SELL" contract	Plan production	"BUY" contract
Ownership	OWN the commodity	OWE delivery	OWE input item to operation	OWN the right to receive input
Terminology	LONG the cash	SHORT the futures	SHORT the cash	LONG the futures
Offset to initial position	Cash sale of commodity	"BUY" contract	Buy cash commodity	"SELL" contract

Figure 2-10: Summary of Terminology and Futures Positions for Hedge Selling and Buying.

Margins: Settling the Accounts. There are two kinds of margins: initial and maintenance. Before a producer can enter into a futures contract, he or she must make a deposit of "good faith" money to establish financial integrity and guarantee performance. This deposit is similar to the house buyer's down payment when he or she signs a contract with a builder. In commodity trading the payment is called the "initial margin deposit."

The second type of margin is maintenance margin. When the market moves against a futures position, a deficit occurs in the brokerage account. The broker makes a margin call to the producer, requesting an additional deposit of maintenance margin to cover the withdrawals and to maintain a minimum balance. In order to continue holding the contract position, the producer must send the deposit within a short time or else the broker has the authority to cancel the contract on the farmer's behalf.

With a perfect hedge, the parallel fluctuation of cash inventory value and futures account balance produces offsetting gains and losses that fix a preestablished price at the termination of the hedge. Day by day, however, from contract initiation until the hedge is covered, the inventory fluctuation is a non-cash change in value, but the futures account is settled in cash every day. Price change after initiating a futures position produces a cash surplus or cash deficit at the end of each day in the brokerage account. Daily settlement is made by the commodity brokerage firm in the producer's margin account. If the market price goes down after a selling hedge is made, the margin account will receive deposits each day equal to the price change from the previous day. The opposite situation occurs when the market price goes up after a futures contract is sold. In this case, payments must be made each day from the margin account.

A cash surplus in a brokerage account is often seen as "profit" for the farmer-hedger. This is "found money"; it came so easily. The hedger did not have to work for it. All the hedger had to do was sell a futures contract and watch prices go down. The whole experience creates a false euphoria about what has actually happened.

Conversely, cash deficits or "margin calls" generate personal agony. Obviously, hindsight now tells the hedger that he or she sold crops or livestock at too low a price. Second, cash needed to meet margin calls can exceed available funds. Due to disgust over hedging at "low" prices, or inadequate financing, the farmer cancels the futures position at a loss. Invariably, price will turn the other way and begin to fall; but it is too late. The price protection the farmer sought is gone. Sold-out hedgers end the experience with a realized low price for their production and a loss in their futures account. They forever tell the story of "losing their shirt" in the futures market and repeat a vow never to "gamble" again by "speculating" in commodity futures. They are actually saying that they do not understand the purpose of futures trading nor the mechanics of the system. To use futures effectively, farmers need to understand the offsetting gains and losses that are occurring in two markets due to parallel movement in cash and futures prices.

Hedged-to-Arrive Contracts

Hedged-to-arrive contracting achieves the opposite result of a basis contract. With a basis contract, the producer locks in the basis level and continues to be exposed to price level fluctuation in the futures market. By contrast, the hedged-to-arrive contract fixes the price level, but the producer is exposed to basis fluctuation. Only the *price level is fixed* by hedged-to-arrive contracts at the time the contract is made. Delivery, title transfer, settlement, and fixing the basis are delayed until later.

Hedged-to-arrive contracts produce the same results as hedging with futures; the difference between the two arises from who handles the details. With hedged-to-arrive contracting, the elevator assumes responsibility for placing the futures position and absorbs the costs normally associated with hedging. These costs include margin calls, interest on margin deposits, brokerage fees, and associated management expenses. A farmer-hedger using futures would incur these costs and obligations personally.

The primary advantage of this method for the elevator operator is the potential for increased volume of business. While the elevator operator may give up some income from basis appreciation, increasing business volume and spreading fixed costs are definite advantages to the operator. For the grain seller, the hedged-to-arrive contract offers the price protection advantages of hedging without sacrificing the potential benefit from basis improvement. Because the hedged-to-arrive contract is settled in cash at time of delivery, there is little business risk. And the seller avoids margin costs and commission fees associated with hedging.

Elevator operators may allow the seller to roll forward the contract obligation. For example, the elevator may allow the farmer to sell July futures contracts to cover a harvest time delivery. This allows the seller to take advantage of present price spreads bid into the market. The seller would roll the futures position into another futures month as many times as desired, until delivery time is reached.

Since the hedged-to-arrive contract is predicated on the farmer's planned delivery period, a future's price may be established up to 18 months in advance of actual delivery. This greatly increases the producer's opportunity to select a profitable price by extending the market season far into the future. By understanding month-to-month spreads in futures contracts and basis appreciation, a producer can use this marketing tool to capture basis returns as well as price level protection.

Agricultural Options[4]

In October, 1984, trading of agricultural options was reinstated after a 50-year ban. They provide another marketing tool for farmers.

In contrast to a futures contract, an option is a contract that conveys the right, *but not the obligation,* to the option buyer to buy or sell the underlying commodity (either the physical commodity or the futures contract) at a specified price (the strike price) on or before a specified date of expiration. The person who grants the option buyer this right to buy or sell is known as the option writer. The writer receives a non-refundable lump-sum payment, called a premium, as compensation.

To illustrate the components common to all options, consider a call option that conveys the right to purchase a tract of farmland. Mr. Sellit offers Mr. Buyit the right to buy 100 acres at $1,000 an acre any time between now and July 1. For this right, Mr. Buyit pays Mr. Sellit an option fee of $2,500.

In options terminology, the specific 100-acre tract is the underlying commodity. The $1,000-per-acre purchase price is the strike price. July 1 is the expiration date. And the $2,500 paid by Mr. Buyit is the premium. At any time prior to the expiration date, Mr. Buyit can buy the underlying commodity at the option strike price, but Mr. Buyit is not obligated to do so. That is, Mr. Buyit can exercise the option if he chooses. If Mr. Buyit decides not to exercise, he forfeits his premium and allows the option to expire worthless. Or Mr. Buyit might sell his option right to someone else. In any event, Mr. Sellit keeps the $2,500 premium.

In using options in farmer marketing, none of the business events of delivery, title exchange, or settlement is fixed by the transaction. Merely the right to receive either a short or long futures position is conveyed, should the buyer choose to exercise the option. As the name suggests, that right does not impose an obligation. Hence, options allow a farmer to fix a favorable price without being obligated to accept it. If price improves, the option buyer can walk away from the original option and capture the better price being offered by the cash market. In that event, the maximum loss to the option holder is limited to the original option premium paid.

There are two types of options: a PUT option and a CALL option. A put option conveys to the holder the right to *sell* at the strike price on or prior to the expiration date. A farmer can "put" the crop on the market at the stated price. A call option conveys the right to *buy* at the strike price on or prior to the expiration date. A feeder can "call" feed corn, for example, from the market at a stated price.

A farmer who wishes to protect the value of the growing soybean crop can purchase a PUT option as price insurance. A soybean option designated as a "Nov. 750 put" gives the option buyer the right to sell (go short) a November soybean futures contract (5,000 bushels) at a strike price of $7.50 a bushel at any time prior to expiration, regardless of what the futures price is at the time. Purchasing put options provides a means of obtaining protection against declining prices without giving up the opportunity to profit from a price increase. In essence, options set a floor price. The expected price is calculated by subtracting the option premium and expected basis from the strike price ($7.50 strike, − $0.75 premium, − $0.25 anticipated basis, = $6.50 anticipated cash price floor). If the futures price level declines to $6.50 just prior to expiration, the option could have a value of $1.00. Conversely, if the futures price rises to $8.50 prior to expiration, the option would be worthless and the farmer would let it expire without exercising it. The right to sell soybeans for $7.50 when the market is offering $8.50 makes the option worthless.

The buyer of a futures option, unlike the buyer of a futures contract, does not have to worry about daily margin calls. The buyer must instead pay an option premium at the time of purchase. Thus the option holder can weather a period of adverse price moves in the expectation that the position will eventually become profitable. If this expectation remains unfulfilled at the time of expiration, the buyer has the alternative to "meet a margin call" by paying another premium to buy another option. Options are, therefore, not cost-free. Using them to hedge production over a period of time may result in higher costs than meeting margin calls on futures contracts.

SUMMARY

Various marketing tools allow producers to effect different marketing decisions. Different tools enable them to transact each business event at the most opportune

Time Frame	Tool	Pricing		Delivery	Title Exchange	Payment	Speculation			
		Price Level	Basis Level				Price	Basis	Quality	Quantity
Prior to harvest	Cash Contract[1]	Fixed	Fixed	Later	Later	Later	No	No	Yes	Yes
	Futures Contract[1]	Fixed	Open	Later	Later	Later	No	Yes	Yes	Yes
Harvest and post-harvest	Cash Sale[1]	Fixed	Fixed	Now	Now	Now	No	No	No	No
	On-farm Storage	Open	Open	Later	Later	Later	Yes	Yes	Yes	No
	Commercial Storage	Open	Open	Now	Later	Later	Yes	Yes	No	No
	Delayed Pricing	Open	Open	Now	Now	Later	Yes	Yes	No	Yes[2]
	Delayed Payment[1]	Fixed	Fixed	Now	Now	Later	No	No	No	Yes[2]
	Basis Contract	Open	Fixed	Now	Now	Partial now; remainder later	Yes	No	No	No
	Cash Sale Plus Buy Futures[1]	Open	Fixed	Now	Now	Now on cash; later on futures	Yes	No	No	No

[1]Options available to livestock producers to sell feeder cattle, fat cattle, and hogs.
[2]Payment is at risk.

Figure 2-11: Summary of Marketing Tools and Their Respective Business Transactions.

time. Pricing, delivery, title transfer, and settlement need not occur simultaneously. Through a judicious use of these tools, each of the business events can be transacted at the most opportune moment over a broad span of time.

Distinguishing between the market price level and basis price level is important in determining which tool to use at the appropriate time.

The market pricing alternatives available to farmers and their respective pricing commitments are summarized in Figure 2-11. Grouping marketing tools under two kinds of strategies should help explain what they accomplish. For example, to fix price level, a farmer might use a cash sale, a forward contract, or a hedge. To speculate on price level, a farmer could store unpriced grain, use the delayed pricing contract, use the basis contract, or sell cash/buy futures. These four alternatives will hold grain for speculation. The proper time to use each tool will be discussed as part of strategy considerations in Chapter 6.

Questions for Study and Discussion:

1. Name and discuss the four business events of commodity transactions.
2. Discuss the difference between market price level and basis price level.
3. List the marketing tools of agricultural marketing. What price is fixed by each tool and what is left open to fluctuation? What speculation is removed with each tool?
4. Why are grain moisture considerations important in cash sales?
5. Why is delayed pricing a risky marketing tool?
6. Explain the reluctance of farmers to forward price production prior to harvest. Is the reluctance justified?
7. How does futures contracting differ from other tools as a marketing device?
8. What two methods may be used to satisfy a futures contract obligation?
9. What is the purpose of margins in futures trading?
10. Explain parallelism of cash and futures markets. Assume various prices and calculate gains and losses from cash and futures positions.

Exercises for Using a Standard Grain Shrinkage Table:

1. Compare value of shrink (grain price times percent shrinkage, from Table 2-3) against the discount charged by your local elevator. For example, 21 percent corn dried to 15.5 percent has 7.01 percent true shrinkage worth 21 cents per bushel on $3.00 corn ($3.00 x .070). By comparison, an elevator might discount 21 percent moisture grain by 44 cents.
2. Calculate the cost of over-drying corn. How much value loss occurs by taking grain from 15.5 percent to 13 percent? Add to the value loss the extra fuel cost required for extra drying. The sum of both costs is the cost of over-drying.

Table 2-3: Grain Shrinkage Chart (Standard Moisture Table)[1]

Initial Moisture	13.0%	13.5%	14.0%	14.5%	15.0%	15.5%	16.0%	16.5%	17.0%	17.5%	18.0%	18.5%	19.0%
Percent	. . . . (percentage of shrinkage) . . . .
15.5	3.37	2.81	2.24	1.67	1.09	—	—	—	—	—	—	—	—
16.0	3.95	3.39	2.83	2.25	1.68	1.09	—	—	—	—	—	—	—
16.5	4.52	3.97	3.41	2.84	2.26	1.68	1.10	—	—	—	—	—	—
17.0	5.10	4.55	3.99	3.42	2.85	2.28	1.70	1.10	—	—	—	—	—
17.5	5.67	5.12	4.57	4.01	3.44	2.87	2.29	1.70	1.11	—	—	—	—
18.0	6.25	5.70	5.15	4.59	4.03	3.46	2.88	2.30	1.71	1.11	—	—	—
18.5	6.82	6.28	5.73	5.18	4.62	4.05	3.48	2.90	2.31	1.72	1.11	—	—
19.0	7.40	6.86	6.31	5.76	5.21	4.64	4.08	3.50	2.91	2.32	1.72	1.12	—
19.5	7.97	7.44	6.90	6.35	5.79	5.23	4.67	4.10	3.52	2.93	2.33	1.73	1.12
20.0	8.55	8.01	7.48	6.93	6.38	5.83	5.27	4.70	4.12	3.54	2.94	2.35	1.74
20.5	9.12	8.59	8.06	7.52	6.97	6.42	5.86	5.30	4.72	4.14	3.55	2.96	2.36
21.0	9.70	9.17	8.64	8.10	7.56	7.01	6.46	5.89	5.32	4.75	4.16	3.57	2.97
21.5	10.27	9.75	9.22	8.69	8.15	7.60	7.05	6.49	5.93	5.35	4.77	4.19	3.59
22.0	10.84	10.33	9.80	9.27	8.74	8.19	7.65	7.09	6.53	5.96	5.38	4.80	4.21
22.5	11.42	10.90	10.38	9.86	9.32	8.78	8.24	7.69	7.13	6.57	5.99	5.40	4.83
23.0	11.99	11.48	10.97	10.44	9.91	9.38	8.84	8.29	7.73	7.17	6.60	6.03	5.44
23.5	12.57	12.06	11.55	11.03	10.50	9.97	9.43	8.89	8.34	7.78	7.21	6.64	6.06
24.0	13.14	12.64	12.13	11.61	11.09	10.56	10.03	9.49	8.94	8.38	7.82	7.25	6.68
24.5	13.72	13.22	12.71	12.20	11.68	11.15	10.62	10.09	9.54	8.99	8.43	7.87	7.30
25.0	14.29	13.79	13.29	12.78	12.26	11.74	11.22	10.68	10.14	9.60	9.04	8.48	7.91
25.5	14.87	14.37	13.87	13.37	12.85	12.33	11.81	11.28	10.75	10.20	9.65	9.09	8.53
26.0	15.44	14.95	14.45	13.95	13.44	12.93	12.41	11.88	11.35	10.81	10.26	9.71	9.15
26.5	16.02	15.53	15.03	14.54	14.03	13.52	13.00	12.48	11.95	11.41	10.87	10.32	9.76
27.0	16.60	16.11	15.62	15.12	14.62	14.11	13.60	13.08	12.55	12.02	11.48	10.93	10.38
27.5	17.17	16.69	16.20	15.71	15.21	14.71	14.20	13.68	13.16	12.63	12.09	11.55	11.00

[1]Formulas: (1) Shrinkage = 100.0% − ($\frac{\text{\% dry matter in wet grain}}{\text{\% dry matter in dry grain}}$ × 100) + 0.5% handling shrink.

(2) Value of shrink = Price basis grade × shrinkage.

(3) Returns to drying = discount − value of shrinkage.

Table 2-4: Grades, Grade Requirements, and Grade Designations for Soybeans

The following grades, grade requirements, and grade designations are applicable under these standards:

				Maximum Limits of —			
Grade	Minimum Test Weight per Bushel	Moisture	Splits	Damaged Kernels		Foreign Material	Brown, Black, and/or Bicolored Soybeans in Yellow or Green Soybeans
				Total	Heat-Damaged		
	(lb.)	. . . . (%) . . . .
U.S. No. 1	56.0	13.0	10.0	2.0	0.2	1.0	1.0
U.S. No. 2	54.0	14.0	20.0	3.0	0.5	2.0	2.0
U.S. No. 3[1]	52.0	16.0	30.0	5.0	1.0	3.0	5.0
U.S. No. 4[2]	49.0	18.0	40.0	8.0	3.0	5.0	10.0
U.S. Sample grade	U.S. Sample grade shall be soybeans which do not meet the requirements for any of the grades from U.S. No. 1 to U.S. No. 4, inclusive; or which are musty, sour, or heating; or which have any commercially objectionable foreign odor; or which contain stones; or which are otherwise of distinctly low quality.						

Source: USDA, *The Official United States Standards for Grain* 8.3.

[1]Soybeans which are purple mottled or stained shall be graded not higher than U.S. No. 3.

[2]Soybeans which are materially weathered shall be graded not higher than U.S. No. 4.

Table 2-5: Grades, Grade Requirements, and Grade Designations for Wheat

The following grades, grade requirements, and grade designations are applicable under these standards:

Grade	Minimum Test Weight Per Bushel (pounds)		Percent Maximum Limits of—					Wheat of Other Classes[3]	
	Hard Red Spring Wheat or White Club Wheat	All Other Classes and Sub-classes	Heat-Damaged Kernels	Damaged Kernels (Total)[1]	Foreign Material	Shrunken and Broken Kernels	Defects (Total)[2]	Con-trasting Classes	Wheat of Other Classes (Total)[4]
U.S. No. 1	58.0	60.0	0.2	2.0	0.5	3.0	3.0	1.0	3.0
U.S. No. 2	57.0	58.0	0.2	4.0	1.0	5.0	5.0	2.0	5.0
U.S. No. 3	55.0	56.0	0.5	7.0	2.0	8.0	8.0	3.0	10.0
U.S. No. 4	53.0	54.0	1.0	10.0	3.0	12.0	12.0	10.0	10.0
U.S. No. 5	50.0	51.0	3.0	15.0	5.0	20.0	20.0	10.0	10.0

U.S. Sample grade .... U.S. Sample grade shall be wheat which:
(1) Does not meet the requirements for the grades U.S. Nos. 1, 2, 3, 4, or 5; or
(2) Contains a quantity of smut so great that 1 or more of the grade requirements cannot be determined accurately; or
(3) Contains 8 or more stones, 2 or more pieces of glass, 3 or more crotalaria seeds (*Crotalaria* spp.), 3 or more castor beans (*Ricinus communis*), 4 or more particles of an unknown foreign substance(s), or 2 or more rodent pellets, bird droppings, or an equivalent quantity of other animal filth per 1,000 g of wheat; or
(4) Has a musty, sour, or commercially objectionable foreign odor (except smut or garlic odor); or
(5) Is heating or otherwise of distinctly low quality.

Source: USDA, *The Official United States Standards for Grain* 1.7.
[1] Includes heat-damaged kernels.
[2] Defects (total) include damaged kernels (total), foreign material, and shrunken and broken kernels. The sum of these 3 factors may not exceed the limit for defects.
[3] Unclassed wheat of any grade may contain not more than 10 percent of wheat of other classes.
[4] Includes contrasting classes.

Footnotes for Chapter 2:

(1) USDA, *The Official United States Standards for Grain* (Washington, D.C.: USDA, Jan., 1978).

(2) A complete analysis of the economics of drying corn and soybeans is presented in Gene A. Futrell and Robert N. Wisner, eds., *Marketing for Farmers* (Doane Information Services, 11701 Borman Drive, St. Louis, Mo. 63146), 1982.

(3) Standard Federal Tax Reports, "Tax Accounting Installment Method Prepaid Income," Vol. 4 (Chicago: Commerce Clearing House, Inc., 1985), paragraphs 2887L.025, 2887P.01.

(4) Randall E. Sheldon and Jin W. Choi, *Agricultural Options* (Chicago: Chicago Board of Trade, Jan., 1983).

Chapter 3

The Futures Markets: Focal Point in Commodity Marketing

FUTURES MARKETS 56
   Historical Evolution of Grain Marketing 56
   Exchanges 58
   What Is a Commodity? 59
ROLES OF FUTURES MARKETS 61
   Futures Trading and Price Discovery 61
      How Trading Takes Place 63
      Price Quotations 65
      The Clearing Corporation 68
      Opening an Account 69
      Tax Treatment of Futures Transactions 69
      Margins 70
      Types of Orders 70
   Risk Transfer 72
      Risk and Uncertainty 72
         Production and Product Risk 73
         Price Risk 73
      Hedging 74
         Who Uses Short Hedges and How 74
         Who Uses Long Hedges and How 74
         Selling Hedge 75
         Buying Hedge 77
      Financing Price Protection 78
      Types of Hedging 79
      Hedging Hazards 80
   Investment Medium 81
      Participants in Futures Trading 82
      Attractions of Speculating 85
      Trading Units 85
      Trading Price Fluctuations 87
      Regulation of Futures Trading 88
      Farmer Trading 89
      Rules for Speculating 90
      Is Futures Trading Gambling? 91
SUMMARY 92

The Futures Markets: Focal Point in Commodity Marketing

FUTURES MARKETS

Historical Evolution of Grain Marketing[1]

Agricultural commodity marketing, as we know it, did not just suddenly happen; it evolved over a considerable period of time along with the industrial and financial development of the nation. Removal of the Indian threat after the War of 1812 opened the Northwest Territory to rapid settlement. Pioneers found vast and rich resources that yielded bountiful production beyond the needs of the pioneer family. As they sought to sell the surplus, farmers brought their wagonloads of grain across the prairie on plank roads to the log cabin village of Chicago. Canal boats and barges loaded with grain converged on the town via the river system. When navigating the Great Lakes was possible, vessels crowded the Chicago harbor to take grain east for milling and processing, or to connect with boats on the Erie Canal for export.

After reaching Chicago, the crossroads of supply and demand, the farmer still had to haul the grain from store to store until a buyer could be found. Sometimes the farmer found willing buyers, sometimes not. During harvest, wagon lines were often 20 miles long. Excess grain was often dumped on the ground or in inadequate storage facilities. Spoiled grain was hauled on to Lake Michigan and dumped. Yet before the next harvest, supplies were short again. The farmer was at the mercy of severely fluctuating prices; either too much grain was received, dropping prices drastically for latecomers, or deliveries were too small, causing prices to skyrocket. Prices of consumer goods, such as bread, flour, and meal, rose and fell with the tide of raw materials. Sometimes products disappeared from the market entirely. Merchants safeguarded themselves against price change by offering less and less for raw commodities and charging more and more for the finished product.

With an ever-increasing volume of grain pouring into Chicago, old trade practices became inadequate. Standard grades had not been established. Lack of standard bushel weights fostered unfair practices in measuring. Storage facilities were inadequate. The only contracts were receipts from warehouses for grain of dubious quantity and quality. Bitter disputes often flared between buyer and seller.

Out of this chaos, one thing became clear — expanding production required a large, liquid, year-round central market for sellers and buyers of the new wealth of the prairie. The Chicago Board of Trade was founded in 1848 by 82 businessmen as a cash market. It was set up as a central meeting point in town for buyers and sellers. Trades could be made on the spot, hence the name "spot market."

The stated purposes of the Board of Trade were "to maintain a commercial exchange; to promote uniformity in the customs and usages of merchants; to inculcate principles of justice and equity to trade; to facilitate the speedy adjustment of business disputes; to acquire and disseminate valuable commercial and economic information; and generally to secure to its members the benefits of cooperation in the furtherance of their legitimate pursuits."[2]

One of the first activities of the Board of Trade was to appoint a committee to determine the quality of commodities being handled by members. Standardized definitions of grade quality, accepted everywhere in the world, were developed. Inspectors were appointed to grade incoming grain in conformity with Exchange standards. The method of buying grain by measured bushels was replaced by a weight system, and a weighing department was inaugurated to assess accuracy and integrity. In 1859, the state of Illinois legislatively set quality and quantity standards for grain trading. A governmental agency employed unbiased inspectors to make judgments on quality and enforce the standards. At this point, a grade designation such as U.S. No. 2 yellow corn meant the same throughout the trade, and grain could be bought and sold by description.

By 1860, forward contracting for future delivery of grain came into existence. Forward contracts were a vehicle for the seller and buyer to agree upon quality, quantity, and price for delivery of a commodity at a later date. Merchants located downstate along the Illinois River would buy surplus grain from local farmers at harvest, store it in their facilities, and forward contract its later sale to Chicago merchants. Forward contracts became the security instrument for financing grain held in storage.

Later, when river merchants brought grain to town, there was often a surplus and buyers would not honor the agreement. If there were a short supply and high price, merchants would default. The story is told of grain merchants going to Chicago with grain on the back of a wagon and a shotgun on the front. They would find the buyer and ask, "Which do I unload first?" Solution to the default problem was achieved with earnest money deposits by both buyer and seller, placed with a disinterested third party until delivery was made. These were the first margin deposits. Forward contracting, as a method of pricing, developed rapidly after adoption of this practice. Grain merchants were the first hedgers. They locked in a price to be received at time of delivery irrespective of whether cash market price rose or fell in the interim. By knowing their selling price, grain merchants carried less risk. They were willing to bid a relatively higher price to farmers for their grain.

As the practice of contracting for forward sales expanded, some buyers found that they had overbooked, while others were short of commitments. An active market for trading the forward sale contracts developed in order to match needs with availability; this practice laid the cornerstone of a futures contract market.

Trading in forward contracts led to two other practices. First, a clearinghouse was established to match buyers and sellers. In the past, merchants bringing grain into town to satisfy forward commitments had difficulty locating the buyer, since the forward contract might have been sold two or three times. The seller would be sent from place to place to find the holder of the contract. The clearinghouse provided a centralized clearing point to accept grain from the seller and match deliveries with contract holders. Warehouse receipts, as evidence of title, also came into being. With them, the grain merchant could deliver the warehouse receipt to the clearing corporation. In turn, the clearinghouse personnel knew who was to receive the grain; they matched the supplier with the buyer.

The second practice that evolved was speculative buying and selling of contracts. Individuals found they could buy excess contracts from an overextended buyer and later resell them to someone who was short of grain. Also, speculators soon learned they could sell contracts to people who needed grain at a later time and subsequently buy the cash commodity at, hopefully, a lower price to fulfill their contract commitment.

Maturity dates of the forward contracts and futures contracts changed gradually. Instead of contracting for delivery in 10, 30, or 90 days in the future, more distant delivery months were gradually agreed upon. Maturity dates for grains generally corresponded with important production or marketing months. July was the first month for cash trading newly harvested winter wheat. September was the harvest month for spring wheat. By December, final crop size was known and most of the demand factors for the year were known. December was also the last month of shipping via the Great Lakes until the spring thaw in March. December was the first month that the new corn crop was available at the terminals. March related to the reopening of Great Lakes shipping. During May, traders could estimate carryover and assess spring planting intentions. July figures reflected rate of use of old crop and provided a view of summer growing conditions. September was the last old crop month.

Soybean trading months evolved somewhat differently. At one time, November was the month the new crop was available, and September was the last old crop month. Harvest patterns changed to make October the time when the bulk of the new crop was available. Consequently, September became a combination of old and new crop trading forces, and July became the end of the old contract year. July proved to be too far from harvest and allowed for extensive supply fluctuation due to summer growing conditions. Finally, August was added as a soybean trade month.[3]

The business practices of forward contracting a future cash transaction, margin deposits to insure financial performance, clearinghouse matching of buyers and sellers, speculative trading, and fixed months of contract delivery paved the way for trading of futures contracts. By 1865, and the end of the Civil War, trading in futures contracts had become standard business practice. With forward contracts, every detail of delivery, quality, quantity, and price was negotiable. With futures contracts, all but one variable, price, was standardized. It was negotiated in open auction.

Two points can be gleaned from this historical perspective. First, it took many years for the futures market and its business operating practices to evolve. Implementation of one practice led to development of another. Second, futures markets became important to farmer-sellers on one end of the food chain — and consumer-buyers on the other end. Futures pricing tended to minimize price fluctuation by shifting the business practice of matching producers' supplies to the needs of users from trade people into the hands of specialists. Trading in grain movement had much less risk of financial loss. Farmers received higher prices; consumers enjoyed lower costs. The system worked more efficiently.

Exchanges

By 1865, the basic structure of the Chicago Board of Trade was as we know it today. Out of the use of forward sales, the futures market, with its specific futures contracts, developed early in the history of the Chicago Board of Trade. The CBOT provided futures markets in wheat, corn, oats, soybeans, soybean oil and meal, iced broilers, plywood, silver, and gold. In 1975, the CBOT launched the first financial instruments futures contract in Government National Mortgage Association (GNMA) mortgage-backed certificates. In 1977, two additional interest rate-sensitive markets, U.S. Treasury bond futures and 90-day commercial paper loan futures, were launched.

Other exchanges developed in other cities representing major transportation centers through which a substantial portion of commodities moved. In 1987, twelve organized commodity exchanges handling agricultural products were operating in the United States. Figure 3-1 lists the exchanges and their locations. The Chicago Board of Trade and the Chicago Mercantile Exchange were the largest, accounting for almost 75 percent of the volume of all commodity futures trading.

The MidAmerica Commodity Exchange differs from the other exchanges and offers particular advantages to the farmer. Traders on the MidAmerica floor watch the price quote boards reporting price action from other exchanges. Their trading among themselves is at premiums or discounts to the prices on the other exchanges, and their contract sizes are smaller. Their activity is not concerned with price discovery since they reference their trades to a quoted price, but they provide market flexibility in contract size. MidAmerica offers 1,000-bushel contracts in corn, soybeans, and wheat which are one-fifth the size of grain contracts on CBOT. It also offers a 15,000-pound live hog contract (about 70 head) and a 20,000-pound cattle contract (about 20 head). These are one-half the size of similar contracts traded on the Mercantile Exchange. The advantage for producers is that the mini-contracts offer the opportunity to trade in smaller units. A smaller farmer is not eliminated from market participation. Larger operators can more closely tailor market positions with actual quantities of production in 1,000-bushel units compared to 5,000-bushel units. Finally, farmers can execute their pricing decisions in more separate transactions with smaller units.

Commodity Exchanges conform to the definition of a market: They provide a place for trading, and they facilitate exchange. The exchanges are organized, for the most part, as not-for-profit corporations of member-traders authorized to buy and sell futures contracts for the public. Each exchange has a trading floor where buyers and sellers meet; a governing board sets and enforces rules of orderly trading; and a clearinghouse facilitates trading and delivering commodities. The organization of exchanges has been compared to a football game by T. A. Hieronymus: "The commodity exchanges provide the playing field and equipment; they write the rules; and they act as referee, head linesman and field judge."[4] But they do not handle the football. The players in the action are the hedgers and speculators.

What Is a Commodity?

The list of commodities traded on exchanges is extensive, diverse, and subject to change. Commodities are natural products, agricultural and manufactured products, currencies and financial instruments. A list at the end of this chapter provides detailed information on actively traded commodities and their futures contract specifications.

In order to qualify for futures trading, a commodity must possess the following characteristics:[5]

1. Chicago Board of Trade (CBOT)
   141 W. Jackson Boulevard
   Chicago, Illinois 60604

2. Chicago Mercantile Exchange (CME)
   and International Monetary Market (IMM)
   30 S. Wacker Drive
   Chicago, Illinois 60606

3. Coffee, Sugar & Cocoa Exchange (CSCE)
   Four World Trade Center
   New York, New York 10048

4. Commodity Exchange, Inc. (CMX)
   Four World Trade Center
   New York, New York 10048

5. Kansas City Board of Trade (KCBT)
   4800 Main Street, Suite 274
   Kansas City, Missouri 64112

6. MidAmerica Commodity Exchange (MACE)
   141 W. Jackson Boulevard
   Chicago, Illinois 60604

7. Minneapolis Grain Exchange (MGE)
   400 S. Fourth Street
   Minneapolis, Minnesota 55415

8. Chicago Rice and Cotton Exchange (CRCE)
   444 W. Jackson Boulevard
   Chicago, Illinois 60606

9. New York Cotton Exchange & Assoc. (NYCE)
   Four World Trade Center
   New York, New York 10048

10. New York Futures Exchange (NYFE)
    20 Broad Street
    New York, New York 10004

11. New York Mercantile Exchange (NYME)
    Four World Trade Center
    New York, New York 10048

12. Philadelphia Board of Trade (PBOT)
    1900 Market Street
    Philadelphia, Pennsylvania 19103

Figure 3-1: Futures Markets, United States.

1. It must be a homogeneous product such that every lot is the same and units are interchangeable.
2. It must be susceptible to standardization and grading so that well defined, uniform grades become essentially homogeneous.
3. Supply and demand must be large enough so that no single person can control flow in and out of the market.
4. Supply to the market must flow naturally and freely without artificial restraints from government or private agencies.
5. Supply and demand must be uncertain and constantly changing. If supply and demand were both certain, prices could be adjusted without an organized market. However, if supply and demand are large and both are uncertain and subject to wide fluctuations from year to year and season to season, then the forces of supply and demand will be in a constant state of flux. This interplay of uncertain economic forces produces the constant price fluctuations which must exist in a successful futures market.
6. The commodity must be storable in order to allow for delivery on futures contracts that are due months after harvest or production. Any item which deteriorates rapidly would not meet this condition.

Commodities that are not adaptable to futures trading include (1) perishable commodities such as fresh fruits and vegetables, (2) manufactured goods whose supply can be controlled and whose styles change, and (3) goods where supply has been monopolized by a few producers.

Obviously, livestock and livestock products do not satisfy all of these requirements. They are perishable and cannot be easily stored. Futures market trading in these commodities creates a different set of conditions and requires that a farmer or rancher gain a specialized understanding of the market for such goods. The peculiarities of futures trading in livestock will be discussed in Chapter 9.

ROLES OF FUTURES MARKETS

Commodity futures markets have three major roles to fulfill:

1. Price Discovery
2. Risk Transfer
3. Investment Medium

Exchanges provide the physical facilities and business framework for buyers and sellers to conduct trading. Worldwide demand and supply conditions flow into the trading pits and are interpreted by traders. Prices are discovered by the interaction of buyers and sellers.

Two types of people participate in the process — hedgers and speculators. Hedgers seek to transfer the effects of price change from their business, and speculators seek to profit from price change. Thus, roles of risk transfer and investment medium are fulfilled.

Futures Trading and Price Discovery[6]

What an item is worth is an age-old question of trade. Sellers obviously would like price to be high; buyers want it to be low. Somehow a mutually agreeable compromise must be reached so that both are satisfied with the transaction. In an immediate cash trade, value of an item becomes what the buyer is willing to pay and the seller is willing to take. At the time, both the Indians and the Dutch were obviously satisfied with their trade of $24 worth of trinkets for Manhattan Island. But values do change.

In a barter economy, price is established by haggling. In a more complex economy it is established by the buyer's willingness or reluctance to part with his or her money in trade for the item and the seller's willingness to take the money in exchange for the goods.

Assessing a futures price is more complex than assessing a cash price. The cash price reflects "what is" today, but a futures price reflects market conditions as traders anticipate they will be. Because of the time lag between pricing and delivering the commodity, and the uncertainties of events in the interim, wide differences of opinion always exist about the "correct" futures price. The differences of opinion and unpredictable events insure that there will be plenty of buyers (who think price will go up) and sellers (who think it will go down) willing to participate in a futures market.

A trader in the futures market operates something like this: (1) The trader determines what he or she believes is the "correct" price for a commodity in some delivery month in the future, based upon knowledge and interpretation of expected demand and supply conditions. (2) If the current price of the futures contract the trader intends to trade is higher than the expected "correct" price, he or she "sells" (goes short) a contract obligation to deliver the commodity at the future date. The trader intends to purchase the commodity to satisfy the contract after the price has come down to the "true" supply-demand price. (3) If the current price is lower than the expected "correct" price, the trader "buys" (goes long) a futures obligation with intentions of selling it when the price rises to the "correct" level. If the trader's judgment is correct and price changes as expected, he or she profits from the difference between the original and final prices. This profit represents a loss to some other trader whose judgment was wrong.

The actions of traders lead to establishment of the "true" supply-demand price for the commodity as they express their collective judgments. The trader who sells the futures contract, believing that the current price is too high, tends to lower price in the direction that it should go if this judgment is correct. The action of the trader who buys the futures contract tends to raise the price toward what it should be if this judgment is correct. The buying and selling transactions do not balance each other out, even though every sale has a corresponding purchase. Obviously, if they did, the price would not change. For a trade to cause a change in price, that trade must be a reflection of the weighted judgment of market participants on their willingness to buy or sell. If buyers are more aggressive than sellers, the price will move upward; if sellers are more aggressive, the price will move downward. In terminology of the market, bear and bull markets develop in response to traders' actions. Bull markets are made by aggressive buyers ("bulls") who are willing to charge on to higher and higher prices. Bear markets are made by sellers ("bears") who want to retreat into hibernation away from stronger price action. An aggressive buyer must entice a reluctant seller to enter into a transaction by offering a slightly higher price than presently exists. This action thus bids up price. A seller entices a reluctant buyer into a trade by offering to sell at a slightly lower price. Thus, this action tends to lower price. Either up or down, a price direction develops from a trader's willingness to execute a trade at a price different from the previous one.

After the original "open commitments" of speculative traders have been made, some of them hold "long" positions; that is, they have bought futures commitments that have not been offset by a counterbalancing sale. They are committed to take delivery of the commodity. Other traders are "short" in that they have made commitments not yet offset by a counterbalancing buy. They owe delivery of the commodity. The circumstances under which the traders are willing to dispose of their current holdings intensify the direction of price action once a trend begins to develop. As supply and demand conditions unfold and point to higher prices, the natural reaction of the long-position traders is to maintain, or possibly increase, their open commitments. There are always traders willing to sell, but only at a higher price. How much higher depends upon judgments by both buyers and sellers of the eventual effects of a new development on supply and/or demand conditions. As a stronger market unfolds, the reaction of short sellers is to cancel their previous sale by purchasing a corresponding amount as quickly as possible. Since the short is more anxious to buy than someone else is to sell, the price of futures contract is pushed upward further by the decision of shorts to offset their original. Thus, as price direction is confirmed, it tends to feed upon itself.

More eager buyers than willing sellers will raise prices, transforming some sellers into buyers. More willing sellers than buyers will lower prices, changing some buyers to sellers. Judgments expressed by buyers and sellers cause the current market price of the futures contract to approach the price it should attain, based upon currently available information relating to expected supply and demand conditions in some future delivery month. Thus, a price at which trades are made is discovered.

In the process of price discovery, the commodity exchanges themselves do not buy or sell, nor do they set prices. The role of the exchanges is like a thermometer which does not influence temperature but merely records it. As hot weather sends the thermometer reading up and cold weather sends it down, strong demand sends price up and heavy supply sends price down. "Through the commodity exchanges, supply, demand, and other price making forces are translated into a single figure — into a price."[7] An exchange does not set price but simply provides facilities where collective judgments of buyers and sellers are expressed as price.

How Trading Takes Place[8]. The most visible aspects of exchanges are the exchange floor and the trading pits where trading takes place. Less visible, but as important, is the communication system that links brokers throughout the world with traders in the pits. The commodity trading floor is a physical place, but it represents a worldwide market. Up-to-the-second reports of exchange prices are available across the land on video equipment, and most newspapers print daily price quotations of futures market trading.

A picture tour of the Chicago Board of Trade can serve to illustrate the exchange as a market place. Figure 3-2 is a sketch of the trading floor where members of the exchange buy and sell futures contracts in open, competitive bidding. Trading takes place in "pits," which are raised octagonal platforms with descending steps. The shape permits all buyers and sellers to see all others. Traders stand in the pit for the commodity they wish to trade, on steps or in groups according to the delivery month in which they trade.

Trading is officially signaled each business day by a gong from the opening bell. No futures trades may be made before it sounds the official opening, nor can

Figure 3-2: The Chicago Board of Trade, Trading Floor. Source: Besant, *Action in the Marketplace* 6. Used by permission.

any be made after it rings the close. During trading hours the trading floor is a beehive of activity as traders yell and signal their bids and offers.

At any moment there are thousands of buyers and sellers of futures contracts participating in the market, and there are even more potential participants. Anyone can participate. Contract orders from outside the exchange are relayed through brokers to member traders ("locals") on the floor. Batteries of telephones are located near the pits to handle the thousands of calls from commercial traders and brokerage firms. As orders are received by phone operators, they are time-stamped and rushed by messengers (runners) to traders in the pit. Exchange Rules and Regulations require pit traders to use open outcry to present bids and offers so that anyone who wishes to take the opposite side of the trade may do so. In addition, they use hand signals to clarify their verbal outcry. Position of the palm shows whether they are buying or selling. Traders hold the palm facing themselves when they want to buy and facing away when they wish to sell. Vertical fingers and arm positions indicate the number of contracts the trader wishes to buy or sell. For grains each vertical finger represents one contract of 5,000 bushels. Finger signals with the hand held horizontally relay the fraction of a cent that the trader bids or offers relative to the last trade as posted on the electronic quote boards. Figure 3-3 shows the one-quarter to full-cent designations of the finger positions in grain trades.

Once a trade has been confirmed with a nod, each trader lists the completed transaction on respective trading cards and the multi-part order form. A carded trade shows the number of contracts involved, the commodity, the maturity month of the contract, the price, the name of the clearing firm, the initials of the other broker, and an alphabetic symbol that indicates the half hour in which the trade was executed. The broker jots down the price and other pertinent information on the order form that is tossed from the pit to the messenger for return to the phone desk. The information is then relayed to the office in which the order originated so the customer can be advised that the trade was completed. The whole process often takes less than two minutes between the time of placing the order until confirmation. The ease and speed of transaction reflect the liquidity of the market. Commodity markets lose their effectiveness if they cannot execute trades quickly and at slight price deviations.

Price Quotations. Above each trading pit is a raised platform where two or more Exchange employees (market reporters) constantly watch the changing price picture. As trades are executed at new prices, a reporter records the new price and transmits it to the central quotations computer. The computer flashes the new price on the quotation boards at the exchange and through the communications network to the world. Prices are posted immediately and constitute "the market" until another trade is made at a different price.

Newspapers expand the availability of market information by printing commodity quotations daily. A report from *The Wall Street Journal* in Figure 3-4 summarizes a day's trading activity. Similar commodities are grouped together under headings of grains, livestock, food and fiber, metals and petroleum, wood, and other categories. A description of each commodity indicates its name, where it is traded, the number of units, and the unit price designation. The month of maturity of the futures contract is noted on the left column. "Open" is the agreed price for the first transaction of the day. "High" is the highest price of any transaction during the day; "Low" is the lowest trading price of the day. "Settle" refers to the

Figure 3-3: Hand Signals Used by Traders to Flash Bids and Offers. Source: Besant, *Action in the Marketplace.* Used by permission.

Figure 3-4: Daily Commodity Price Quotation. Source: *The Wall Street Journal*. Reprinted by permission of Dow Jones and Co., Inc. All rights reserved.

settlement price, which is the last trade, if there was only one. If frenzied trading at the closing bell creates a closing range of prices, officials of the exchange determine a settlement price, usually the midpoint of the trading range. "Change" refers to the increase or decrease in settlement price from the previous day's settlement price. "Lifetime high-low" reports the extremes in price at which a specific contract has been traded since its introduction. "Open interest" listed in the right column reports the number of contracts which have been initiated and are still in force for each contract month. "Est. vol." indicates approximately how many contracts were traded during the trading day reported. "Vol. Tue." (for example) gives the accurate count of previous day's trades. "Open interest" reported at the bottom of the table is the total number of open contracts for all months. The last number, either positive or negative, indicates the change in open interest from the previous day.

The Clearing Corporation. The Clearing Corporation of the exchange handles the mechanics of matching trades between buyers and sellers.[9] Each trader turns endorsed orders of completed trades over to the clearing firm, which in turn reports the trades to the Clearing Corporation. The Clearing Corporation substitutes itself as the opposite party to each transaction. It becomes the buyer to every seller and the seller to every buyer and is, in effect, a party to every transaction. What was initially a sale by A to B becomes a sale by A to the Clearing Corporation. B's purchase becomes a transaction with the Clearing Corporation rather than with A. Because of this substitution of parties, it is not necessary for the buyer (or seller) to find the original seller (or buyer) when he or she decides to offset his or her position. The trader merely executes an equal and opposite transaction, usually with an entirely different opposite party, to end up with a net zero position. The Clearing Corporation serves as the intermediary. By rule, all members of the exchange must clear their trades on the day the trade is made and settle disputes before trading resumes the next day.

The Clearing Corporation guarantees financial performance of all contracts traded and cleared. Financial integrity of futures trading is thus insured by the Clearing Corporation. Each trader must deposit original margins with the clearinghouse and maintain a minimum in the event of adverse price fluctuations. At the close of the trading day, all settling of accounts is done with the Clearing Corporation; brokers do not make settlement with each other. If wide fluctuations in price occur during a trading day, the Clearing Corporation can call for additional margins throughout the day without waiting for routine settlement. Thus, a trader is not concerned about possible financial embarrassment of other traders with whom trades have been made. Every trader's position is with only one party, the Clearing Corporation, which is financially responsible. The financial position of the Clearing Corporation is assured (1) by the original margin deposits on each contract; (2) by guaranty funds deposited by each clearinghouse member firm; (3) by the surplus reserve of the clearinghouse; and (4) by the joint financial responsibility of all Clearing Corporation members based on their proportion of the total business done over a period of time. "Even in periods of heavy trading and wide fluctuations, original margins alone have almost always been adequate to cover demands made on commodity clearing houses. No semblance of strain has ever been placed on their resources. Not a penny has ever been lost as a result of a commodity clearing house being financially unable to meet its obligations."[10]

Opening an Account. Commodity exchanges are generally membership associations. Only members owning seats on the exchange are permitted to participate in the auction on the trading floor. The largest participation in trading, however, comes from individuals outside the exchanges. Futures commission merchants represent these individuals on the exchange floor. They solicit customers from the public and maintain the commodity accounts of these customers.

The first step by an individual toward using the futures market as a marketing tool is to select a brokerage firm, to select an account executive (broker), and to open an account with the firm. Many highly visible financial firms offer commodity brokerage services. A reply to newspaper advertising, public meetings, or trade literature will elicit responses from them. A review of their advertising, market letters, and services available gives a feeling for how they operate. Most emphasize the quality of their research departments and their ability to fill orders quickly and efficiently. Since all of them are highly regulated and closely supervised by the exchange and the government regulatory authority (the Commodity Futures Trading Commission), there is little difference between their operating procedures. But farmers may prefer to deal with firms which specialize in agricultural commodities.

Brokerage houses charge commission fees for the service they perform. These fees ($25 to $100) cover both the initial and liquidating trades — a "round turn" trade. On March 6, 1978, commission rates became negotiable rather than fixed. Cut-rate brokerage firms offering fewer services, at lower rates, have sprung up as a result. Some of them advertise rates as low as $15 per "round turn" trade.

Selecting an account executive (broker) is probably more important than selecting the firm. He or she is the individual who will provide day-to-day contact with the market for the customer. Account executives are usually called brokers, but it should be remembered they are distinctively different in function from floor traders who are also referred to as brokers. Account executives (brokers) come in all shapes, sizes, colors, and philosophies. Some are highly aggressive in providing a continuous flow of market tips and information; others are reserved in their recommendations. Personal compatibility and acceptance of customer objectives are important traits between the customer and broker. They should enjoy working together.

After choosing a brokerage firm, the customer fills out information forms and signature cards required by the firm to open an account. The customer must specify whether he or she is opening a hedge or speculative account. These accounts have different margin deposit requirements and tax implications. Hedging account margins are typically half the amount required for speculative accounts.

Tax Treatment of Futures Transactions. Whether a futures market participant is a hedger or a speculator affects his or her federal income tax status. Transactions that reduce price risk of commodities used or produced in a taxpayer's business are considered, for tax purposes, legitimate business transactions. The costs of such transactions, including losses sustained, are deductible in full amount as a business expense. Likewise, profit realized from hedging in futures contracts must be included in the gross income of the taxpayer. Gains or losses in futures positions are thus treated as ordinary income for hedgers. Proof that trades were for hedging rests with the trader. Even though a trader identifies the transaction as a hedge, the trader must be able to prove that the purpose for entering the market was for price protection on commodities that were part of the business operation. Accurate inventory records should be maintained to show the actual cash position at the time of the hedge.

Commodity futures contracts traded for speculative purposes are considered capital assets for income tax purposes.[11] Speculative transactions produce either short-term or long-term capital gains or capital losses. At the end of the year, gains or losses are allocated 40 percent to short-term and 60 percent to long-term categories. Commodity gains and losses are added to gains or losses in other investments for tax calculation purposes. Net gains from commodity speculation must be included in full as gross income, but a speculator is limited to a total capital loss deduction (from all investments) of $3,000 per year. Losses greater than $3,000 may be carried forward to later tax years. At the end of the tax year, open traders are "marked to the market" to eliminate the effect of unrealized losses or gains of switching income between tax years.

Margins. An initial margin deposit must be made with the brokerage firm before a trading order can be placed. Margins on commodity trades are money deposits by the customer to guarantee financial integrity. Inasmuch as the brokerage firm must deposit and maintain cash margins with the Clearing Corporation, the brokerage firm protects itself by securing original margin deposits from the customer. The customer is asked to place from 5 percent to 20 percent of the total value of the commodity contract involved as original good faith money. Minimum margins, as set by the various exchanges, vary in amount between commodities in accordance with prevailing price and price volatility.

Brokerage firms usually require a margin deposit larger than the exchange minimum. The margin money is deposited in a segregated fund at the brokerage firm's bank. A segregated account means that the money will be used only for margin and not for expenses of the brokerage firm. Most firms do not pay interest on the account balance to the customer, but some do offer money market or T-Bill accounts.

After the original margin is deposited and a trade is made, a minimum margin equity must be maintained in the event of adverse price fluctuation. The broker reserves the right to call for additional margin, called maintenance margin, at any time to maintain the customer's minimum equity. The initial margin is usually sufficient to cover small fluctuations, but the trader must be prepared with liquidity to satisfy a call for additional margin should the market move against his or her position. Otherwise the broker has the right to close out the position on behalf of the customer. The broker also reserves the right to raise margin requirements to the customer if the price changes suddenly or drastically and the broker feels a higher margin would be more consistent with the general safety of the account.

On the other hand, if the customer's transaction shows a profit, the customer can ask for and receive the surplus balance in cash from the broker without closing out the futures position. Usually, the customer should leave only enough cash in the account to maintain the current minimum margin requirements of the broker. The surplus should be withdrawn and placed in an interest-bearing account.

Types of Orders. A market participant can place an order for a futures transaction with the broker in several ways. Usually, a market participant places a "market order" which instructs the broker to execute the trade at the most advantageous price obtainable at the moment the order reaches the pit. The customer will not know the execution price until it is confirmed by the broker.

A "limit order" instructs the broker to buy or sell only if the market price reaches a certain price. The order will not be filled until the price reaches the limit specified by the customer. A trader who gives a limit order to sell 5,000 bushels of May wheat at $4.50, for example, is saying in effect that he or she does not want to sell unless the price is at least $4.50. Such an order is given when the market is trading below $4.50. To protect against the event that the market may have exceeded the $4.50 price by the time the order arrives, the customer may add an "OB" (or better) designation. It insures that the trade will not be filled at a price lower than $4.50 and will be filled above that level if the market is strong. Limit orders are usually considered good only during a specified trading session, but an order may be marked "GTC" (good till canceled) to indicate that it is a standing order until canceled by the trader.

"Stop-loss" orders are used by traders who want to limit their risk or stop their losses. For example, a trader who is "long" at $4.20 may enter a stop-loss sell order at $4.15. The trader hopes that the market will rise; but if it slumps instead, he or she does not want to hold the position if the market declines below $4.15. By the same token, a trader who is "short" may enter a stop-loss order to buy if the market advances to the price level specified. Stop-loss orders are entered by traders who want to pre-set how much loss they are willing to incur before they will cancel their position, but they are not close enough to the market action themselves to be ready to place market orders when their price is reached. A stop-loss order does not guarantee, however, that the order will be filled at the price specified. When a market is moving fast, a stop-loss price may be reached and passed before the floor broker can execute the trade. The price may jump over a sell-stop or fall right through a buy-stop. Accordingly, the broker cannot be held responsible for filling stop-loss orders. Since such an order immediately becomes a "market order" once the stop-loss price has been reached, the sale may be made at a price lower — or the purchase higher — than the order. One of the best uses of stop orders is to initiate a position at a particular price the customer wants rather than what the market is offering. There is no assurance the order will be filled at the objective price, but it usually will be within some reasonable range. It gives the trader control over the price at which to enter the market.

By adding "MIT" (market if touched) to a limit order, the customer specifies that the order will have the effect of a market order when the limit is reached or touched even though price may move away quickly. This order can be useful in a thinly traded market that has a particular price objective as defined by some method of analysis. The customer wants an order filled near the objective price, but executing the order is more important than exact price. If trading touches the stated price but quickly moves away, the broker can fill it as a market order as soon as possible.

"Spread" orders initiate simultaneous long and short positions. The trades may be placed in related months of a single commodity, in products of a commodity, or between related commodities. For example, a spread order would buy one contract month (December corn) and sell another contract month (March corn). It might specify buying soybeans and simultaneously selling soybean oil and soybean meal. A related commodity spread might buy live cattle and sell live hogs. Traders place spread positions, because they expect the relative prices to change. They feel more confident predicting the difference between prices than predicting their price level. Whether the general price level moves up or down, the spread position will show a profit if the relative prices converge or diverge in the direction the trader expects.

A spread order is useful for farmer-hedgers who are rolling their positions forward out of a maturing contract month into another more distant one. Sometimes they prefer a maximum spread between the two contracts; sometimes they prefer a minimum. Rather than placing one market order to buy and another to sell, they specify a spread they want to achieve. For example, the order would state, "buy Feb. hogs and sell July hogs at a $2.50 premium on the Feb." This order instructs the broker to execute the trades whenever the spread between the two contracts reaches the stated difference.

A customer should never assume that an order has been filled until an oral confirmation or paper from the broker is received. A ticker-tape or a price-reporting board can run several minutes behind and does not determine whether a trade was executed or not. Until the customer receives confirmation, he or she should never re-enter an order against that position. Paper confirmations always follow oral confirmations. The written confirmation of a trade should be checked for accuracy in type of commodity, month, trading price, and quantity of contracts. Any discrepancy should be reported to the brokerage firm immediately, regardless of whether there was a profit or loss associated with the error. If the customer does not receive written confirmation of a trade following an oral confirmation, the customer should contact the account executive immediately to correct the oversight.

All mechanical details and records of a transaction are handled by the brokerage firm for its trading customers. For every trade made, the customer should receive a confirmation, and for every close-out a profit and loss statement showing the financial results of the transaction. In addition, the broker should send the customer a monthly statement showing the ledger balance, the open positions, the net profit or loss in all contracts liquidated since the date of the previous statement, and the net unrealized profit and loss on all open contracts figured at the market settlement price on the last business day of the month.

Risk Transfer

Futures markets transfer price risks from those who wish to avoid them to those who make it their business to assume those risks. Use of futures contracts permits hedging to minimize the effects of price change. Futures markets are thus an important tool for risk management.

Risk and Uncertainty. These are inherent with producing and owning agricultural products. Risks can be classified into two types:

> Physical — quality and quantity variation in commodities as affected by natural conditions
>
> Price — rising or declining value of production or inventory due to changing economic conditions

Farmers have faced the threat of insects, disease, and weather for centuries, and they continue to expose their production efforts to these risks. But modern farmers have also become production masters. When Columbus came to America, approximately 1 million Indians lived in the area that was to become the United States. The largest portion of their energies was directed toward providing food for their families. As America's original farmers, they fished, hunted, and raised crops of corn, yams, tomatoes, beans, and squash. These Indians barely subsisted, even though most of their time was spent gathering food. Today, there are about three times as many farm workers in the U.S. as there were Indians in Columbus's time. Today's farmers provide 246 million Americans with an infinite variety of quality food and have enough left over to be the world's largest food exporter. Two centuries after Independence, the U.S. became the world's breadbasket.[12] Farmers have become experts at overcoming production risk.

Economic conditions of the 1970's and 1980's added a new dimension that farmers had to cope with: price risk. Wide fluctuations in commodity prices mean a potentially wide margin for profit or loss. When a commercial family farm of 600 acres faces gross income fluctuations of $60,000 annually due to price variability, a new form of risk threatens financial security and survival on the farm. Successful production alone is not adequate for business success; profitable marketing is also required. The risk challenge of the 1990's is not production risk, but price risk.

Production and Product Risk. Participants in the food chain from production through marketing face the possibility of loss from natural hazards that may destroy or damage the production or inventory on hand. Each firm accepts many of these risks as part of the production process and accounts for them as a cost of production. However, to the extent possible, many farmers transfer physical risk to an insurance company that specializes in accepting risk for a fee. Buying insurance is often more economical for an individual than attempting to provide for his or her own protection. Farmers are accustomed to purchasing coverage for stored commodities against fire, wind, and other natural hazards. Crop insurance permits farmers to protect anticipated production.

Another risk every producer and marketer faces is physical deterioration of stored commodities. A rapid change in temperature, a breakdown in equipment or facilities, and moisture or insect damage to stored commodities are risks that cannot be shifted. This threat of loss is one of the factors that increases the cost of owning perishable commodities.

Price Risk. The combination of inventory ownership and uncertain price creates the need for price risk management. Value loss, arising from price change in agricultural products, may originate from many sources. An elevator operator who purchases large amounts of grain for storage in the fall, at what he or she considers low prices, may face a highly unfavorable spring market because of unforeseen declines in general business conditions. An untimely frost early in the fall or late in the spring may greatly increase the value of current orange juice stocks by sharply reducing the growing crop. A series of rains may make or break the wheat crop, greatly affecting the fortunes of wheat farmers producing a crop or holding large inventories of grain in storage.

Various marketing devices either minimize price risk or shift it from one person to another. Vertical integration of production and marketing reduces or transfers risk. Poultry production contracts by farmers typically specify a fixed return and thereby transfer risk of changing broiler prices to the integrator. The integrator in turn may attempt to obtain control over the kind and amount of production coming forth to reduce supply risks. Selling in advance of product availability eliminates price risk. Farmers can forward contract anticipated production of grain with their elevator for fall delivery. Elevators, in turn, may book delivery later in the season from their storage with a terminal elevator or exporter.

Government price-support programs transfer price risk from producers to taxpayers. Non-recourse loans and deficiency payments establish minimum prices for farmers who participate in government programs. Government purchase, sale, or storage of commodities also helps reduce variation in available market supplies and thereby reduces price volatility. Since price fluctuations can also arise from inadequate or inaccurate information, government efforts that improve the gathering and dissemination of market news and statistics help reduce this risk. Equally informed buyers and sellers can more closely negotiate a true price.

Each of these devices helps reduce farmer price risk by either minimizing the risk or transferring it to someone else.

The futures market provides the mechanism for shifting price risk to individuals outside the commodity production and marketing chain. Buying or selling futures contracts allows producers, owners, and buyers of commodities to hedge against price change. The effects of price change are accepted by speculators.

Hedging. Hedging is defined as taking opposite positions in two different markets with equal quantity commitments so that losses in one market are offset by gains in the other. The two markets are the cash market and the futures market.

Hedging provides protection against price change for both sellers and buyers of commodities. The owner of a physical commodity hedges against price change by "selling" futures contracts equivalent to his or her production or inventory of the same commodity. This is called a "short" hedge. It is used by farmers, elevators, processors, and exporters. They "own" the commodity and they "owe" delivery of it against a futures contract. Commodity buyers "buy" futures contracts to protect against a possible advance in price of the commodity not yet owned but needed to fill feeding, processing, or export commitments. This is a "long" hedge. Long hedgers "own" the right to receive delivery of the commodity, but they "owe" the physical commodity to their production process or buyer.

Who Uses Short Hedges and How. Hedging is not an automatic operation with a stereotyped procedure. Each farm or firm hedges on the futures market in a manner that corresponds to its respective cash position. Following are some of the trade interests who use short hedges when they wish to be protected against possible market price declines:

    a. The grower, who sells futures contracts against crops or livestock in production.

    b. The farmer-merchandiser, who sells futures contracts against grain in storage.

    c. The grain elevator operator, who sells futures against current purchases from farmers, while awaiting an opportunity to sell to consuming, processing, or export interests.

    d. The exporter, who sells futures against purchases of commodities.

    e. The processor, who sells futures against purchases of actual commodities pending resale in the form of a manufactured product.

    f. The manufacturer, who sells futures against unsold manufactured products.

Who Uses Long Hedges and How. Producers, shippers, importers, manufacturers, and others use the futures market to place "long" hedges when they wish to be protected against possible price advances in commodities they intend to buy. Following are some of the trade interests who use the "long" hedges:

   a. The livestock producer, who plans to purchase grain and feed supplements.
   b. The cattle feeder, who intends to buy feeder animals.
   c. The processor, who buys futures against possible price rise in required raw materials.
   d. The importer and merchant, who makes forward sales for specific shipment or delivery and buys futures as protection against market changes pending the purchase of the actual product.

Selling Hedge. One way to understand the "equal but opposite" requirement of hedging is to trace the results through examples. Consider a farmer who expects to produce 100,000 bushels of corn for November delivery. About mid-July the price level is high enough to cover his production costs and yield a reasonable profit. The producer contacts the local country elevator to find that the current cash price quoted for new crop corn is $3.30 per bushel.

Cash Market	Futures Market
July 19	
Price of New Crop Corn $3.30/bu. 100,000 bu. of expected production	No Position

The producer does not yet have the corn harvested. However, since the quoted price meets a pricing target, the opportunity presents itself to eliminate the risk of declining price between now and the time he expects to deliver corn. Setting a hedge will protect the corn crop against price change before harvest. The corn grower contacts a commodity account executive and places a short hedge in the December futures (currently trading at $3.55) per bushel by selling 20 futures contracts. As a hedger, the farmer has an account which now looks like this:

Cash Market	Futures Market	Price Spread (Basis)
July 20	July 20	
New Crop Corn $3.30/bu.	Sell 20 Dec. Corn Contracts @ $3.55/bu.	$0.25 under Dec.
100,000 bu. of expected production	100,000 bu.	

Let's move forward in time to November, when the corn is ready for delivery. To lift the hedge, the farmer simultaneously offsets the futures obligation by buying 20 December futures contracts and sells the corn for cash by delivering it to the local elevator. The hedger could have fulfilled the short futures obligation by delivering corn against it, but for a corn grower in central Minnesota it is usually more convenient to buy back the futures and sell the cash corn locally. In practice, few futures contracts are settled by delivery.

If the price of corn declined, as the farmer expected, the cash price was protected because the same market forces affected both the cash and futures prices of corn. If both declined by an equal amount, meaning that the basis remained the same, the farmer would realize the $3.30 per bushel cash price he was trying to protect.

The accounting follows:

Cash Market	Futures Market	Basis
July 20 New Crop Corn $3.30/bu.	Sell 20 Dec. Corn Contracts @ $3.55/bu.	$0.25 under Dec.
Nov. 20 Sell 100,000 bu. Cash Corn @ $3.14/bu.	Buy 20 Dec. Corn Contracts @ $3.39/bu.	$0.25 under Dec.
Result: $0.16/bu. loss in value	$0.16/bu. gain on futures	Change $0.00

Cash sale price $3.14
Futures gain   +0.16
Realized sale price $3.30

The hedger realized $0.16 per bushel less in the cash market than the desired price of $3.30 per bushel. However, he recovered that $0.16 from the gain in the futures market position. This represents a perfect hedge, with both prices declining exactly the same amount. They moved parallel to each other over time; i.e., no change in basis.

The price levels might have risen following the placement of the hedge. What would happen if both cash and futures prices rose by equal amounts of $0.30 per bushel? The accounting would look like this:

Cash Market	Futures Market	Basis
July 20 New Crop Corn $3.30/bu.	Sell 20 Dec. Corn Contracts @ $3.55/bu.	$0.25 under Dec.
Nov. 20 Sell 100,000 bu. Cash Corn @ $3.60/bu.	Buy 20 Dec. Corn Contracts @ $3.85/bu.	$0.25 under Dec.
Result: $0.30/bu. gain	$0.30/bu. loss	Change $0.00

Cash sale price $3.60
Futures deficit   −0.30
Realized sale price $3.30

The farmer had a deficit of $0.30 per bushel on the futures market transaction because the December futures were repurchased at a higher price. However, since the cash price of corn also increased by $0.30 per bushel, the cash market gain offsets the futures market loss. The important result is that the corn grower received the desired $3.30 per bushel for the corn.

Obviously, the farmer would have received a higher net price by not hedging. However, when the hedge was initiated on July 20, the farmer was satisfied with the $3.30 per bushel price and was concerned that a drop in price would have serious financial consequences. Therefore, the hedge protected the profit margin that the $3.30 per bushel net price offered. The 30 cents per bushel foregone was the "cost" of insuring a $3.30 price.

Buying Hedge. The buying hedge fixes input costs by protecting against a possible rise in the commodity's cash price. When a hedger places a "long" or buying hedge, the first transaction in the futures market is a "buy." The futures market "purchase" is a temporary substitute for the actual purchase of the cash commodity the long hedger intends to acquire at a future date.

Soybean meal is an important feed ingredient in broiler production. Therefore, poultry producers face significant risks from rising feed costs. Producers are able to control part of their feed costs by placing "long" hedges in soybean meal futures.

Suppose that in early October a broiler producer projects soybean meal needs for January to be 100 tons. Cash meal is currently selling at $238 per ton. The producer likes the price but has no storage space. Concerned about rising meal prices, she decides to hedge by buying a single contract (100 tons) of January soybean meal. On October 4, the January soybean meal futures are trading at $243 per ton. The accounting follows:

Cash Market	Futures Market	Basis
Oct. 4		
Need 100 tons of meal in Jan. Price quotes @ $238/ton	Buy 1 Jan. soybean meal contract @ $243/ton	$5 under Jan.

Assume that during the intervening months there was a greater than anticipated demand or a smaller than expected supply, causing soybean meal prices to rise. On December 30, the poultry feeder buys meal to meet feed needs and lifts the hedge.

What happened if both soybean meal cash and futures prices rose by $8 per ton? The accounting looks like this:

Cash Market	Futures Market	Basis
Oct. 4		
Soybean meal quoted @ $238/ton	Buy 1 Jan. meal contract @ $243/ton	$5 under Jan.
Dec. 30		
Buy soybean meal @ $246/ton	Sell 1 Jan. meal contract @ $251/ton	$5 under Jan.

Result: $8/ton higher cost (loss)	$8/ton gain	Change $0.00

Cash purchase price $246/ton
Futures gain (lower cost)   −8
Actual price $238/ton

In this example, the futures market gain exactly offsets the cash market loss (increased cost) because the basis was the same, $5 under January, when the hedge was placed and lifted. The poultry feeder would have paid $8 more per ton without the hedge.

What would happen if both cash and futures prices fall by an equal amount? The accounting follows:

Cash Market	Futures Market	Basis
Oct. 4 Soybean meal price quoted @ $238/ton Need 100 tons of meal in Jan.	Buy 1 Jan. soybean meal contract @ $243/ton	$5 under Jan.
Dec. 30 Buy soybean meal @ $233/ton	Sell 1 Jan. soybean meal contract @ $238/ton	$5 under Jan.
Result: $5/ton lower cost (gain)	$5/ton loss	Change $0.00

Cash purchase price $233/ton
Futures deficit (higher cost)   +5
Actual purchase price $238/ton

Without a change in basis, the futures deficit is exactly offset by the cash gain. In this situation the unhedged poultry feeder would have paid less. However, since the objective was to fix input cost, hedging fulfilled that objective.

These are examples of perfect hedges — the basis is the same the day the hedge is initiated and lifted. These examples are used to illustrate how hedging reduces price fluctuation risk. The price level may vary considerably but, as long as basis does not change, the price objective will be realized through hedging.

Hedging fixes a price. The opposite positions in the cash and futures markets eliminate the effects of price changes, both up and down. It protects the hedger against adverse price movement, but at the same time, it also eliminates the opportunity to enjoy favorable price changes.

Financing Price Protection. Where does the money for price protection for the hedger come from? Every buyer of a futures contract must have a seller, and every seller must have a buyer. Futures trading is a zero-sum game: If someone profits from a futures trade, someone else must lose.* For a short hedger who

---

*Actually, it is a negative-sum game because of transaction costs and commissions.

judges price to be high and likely to go lower, there may be a long hedger who believes it is low and likely to go higher. One enters the market as a seller and the other, a buyer. If the price level declines, the short hedger enjoys the protected high price, and the long hedger incurs higher input cost than later markets offer. If price level increases, the short hedger pays the piper by having foregone the opportunity of higher price, and the long hedger enjoys the protected low price.

Seldom does a farmer offering 20 corn contracts as a hedger-seller match with a corn processor buying 20 contracts as a hedger-buyer. Most likely, the other party to any hedge trade is a speculator. For the short hedger in a declining market, the speculator finances the price protection by suffering losses in the futures position. In a rising market, the short hedger pays the speculator the extent of the price rise in the futures market. Only the speculator profits or loses from being correct or incorrect in judging direction of price change. The hedger receives a fixed price, neither profiting nor losing from the hedge position. The changing value of the hedger's inventory offsets the gain or deficit in the futures portion of the hedge. For a short hedger, a falling market generates declining inventory values that are offset by gains in futures. In a rising market, a deficit in futures position is offset by gains in inventory value. In the end, when the hedge is canceled by selling cash and simultaneously buying futures, the net price to the short hedger will be what he or she had decided would be satisfactory when the hedge position was initiated. Selling hedgers give up the opportunity to enjoy higher price in return for protection from experiencing lower price. Buying hedgers give up the opportunity to buy inputs at lower price in return for protection against having to pay higher price.

Hedging fixes a price (ignoring basis change) that will be realized eventually. Hedging enables the producer to shed the effect of price change he or she would otherwise have to carry; an uncertain outcome is replaced by a certain one for the hedger. The speculator either profits or loses from being right or wrong in his or her position; speculators do not have the offsetting inventory value of the commodity. Speculators finance the hedgers' price protection.

Types of Hedging. Hedging can be categorized into three types: routine hedging, selective hedging, and carrying-charge hedging.

Routine hedging is a common practice in the U.S. grain and cotton trade. It "covers" the trader's position in the cash market by executing offsetting transactions in the futures market. Routine short hedging is attractive to an intermediary, elevator, processor, or manufacturer whose main business activities require holding stocks in inventory. They do not wish to risk business profitability in the face of uncertain price movements. Routine hedging assumes no need for forming judgment on price movement. When inventory is on hand, it is automatically hedged, either in futures or in forward contracts. Elevators which buy grain from farmers at a price, add on an operating margin, and contract sale at the higher price essentially practice routine hedging by contracting. Banks encourage routine hedging when they require that borrowers' inventory be hedged so that their solvency is not impaired by a drastic fall in market price.

Selective hedging differs from routine hedging because the trader exercises judgment about expected price changes. A selective short hedger hedges only when prices are expected to fall. A selective long hedger hedges only when he or she expects a rise in price. According to the confidence in one's market judgment, the extent of one's cash commitment, one's ability to withstand losses, and the likely cost of hedging, one may hedge all or only part of the cash commitment.

The selective hedger sometimes takes hedging action to reduce exposure to price risk and sometimes does not. The selective hedger's decisions on when and how much to hedge depend heavily on personal price expectations. This type of hedging should be termed "selective hedging/selective speculating." The secret to being successful in a selective hedging program is having the tools and/or insight to make profitable decisions as to when to hedge and when to be unhedged.

A selective hedger enters a futures position for the same reasons as a routine hedger. However, the selective hedger is not committed to that position unalterably until the cash market risks have been satisfied. A $3.00 December futures price may be very attractive for a farmer planting corn in May. However, dry weather that pushes prices to $3.30 by mid-July may encourage the hedger to reevaluate the original position. If the farmer anticipates that a smaller corn crop could push prices to $4.00, he or she would lift the hedge at $3.30 in mid-July. The net effect, if correct, is $3.70 corn in a $4.00 market. Neighbors who are cash market speculators enjoy $4.00 corn prices because they were not hedged.

Both routine and selective hedging represent responses to uncertainty about changes in market price. They attempt to fix price level and reduce uncertainty. The third type, carrying-charge hedging, differs from the first two types because it is not a response to market uncertainty. Carrying-charge hedging takes advantage of expected favorable changes in the relative prices of the futures market and the cash market. Short hedging of this type is used by those who expect the spread between two prices (basis) to narrow so that the overall result of the hedged transaction will be profitable. A profitable outcome can be anticipated when the hedger expects the favorable change in basis to exceed the costs of carrying inventory over the period involved. In carrying-charge hedging, the hedger reacts to prospective changes in the basis, rather than the expectation of changes in price level. This hedger does not care whether price level rises or falls; the bet is that basis will narrow. Carrying-charge hedging is used to capture a storage profit. It is a common practice among grain elevator operators. It should be practiced by farmers who own on-farm storage facilities. Its implementation will be developed in Chapter 7.

Hedging Hazards. Hedging with futures contracts can be hazardous. Some hazards that producers face are:

1. Changing basis,
2. Failure to distinguish between hedging and speculating,
3. Hedging too early in rising markets or too late in declining markets,
4. Inadequate liquidity to finance margin calls,
5. Production shortfalls.

In the hedging illustrations outlined above, it was assumed that the price difference, or basis, between cash and futures remained the same at all times. This is not usually the case. The difference may narrow or widen, as one price changes relative to the other. Such disparity prevents price protection through hedging from being 100 percent perfect at all times. Hedging with futures permits fixing price level but does not fix basis. Changing basis is a variable that can either improve or reduce the price protection of a hedge.

Failing to accurately distinguish between hedging and speculating is a hedging hazard. The definition of hedging requires **equal but opposite** positions in two markets to counterbalance the effect of price change. Without fulfilling both requirements, a commodity user or buyer is a speculator. A farmer who places seed in the ground or livestock in pens is speculating in the cash market that he or she can sell the output at a profit. The farmer who sells coffee futures or buys silver is a futures market speculator. A hedger who makes a cash sale but does not cancel the futures position is speculating in futures. A producer who buys futures while owning cash commodities is a "Texas hedger." Texas hedgers have two long positions (one in the cash, one in the futures) and are doubling the effect of price change rather than removing it. A commodity producer or user is not perfectly hedged unless there is a futures market position equal to, but opposite, the cash market position. This requirement is a serious impairment to farmer use of futures in hedging growing crops or anticipated livestock production. The growing production is not yet on hand; actual production may not equal expected (hedged) production.

At times profits are increased by hedging; at other times they are reduced. A seller who hedges before the market top or a buyer who hedges before the market bottom will receive less benefit from a hedge than one established at the market extreme. If price level rises after a hedge is initiated, a seller will receive a lower net price than the market later offers. If price level declines after a buying hedge is established, the net cost will be higher than that of not hedging. Without routine hedging, the hedger must decide before the fact that he or she will be satisfied with the price offered by the hedge. The hedger gives up the opportunity to realize a more favorable price for the comfort of avoiding an unfavorable one.

Cash liquidity to meet margin calls poses a hazard for hedgers. Since only one of thousands of market participants will place hedges at the market top or bottom, all others must be ready with cash to finance holding their position if the futures move against them. At the time of the cash sale, a more favorable cash price will cover futures account deficits, but the futures market will not wait for the cash sale to occur. It settles accounts every day and does not finance negative balances. The role of the banker in providing market liquidity for a hedger is important and will be discussed later.

Hedging sometimes fails the farmer because of yield fluctuation. A farmer who hedges corn at $3.00 per bushel may see a widespread drought reduce production and push prices to $4.00. Caught by poor weather, anticipated production of 100,000 bushels might be reduced to 70,000 bushels. The $1.00 inventory value gain on only 70,000 bushels does not offset the futures deficit of $1.00 on 100,000 bushels. Hence, farmers are reluctant to hedge 100 percent of their anticipated production without some recourse for exiting their hedge.

Hedging is not a simple process. It requires understanding, planning, patience, and perseverance. Without them hedging can increase risk rather than reduce it.

Investment Medium

Three major types of participants use commodity exchanges.

1. Hedgers, who may be short or long, as noted above.
2. Large speculators: businesspersons who make trading in futures their principal business. These professional speculators are ready at all times to buy or sell futures contracts, provided that the price offers the prospects for profits.
3. Small traders: the "public" whose gambling instincts draw them into speculation in search of a "quick kill."

Hedgers and futures market speculators are in two different worlds. One depends upon the other, but each has its respective objectives.

Futures market speculators neither own the commodity nor expect to take delivery of it. They are willing to assume the risk of price change in the hope of profiting by it. Speculators do not care whether price is high or low; they merely want to be right in gauging the direction of price movement. The speculator is, in effect, the risk bearer who assumes the risks which the hedger seeks to transfer. Speculators are like insurance underwriters. They stand between producers who want the highest price for their production and users who seek the lowest price for inputs. Without them, the farmer anticipating 100,000 bushels of corn for sale for November delivery might not have a market at favorable prices. Farmers do not care whether they sell to a processor, cattle feeder, grain merchant, or speculator. Their single-minded objective is to sell for the highest possible price. But in times of high price, user-buyers may prefer to delay their decision in hope that price will decline. The farmers' search for buyers then shifts to the commodity exchange, where they find speculators who are willing to take the other side of their trade. By providing market liquidity, speculators perform an important economic function.

Participants in Futures Trading. An interesting set of characters exists in the drama of the action on the floors of the commodity exchanges. They were described by Hieronymus by the type of trading in which they specialize.[13]

"*Scalpers* busily inject themselves into as many trades as possible. They stand ready to buy a quarter cent below the last trade or quarter cent above. They follow the principle that an incoming buy order will raise the price slightly, or conversely, an incoming sell order will depress price slightly, but shortly, price will return to the previous equilibrium level. Scalpers are always ready to buy or sell at small price concessions; they bridge the time gap in the flow of orders. They add liquidity to the market and enable orders to be filled without delay.

"*Pit Traders* are grown up scalpers. They come into the pits each day with no positions and leave with no positions. Their objective is to profit from intra-day price changes. They buy when they think price is going up and sell when they think it is going down. They watch the other traders in the pit trying to anticipate what trades others will make before the end of the session. Pit traders want to have a handful of contracts to trade at the close. They collect lots of 5,000 bushel trades and dump them in 100,000 bushel trades. At one time they may trade with the market; at another, against it. They try to sense the mood of the market and buy or sell quickly to get in on the bottom or top of the move and liquidate after the move is made.

"A third group of traders are *Spreaders*. They hope to profit from abnormal price differences between commodities, between delivery months, or between commodity markets. Spreaders may buy December cotton and sell May cotton; they may sell soybeans and buy oil and meal; they may buy Kansas City wheat and sell Chicago wheat; they may sell live cattle and buy live hogs. Spreaders practice arbitrage in hope of profiting from skewed price relationships that move back to normal relative value. In the process they add liquidity to the markets and stability to price relationships.

"*Floor Traders* are professional speculators who trade mostly for their own account. Their positions are flexible. They may trade with the market or against it, hold positions for long periods or short periods. They often take large positions and add to them as they sense they are right about direction of the move.

"*Floor Brokers* are the largest group on the exchange floor. They execute orders that flow through brokerage houses from outside the exchange or are placed with them by floor traders. Each broker holds an individual seat, but it may be financed by a firm, so in essence, they work for the firm. They act as agents and are paid a fee rather than a salary for each contract they execute. They can and do trade for their own account, but they must execute customer orders first. The broker might buy 25,000 bushels for his or her own account and then get an order to sell 500,000 bushels for the firm. The broker must execute for the firm first, then execute his or her own, no doubt after the price has fallen. They have decks of orders tucked in their fingers. Some are sell orders; and some, buy orders, all at different prices. If they make a mistake in executing a trade, the loss is covered out of their pocket. If an order was for 10,000 bushels and the floor broker sold 20,000, the broker makes up the difference if losses result."

For the observer in the visitors' gallery, the activity on the floor resembles a beehive. With hundreds of brokers shouting and gesturing wildly in the pit, each attempting to effect a quick sale or purchase before the price changes, the scene looks more like a yelling convention than an organized market. But they are normal, sane people participating in an orderly arrangement for trading which makes possible the completion of an enormous number of individual transactions within a remarkably short time and in a most effective manner.

In addition to the participants on the floor of the exchange, small trader investors from all walks of life are attracted to the profit potential of commodity speculation.

In order to better understand the nature of market participants, a special study of the corn futures market was made on January 27, 1967. Virtually every occupation was represented among the non-member traders that day — farmers and ranchers, architects, physicians, secretaries, students, plumbers, and barbers. There were 16 homemakers with positions in excess of 100,000 bushels; a large European trader was short 200,000 bushels; and a South Vietnamese trader was long 10,000 bushels. The number of speculators outnumbered hedgers by 8 to 1. Farmers and food intermediaries were among the most active speculators; more than one-third of the long speculative positions were held by farmers.[14]

A later study took a closer look at farmer participation in the futures market.[15] This survey indicated that almost 6 percent of farmers in the U.S. with gross sales over $10,000 traded futures contracts in 1976. Large farmers and grain producers were more active in the market than smaller farmers or livestock producers. Over two-thirds of the participating farmers traded commodities they neither used nor produced; this group was speculating rather than hedging. It was estimated that only 1 percent of the farmers were actually using the futures market for hedging in 1976. Their most frequently stated reasons for not trading futures contracts were (1) not acquainted with how the futures market operates, (2) farm too small, (3) futures market too risky, (4) lack of capital for trading, and (5) limited time to devote to trading. A larger group of farmers used the market in other ways. An estimated one-third of all farmers watched market prices as a barometer of commodity values, even though they did not trade futures contracts.

It is important to note that no commodity exchange or reputable commission house encourages uninformed public speculation. The modern commodity trader is, for the most part, very well informed and a keen student of markets. He or she may well be considered akin to any experienced, resourceful businessperson. The trader's business is to analyze all price-making factors, buying when he or she thinks prices are too low and selling when prices are too high. In addition to foresight, the trader must have courage and resources and know how to limit losses when his or her judgment is wrong.

Two views have been held on the role of the small speculative traders in the futures market. On one hand, they are seen as a disturbing influence in the market, accentuating price swings, and on occasion contributing to wild and disastrous price fluctuations. On the other hand, such traders are seen as a necessary element in the market, since their presence makes it possible for the expert trader and hedger to find traders to take the opposite side of a transaction. This view holds that small traders, through their losses, "supply the income which is necessary to support the continued trading activity of the professional."[16]

Stewart conducted a pioneering research project on the trading behavior of small speculators in grain futures over a nine-year period, 1924-1932.[17] Information on trading activities of nearly 9,000 traders and more than 400,000 individual transactions were analyzed. The study confirmed a number of commonly held opinions on results of speculative trading, but it tended to disprove others which had been widely accepted.

The first conclusion was that the great majority of small speculators lost money in the grain futures market. There were 6,598 speculators with net losses, compared with 2,184 with net profits. About 75 percent of small traders lost money speculating. The total net losses of almost $12 million were six times as great as profits of about $2 million.

The data showed that the great majority had relatively small profits and losses. The average loss for unprofitable traders was $1,812, and the average profit for successful traders was only $945 per trader. The profits of 84 percent of the profitable traders were less than $1,000 each; the losses of 68 percent of the losing traders were less than $1,000 each. Obviously, a large percentage of the traders in the sample operated on a small scale, and many of them discontinued trading before realizing large profits or suffering large losses. At the extreme, the most successful trader had profits of $395,718; the most unsuccessful trader had losses of $413,212, even though four of the first five trades showed profits. The major reason for the high ratio of losses was the small speculator's characteristic hesitation to close out losing positions. An often-quoted maxim for speculative trading is "Cut your losses short and let your profits run." Contrary to this philosophy, speculators in the study exhibited a clear tendency to cut their profits and let their losses run. Losing positions were usually held for longer durations than profitable trades.

The study also confirmed the commonly held impression that the amateur speculator is more likely to be long than short in the futures market. About half of the speculators in wheat and corn held positions only on one side of the market, mostly long. The other half of the traders who held both long and short positions accounted for a larger proportion of the trading, however. From the standpoint of market activity, 63 percent of the positions were long and 37 percent short in wheat futures, but in corn, 58 percent were long and 42 percent were short. Short positions tended to show profits more frequently than long positions.

In the study, people from all walks of life participated in trading, as shown in Table 3-1. Businesspersons were the largest occupational group, and farmers were next. There were 316 persons engaged in the grain business who also carried personal, speculative accounts with the brokerage firm from which the sample was obtained. Among managers of business concerns, 840 were profitable traders, compared to 2,563 losing traders. Persons with occupations "unknown" had the greatest proportion of profitable trades, 32.3 percent. Farmers had the lowest proportion of profitable trades, 21.2 percent. Special knowledge of the commodity being traded seemed to have little effect on the outcome of speculative trading in the group studied.

The study showed the tendency of long speculators to buy on days of price declines and for shorts to sell on price rises. Trading against the current movement of prices was the dominant pattern. This indicates that traders in the sample were predominantly price-level traders. Longs tended to buy when prices fell below levels that they considered proper, and shorts sold when prices advanced above levels that they believed justified.

The results of the study confirm the saying that small traders are usually wrong in their analysis of market direction. But someone has to supply grist for the commodity mill.

Attractions of Speculating. Huge potential profits earned quickly are the greatest lure to speculation. Books have been written on *how I made a million dollars trading commodities.*[18] Speculators must keep in mind that losses occur as quickly as profits. Financial leverage works both ways. In the zero-sum game of futures trading, someone's gain is always someone else's loss. Either the hedger is gaining price protection at the expense of speculators or hedgers are sending speculators reward for their efforts.

In addition to possible high profits, other positive factors attract speculative participation in the market:

1. Financial leverage. A small amount of money controls a larger value. Typically, speculative margin requirements are 5 to 10 percent of the contract value. At $1,200 per contract on corn, it is possible for $24,000 to control the price effects on 100,000 bushels of grain worth maybe $300,000.
2. No long-term dollar commitments. It is possible to be in and out in one day.
3. Choice in the degree of risk assumed. Investments can be made in commodities with relatively large or small fluctuation in price. The speculator can trade 5,000 bushels or 500,000 bushels in a single trade.
4. Profits come as easily from price decline as from price increase. Either short or long positions may be taken, depending on which way the speculator thinks price will move.
5. Commissions are low relative to asset value. Trading fees range from $20 to $100 on a soybean transaction worth perhaps $35,000.
6. Price information is public, and expert advice is available to all traders. Most knowledge of factors affecting prices (politics, weather, etc.) is public, so everyone has an equal opportunity to analyze such factors. Anyone trading the market has access to reams of newsletters, books, and pamphlets providing information and offering advice.

Trading Units. Each commodity has its specific unit of quantity traded. For example, the cotton futures contract on the New York Cotton Exchange is 50,000 pounds (100 bales). The live hog contract on the Chicago Mercantile Exchange calls for 30,000 pounds (about 136 head @ 220 pounds each). A list of contract trading facts will be found at the end of this chapter.

In most cases when placing an order with a broker, the customer does not specify the number of pounds or bushels. A person who places an order to buy "one cotton contract" on the New York Cotton Exchange bears the understanding that it is for a regular contract unit of 50,000 pounds. Placing an order to sell

Table 3-1: Number of Speculative Traders with Profits and with Losses, and Percent with Profits, by Grain and Major Occupational Group

Commodity and Occupational Group	Traders with —		Total	Percentage with Profits
	Profits	Losses		
WHEAT				
Business managers:	(number)	(number)	(number)	(%)
Grain business	83	189	272	30.5
Other	702	1,993	2,695	26.0
Professional	179	498	677	26.4
Semiprofessional	27	64	91	29.7
Clerical	130	349	479	27.1
Farmers	200	697	897	22.3
Manual workers	214	621	835	25.6
Retired	240	576	816	29.4
Unknown	270	509	779	34.7
Total	2,045	5,496	7,541	27.1
CORN				
Business managers:				
Grain business	76	96	172	44.2
Other	577	806	1,383	41.7
Professional	133	221	354	37.6
Semiprofessional	16	20	36	44.4
Clerical	76	159	235	32.3
Farmers	151	261	412	36.7
Manual workers	163	289	452	36.1
Retired	189	280	469	40.3
Unknown	144	271	415	34.7
Total	1,525	2,403	3,928	38.8
OATS				
Business managers:				
Grain business	23	54	77	29.9
Other	223	374	597	37.4
Professional	45	74	119	37.8
Semiprofessional	7	7	14	50.0
Clerical	37	50	87	42.5
Farmers	57	112	169	33.7
Manual workers	53	108	161	32.9
Retired	69	111	180	38.3
Unknown	75	107	182	41.2
Total	589	997	1,586	37.1

(Continued)

Table 3-1 (Continued)

Commodity and Occupational Group	Traders with — Profits	Traders with — Losses	Total	Percentage with Profits
RYE				
Business managers:	(number)	(number)	(number)	(%)
Grain business	22	37	59	37.3
Other	201	303	504	39.9
Professional	36	78	114	31.6
Semiprofessional	2	12	14	14.3
Clerical	33	49	82	40.2
Farmers	43	94	137	31.4
Manual workers	52	88	140	37.1
Retired	62	79	141	44.0
Unknown	46	76	122	37.7
Total	497	816	1,313	37.9
ALL GRAINS				
Business managers:				
Grain business	92	224	316	29.1
Other	748	2,339	3,087	24.2
Professional	185	583	768	24.1
Semiprofessional	23	79	102	22.5
Clerical	121	433	554	21.8
Farmers	218	810	1,028	21.2
Manual workers	212	736	948	22.4
Retired	254	700	954	26.6
Unknown	331	694	1,025	32.3
Total	2,184	6,598	8,782	24.9

Source: Blair Stewart 65.

"one hog contract on the Merc." means that the order is for the Mercantile Exchange unit of 30,000 pounds.

Trading Price Fluctuations. Minimum price fluctuations in many commodities are in one-hundredths of a cent; others are in quarters of a cent. For cotton, the market fluctuates in "points," each equal to one-hundredth of a cent per pound. Therefore, each point equals $5 in the value of the contract, and a fluctuation of a full cent, or 100 points, equals $500 per contract. Livestock contracts also fluctuate in one-hundredths of a cent. Grains are quoted in quarters of a cent per bushel, the minimum fluctuation in most markets. With a full contract representing 5,000 bushels, a fluctuation of a full cent in the price of wheat equals $50 per contract.

In addition to minimum fluctuations, daily trading limits specify the maximum price fluctuation that can occur from the previous day's settlement price. Limits restrict how much price can change from one day to the next. Trading limits on corn are 10 cents per bushel or $500 per contract; wheat, 20 cents or $1,000; soybeans, 30 cents or $1,500 per contract; and cotton, 2 cents (200 points) or $1,000. Live cattle can fluctuate by a maximum of 1.5 cents (150 points) or $600 per contract; live hogs, 1.5 cents (150 points) or $450 each day. It is possible for a commodity to trade limit up and limit down during the same day. The change in value of a contract under those circumstances would be double the limits stated above.

It is also possible for limit price moves to occur on several consecutive days. After three limit days, limits are modified or removed as the exchange governing board deems it necessary in order to allow the severe condition to work itself out in the market. Three days of 30-cent-limit moves on 5 soybean contracts (25,000 bushels) produces a $22,500 change in value of the contracts; three limit moves of 150 points on 10 hog contracts (about 1,400 head) works out to a $13,500 change. The fact that markets can, and sometimes do, change drastically emphasizes the need for market participants to have cash liquidity to meet their financial obligations. Limit day moves often prevent any trading from taking place. The trader on the wrong side of the move cannot cancel the position because no one will take the other side of the offer. As a consequence, margin calls can require lots of money in a short period of time.

Regulation of Futures Trading[19]. Public concern with futures markets, especially the price effects of alleged "excessive" speculation, has resulted in state and federal regulation of these markets. The regulations complement the very extensive self-regulating efforts of the commodity exchanges. Regulation is intended to protect participants and eliminate abuse from unfair practices. But it cannot prevent traders from incorrectly assessing price movement and losing money as a consequence of misjudgment.

The history of futures trading is marked by stories of price fixing, corners, squeezes, and market manipulation. The allegations continue to this day. The markets attract high rollers, because the profits to be won by successful manipulation by "gunslingers" are enormous. The market is fueled by greed, which can be controlled only by competition and by trading rules.

The Grain Futures Act of 1922 was the initial federal regulation of futures markets. In 1936, this law was amended to create the Commodity Exchange Act. These laws granted authority to the federal government to prevent market abuses, price manipulation, cheating, dissemination of false information, and fraud in futures markets. From 1936 until 1975, futures trading was regulated by the Commodity Exchange Authority (CEA), a division of the U.S. Department of Agriculture. Regulatory authority over these markets was shifted to the Commodity Futures Trading Commission (CFTC), an independent governmental regulatory agency, by the Commodity Futures Act of 1974.

The Commodity Exchange Act provides that no organized futures trading may be conducted in any commodities, except on an exchange that is officially designated by the CFTC. Exchanges are required to demonstrate that futures contracts for which they seek trading approval will not be contrary to the public interest. To maintain trading approval, the exchanges must comply with certain stipulations of the Act and the rules and regulations issued thereunder for the protection of customers, who trade through member brokers.

The Act provides that commission merchants trading in regulated commodities must be registered. The registrations must be accompanied by financial statements to demonstrate the financial integrity of the registrant. Registrations expire as of December 31 of each year. Thus, a new registration and new report are required each year. A new registration is also required if any important changes of partnership or ownership occur during the year, as set forth in the rules. Moreover, a new section of the Act requires registration of persons associated with a futures commission merchant who are involved in soliciting or accepting customers' orders. It also requires registration of commodity trading advisors and commodity pool operators. Registration provides a means for detecting and eliminating previously fraudulent operators.

Regulations provide for complete segregation of customers' funds from brokers' funds. When depositing customers' funds, brokerage firms must notify the bank of the nature of the funds by appropriately labeling the account and obtaining an agreement with the bank, waiving any claim or right of offset which the bank might otherwise have against such funds.

The Act gives the Commodity Futures Trading Commission the authority to establish position limits on the number of contracts held by individual speculators (spec limits). The spec limit on net long or short positions for individual speculative accounts has been set at 12 million bushels for wheat, corn, and soybeans. The limit on the maximum net long or net short position which any person may hold or control in cotton on any one contract market has been set at 30,000 bales. These limits upon position and daily trading do not apply to bona fide hedging transactions.[20]

Despite the efforts of regulators, several famous market "corners" and "squeezes" have occurred. A corner results when an individual (or group acting in concert) holds a large portion of the outstanding long positions and also controls most of the cash commodity that could be delivered by short sellers at contract maturity. In liquidating their position as contract maturity approaches, the shorts can either buy back their speculative positions (from a long) or purchase the commodity to deliver against the contract. But in a squeeze, they face the same seller in both cases and must pay the price of the seller. This is reason enough that farmer-hedgers are advised to lift their positions or roll forward their contract obligations before the beginning of a delivery month. It is not wise to get caught in a squeeze.

Extensive regulation has made market manipulation difficult and hazardous. Limits on the quantity of contracts, increases in the number of delivery points, and careful scrutiny of the markets for monopoly behavior diminish the opportunities for manipulation of future markets. On balance, they are probably no more subject to manipulation than other markets.

Farmer Trading. Farmers who become intrigued with commodity marketing face the choice of whether they should become futures market hedgers or speculators. They often fail to make that important distinction and lose the farm before they understand the difference. For example, they accept double jeopardy if they institute a "Texas hedge." Here they own the commodity and buy futures contracts, feeling that the market is going up. If they are wrong and the market declines, they lose on both positions; they suffer from decline in value of their inventory and losses on their futures contracts.

There is one good reason for farmers or ranchers to open a speculative account: It may encourage them to become better students of the market. What they learn from "playing" with $5,000 in a speculative account may cause them to make advantageous sales of their products that could earn $50,000 additional income.

When farmers make a conscientious decision to become speculative traders, they have a new set of rules to learn and apply. "Paper trading" is a good place to start. It provides practice without incurring costs. Writing down the position taken and reasons for it forces the beginner to keep score on gains and losses. In six months' time, a paper trader will determine whether the plan is working and whether he or she has the emotional make-up to be successful. The next step into the real world is to open an account on the MidAmerica Exchange to trade mini-contracts. The initial investment is low, and extent of loss exposure is relatively small. A producer should trade only the commodities he or she actually produced.

Why speculate on futures? For the thrill of it. The thrill encourages farmers to follow the market more closely as they read about the subject and gain understanding of fundamental and technical market forces. The farmer who limits a speculation account to $5,000 is not going to lose much money, but it will provide an outlet to his or her speculative urges. Hedging and speculating are two different worlds. One depends upon the other, but each has its own objectives. Futures trading as a marketing tool should reduce risk, not increase it. As the studies cited show, most small traders lose in their speculative ventures. The objective of farmer speculating is to learn to be a better hedger.

Rules for Speculating[21]. There are a number of basic rules for commodity speculating. An analysis of fundamental and technical conditions is the first step toward successful speculation. Price forecasts must then be combined with a trading plan or strategy which the individual is capable of adhering to and which will generate profits. As a "position" trader, a person must have reasons for his or her trades. By limiting oneself to one or two commodities, learning all one can about them, and studying the technical factors, one can develop a strategy. Emotions should not change the plan; common sense and logic should prevail.

Traders generally believe that profit potential should be large in relation to risk. Traders should specify their profit objective as well as the maximum loss they are willing to sustain. Their own personal preferences will determine if the objective for profit is the maximum potential profit or some lesser figure.

Once the trading guidelines have been determined, the trader should specify the amount of capital to be risked. Although it is implicitly assumed that speculators should be using funds which are totally risk capital, it is still essential to maximize the return from this risk capital. A farmer who chooses to be a futures market trader should open two accounts, one for hedging activities and one for speculating. The two accounts should be kept distinctly separate. Farm production money should not be used for speculation. Trading money should be limited to the amount that can be lost without jeopardizing the business.

Experienced speculators recommend that a great proportion of an individual's capital should not be risked on a single trade. Also, the trader should limit the open positions to as many commodities as he or she can adequately follow, keeping some capital in reserve for additional opportunities. Successful speculators often recommend (1) that additions to an initial position should be made only in the event that profits have already been made and (2) that subsequent commitments should not be in amounts greater than the initial position. Liquidating a position should be based on the original plan. Market conditions may change, however, and it is essential that the speculator maintain some degree of flexibility.

Only the speculator's own experience and preferences can be relied on to assess the desirability of a trade based on profit potential relative to risk, to determine profit objectives and loss limits, and to decide on additions to the original position and final liquidation. Successful price forecasting and trading ultimately depend on the individual's own temperament and objectivity, as well as the economic models and trading plans that have been developed.

In addition to the numerous rules implicit in a systematic approach to speculation, other guidelines, representing the trading knowledge of successful speculators, are helpful. The following list of such guidelines is intended for reference and not necessarily as a recommendation.[22]

1. "Markets in which trading is contemplated should be carefully analyzed. Quick actions based on rumors, tips, etc., should be avoided.
2. "Speculation should not occur in the presence of doubt about the price forecast or without a trading plan.
3. "Speculation should usually follow the market trend.
4. "Speculators should generally avoid attempts forecasting reversal points of trends.
5. "Forecasts of price declines, and consequently, short positions should have the same priority as forecasts of price increases, which call for long positions.
6. "Speculation should be initiated when the potential profit is great relative to risk.
7. "Successful speculation requires limiting losses and letting profits run. Traders should be prepared to accept numerous small losses. A limited number of highly successful trades will offset those losses.
8. "In addition to a careful analysis of commodities, successful speculation requires a well-developed trading plan. This includes limiting the amount of capital on any single trade and maintaining capital in reserve."

A careful approach to the market, coupled with the appropriate temperament and psychology, will increase the probability of successful trading.

Is Futures Trading Gambling? Much of the historical controversy about commodity speculation has related to the confusion of speculation and gambling. Gambling is almost universally condemned as wasteful and as working great evil on the gambler himself or herself. Commodity speculation has often been branded as the biggest gamble of them all.

Commodity trading is a zero-sum game; for every winner there is a loser. Credits in the brokerage account of one trader are covered by debits in the account of another.* However, brokerage account credits and debits may or may not be profits and losses. And it is inaccurate to characterize the zero-sum nature of trading as an unnecessary exercise for shifting money between winners and losers. As illustrated in Figure 2-9, short hedgers accumulate brokerage account credit balances as price level declines. Speculators ante up the money transferred to hedgers. On the other hand, as price level rises, speculators accumulate credit balances as hedgers pay up. Depending upon price direction, the speculator loses or profits. But the hedger neither profits nor loses; hedgers merely fix a price on their commodity inventory. Because the hedger owns inventory in addition to futures position, the value of the inventory fluctuates in the opposite direction of the brokerage account balance. Losses in one account (inventory) are offset by gains in the other (brokerage). Thus the futures market fulfills its role of shifting

---

*Actually, commodity trading is a negative-sum game. Trader A's debits offset Trader B's credits, but brokers collect commissions from both.

the financial effects of price change from owners of commodities (hedgers) to others (speculators) who are willing to risk loss in hope of gain due to price change.

Speculation is like gambling in several ways:

1. Both depend upon uncertainties,
2. Both involve risk of loss for the sake of possible gain,
3. Occurrence of certain events results in losses to one player while other events have the opposite effect,
4. Both may involve careful calculation of probabilities and developing forecasts, or both may be based on pure chance.

Nonetheless, certain essential distinctions can be made in defense of speculation:

1. Gambling involves creating risks that would not otherwise exist; speculation involves assuming necessary and unavoidable risks of commerce.
2. In every futures transaction, the speculator incurs the duties and acquires the rights of a holder of property and thus is an integral part of commerce. Whether the impact of the speculator's activities are good or bad is neither here nor there — they are inevitable and necessary conditions of business practice.
3. Every trade on futures markets has a buyer and seller; the number of contracts bought equal the number of contracts sold. Without speculation, hedging is impossible.

Gambling is the creation of games of chance; futures trading is a mechanism to transfer unavoidable business risk from one who seeks to avoid it to another who specializes in it. Futures trading has a social and economic justification that gambling lacks. Therefore, futures trading is not the same as gambling.

SUMMARY

Futures markets serve three important roles: price discovery, risk transfer, and investment medium. In a capitalistic economy, value is determined by the interaction between buyers and sellers. Governmentally dictated prices do not work. Neither do monopoly pricing practices. A free and open market available to all buyers and sellers more accurately reflects true supply/demand conditions. Futures markets bring buyers and sellers face to face in their quest for price discovery. They also increase the level of pricing information as the results of trading flash worldwide almost instantaneously.

Farmers are the ultimate speculators because of the nature of their business. They commit thousands of dollars to purchase facilities and equipment, seed, fertilizer, herbicides, pesticides, feeder livestock, feed, and supplements. They and their families work all year tilling the soil, feeding the livestock, while wearing out their machinery and facilities. They risk disease, insects, weather, fire, and theft to produce a crop or pen of livestock. When their product is ready for market, there is no guarantee that their efforts will be adequately rewarded. Market price may offer significant profits, or it may fall below cost of production. Futures markets allow farmers (and others) to transfer the effects of price change and guarantee themselves profit before delivering their products to market.

Speculation in commodity markets does not depend upon the existence of futures markets. Speculation is a part of any economy in which economic decisions have to be made in the face of uncertainty of future events. But futures markets do contribute to speculative activities by making it possible for individuals to engage in speculation without having to become engaged in producing, handling, storing, or processing commodities. Specialization in practicing speculation is facilitated by futures markets. Transaction costs and capital commitments are minimized. Standardization of futures contracts enables the speculator to establish either long or short positions in the market with equal ease. Organized exchanges allow speculators to practice their trade efficiently.

The commodity exchanges thus provide a market which is not matched by markets without futures trading: markets for houses, works of art, or antique automobiles. Agricultural producers should learn to use these exchanges efficiently to minimize the producers' speculation rather than lobby for their elimination.

Questions for Study and Discussion:

1. Trace the evolution of grain trading practices that became part of the features of futures trading.
2. What characteristics qualify a commodity for futures trading? Why are some products unsuitable for trading?
3. Discuss the three major roles of futures markets and the significance of each one.
4. What is price discovery?
5. What functions does the Clearing Corporation perform?
6. How does the Internal Revenue Service treat profits and losses from futures trading for both the hedger and the speculator?
7. List the various types of market orders placed by a futures market trader.
8. What types of risk are associated with ownership of agricultural commodities?
9. How may risks be transferred?
10. Give examples of buying and selling hedgers and how they fix a price.
11. Do future market gains and deficits produce profits or losses for hedgers?
12. Discuss the three types of hedging.
13. Explain the hazards of hedging.
14. Who are the participants in futures trading?
15. What are the features that make commodity trading attractive to investors?
16. What is the regulating agency of futures trading, and what does it do?
17. What generalities serve as rules of successful speculation? Why are they important?
18. Is futures trading gambling?

Futures Markets Exercise: Paper Trading

On any day, students in commodity marketing should have a reason to believe the present futures price of a commodity is too high or too low, relative to what its true equilibrium price will be at a future date. "Paper trading" is a technique to test their analytical abil-

FUTURES TRADING FACTS

COMMODITY	NAME OF EXCHANGE	TRADING HOURS (N.Y. time, Mon. thru Fri.)	CONTRACT SIZE	MINIMUM FLUCTUATION per lb., etc.	MINIMUM FLUCTUATION per contract	DAILY TRADING LIMITS* (from previous close)	QUOTATION
ALUMINUM	Commodity Exch., N.Y.	9:30 A.M.-2:15 P.M.	40,000 lbs.	5/100¢	$20.00	5¢	¢/lb.
CATTLE (Feeder)	Chicago Mercantile Exch.	10:05 A.M.-2:00 P.M.	44,000 lbs.	$.00025	$11.00	1½¢	¢/lb.
CATTLE (Live Beef)	Chicago Mercantile Exch.	10:05 A.M.-2:00 P.M.	40,000 lbs.	$.00025	$10.00	1½¢	¢/lb.
COCOA	Coffee, Sugar & Cocoa Exch.	9:30 A.M.-3:00 P.M.	10 tonnes	$1.00/tonne	$10.00	$88.00	$/ton
COFFEE "C"	Coffee, Sugar & Cocoa Exch.	9:45 A.M.-2:28 P.M.	37,500 lbs.	1/100¢	$ 3.75	no limit spot others 4¢	¢/lb.
COPPER	Commodity Exch., N.Y.	9:50 A.M.-2:00 P.M.	25,000 lbs.	5/100¢	$12.50	5¢	¢/lb.
COTTON #2	New York Cotton Exch.	10:30 A.M.-3:00 P.M.	50,000 lbs.	1/100¢	$ 5.00	2¢	¢/lb.
Currencies BRITISH POUND	International Monetary Market of the Chicago Mercantile Exch.	8:20 A.M.-2:24 P.M.	25,000 BP	$.0005	$12.50	$.0500	$/lb.
CANADIAN DOLLAR		8:20 A.M.-2:26 P.M.	100,000 CD	$.0001	$10.00	$.0075	$/Can.$
DEUTSCHE MARK		8:20 A.M.-2:20 P.M.	125,000 DM	$.0001	$12.50	$.0100	$/mark
JAPANESE YEN		8:20 A.M.-2:22 P.M.	12.5 mill. JY	$.000001	$12.50	$.0001	$/yen
MEXICAN PESO		8:20 A.M.-2:18 P.M.	1,000,000 MP	$.00001	$10.00	$.00150	$/peso
SWISS FRANC		8:20 A.M.-2:16 P.M.	125,000 SF	$.0001	$12.50	$.0150	$/franc
CDs (Domestic Bank)	IMM–Chicago Merc. Exch.	8:20 A.M.-3:00 P.M.	$1,000,000	.01	$25.00	no limit	pts. of 100%
CPI Index	Coffee, Sugar & Cocoa Exch.	9:30 A.M.-2:30 P.M.	$1000 × index	.01 or 1/100th of index	$10.00	3 points	index
CRUDE OIL	N.Y. Mercantile Exch.	9:45 A.M.-3:10 P.M.	42,000 gals.	1¢ barrel	$10.00	$1.00	$/bbl.
EURODOLLAR	IMM–Chicago Merc. Exch.	8:20 A.M.-3:00 P.M.	$1,000,000	.01	$25.00	no limit	pts. of 100%
GASOLINE (Leaded)	N.Y. Mercantile Exch.	9:55 A.M.-3:00 P.M.	42,000 gals.	1/100¢ gal.	$ 4.20	2¢	$/gal.
GASOLINE (Unleaded)	N.Y. Mercantile Exch.	9:30 A.M.-3:00 P.M.	42,000 gals.	1/100¢ gal.	$ 4.20	2¢	$/gal.
GNMA MTGES (CDR)	Chicago Board of Trade	9:00 A.M.-3:00 P.M.	$100,000 @ 8%	1/32 pt.	$31.25	64/32	pts. in 32nds of 100%
GOLD	IMM–Chicago Merc. Exch.	9:00 A.M.-2:30 P.M.	100 troy oz.	$.10/oz.	$10.00	$50.00	$ per troy oz.
	Chicago Board of Trade	9:00 A.M.-2:40 P.M.	32.15 troy oz.	$.10/oz.	$ 3.22	$50.00	
	Commodity Exch., N.Y.	9:00 A.M.-2:30 P.M.	100 troy oz.	$.10/oz.	$10.00	$25.00	
	MidAmerica Com. Exch.	9:00 A.M.-2:40 P.M.	33.2 troy oz.	$.025/oz.	$ 0.83	$50.00	
	Winnipeg Commodity Exch.	11:00 A.M.-2:25 P.M.	20 troy oz.	$.10/oz.	$ 2.00	$25.00 (U.S.)	
Grains–Chicago WHEAT, SOYBEANS, CORN, OATS	Chicago Board of Trade MidAmerica Com. Exch.	10:30 A.M.-2:15 P.M. 10:30 A.M.-2:30 P.M.	5,000 bus. 1,000 bus.	1/4¢ 1/8¢	$12.50 $ 1.25	Wheat 20¢ Soybeans 30¢ Corn 10¢, Oats 10¢	¢/bu.
Grains–Minneapolis WHEAT	Minneapolis Grain Exch.	10:30 A.M.-2:15 P.M.	5,000 bus.	1/8¢	$ 6.25	20¢	¢/bu.
Grains–Kansas City WHEAT	Kansas City Board of Trade	10:30 A.M.-2:15 P.M.	5,000 bus.	1/8¢	$ 6.25	20¢	¢/bu.
Grains–Winnipeg BARLEY, OATS, RYE, WHEAT RAPESEED, FLAXSEED	Winnipeg Commodity Exch.	10:30 A.M.-2:15 P.M.	20 tonnes	10¢/tonne	$2.00 (Cdn.)	Barley, Oats, Rye, Wheat $5.00 (Cdn.) Rapeseed & Flaxseed $10.00 (Cdn.)	Can. $ /ton

* Limits expand under certain conditions (except for C.M.E. livestock and K.C. wheat).

FUTURES TRADING FACTS

COMMODITY	NAME OF EXCHANGE	TRADING HOURS (N.Y. time, Mon. thru Fri.)	CONTRACT SIZE	MINIMUM FLUCTUATION per lb., etc.	MINIMUM FLUCTUATION per contract	DAILY TRADING LIMITS* (from previous close)	QUOTATION
HOGS (Live)	Chicago Mercantile Exch.	10:10 A.M.-2:00 P.M.	30,000 lbs.	$.00025	$ 7.50	1½¢	¢/lb.
LUMBER (Random Lengths)	Chicago Mercantile Exch.	10:00 A.M.-3:05 P.M.	130,000 bd. ft.	10¢/1000 board ft.	$13.00	$5.00	$ per 1000 bd. ft.
MUNICIPAL BONDS	Chicago Board of Trade	9:00 A.M.-3:00 P.M.	$1000 × Bond Buyer MBI	1/32nd of point	$31.25	2 points	Bond Buyer MBI
OIL, HEATING #2 N.Y.	N.Y. Mercantile Exch.	9:50 A.M.-3:05 P.M.	42,000 gals.	1/100¢	$ 4.20	2¢	$ per gallon
ORANGE JUICE (FCOJ)	New York Cotton Exch.	10:15 A.M.-2:45 P.M.	15,000 lbs.	5/100¢	$ 7.50	5¢	¢/lb.
PALLADIUM	N.Y. Mercantile Exch.	8:50 A.M.-2:20 P.M.	100 troy oz.	$.05/oz.	$ 5.00	$6.00	$ per troy oz.
PLATINUM	N.Y. Mercantile Exch.	9:00 A.M.-2:30 P.M.	50 troy oz.	$.05/oz.	$ 5.00	$25.00	$ per troy oz.
PORK BELLIES	Chicago Mercantile Exch.	10:10 A.M.-2:00 P.M.	38,000 lbs.	$.00025	$ 9.50	2¢	¢/lb.
POTATOES	N.Y. Mercantile Exchange	9:45 A.M.-2:00 P.M.	50,000 lbs.	1¢/50 lbs.	$10.00	40¢/50 lbs.	$ per 50 lbs.
SILVER	Commodity Exch. N.Y. / Chicago Board of Trade / Winnipeg Commodity Exch.	9:05 A.M.-2:25 P.M. / 9:05 A.M.-2:25 P.M. / 11:00 A.M.-2:25 P.M.	5,000 troy oz. / 1,000 troy oz. / 200 troy oz.	10/100¢ / 10/100¢ / 1¢	$ 5.00 / $ 1.00 / $ 2.00	50¢ / 50¢ / 50¢ (U.S.)	¢ per troy oz.
SOYBEAN MEAL	Chicago Board of Trade	10:30 A.M.-2:15 P.M.	100 tons	10¢/ton	$10.00	$10.00	$/ton
SOYBEAN OIL	Chicago Board of Trade	10:30 A.M.-2:15 P.M.	60,000 lbs.	1/100¢	$ 6.00	1¢	¢/lb.
Stock Index Futures							
MAJOR MARKET INDEX	Chicago Board of Trade	9:15 A.M.-4:15 P.M.	$100 × index	1/8 pt.	$12.50	no limit	
MAJOR MARKET INDEX MAXI	Chicago Board of Trade	9:15 A.M.-4:15 P.M.	$250 × index	5/100 pt.	$12.50	no limit	
NYSE COMP INDEX	N.Y. Futures Exch.	9:30 A.M.-4:15 P.M.	$500 × index	.05	$25.00	no limit	index
S&P 500 INDEX	IMM–Chicago Merc. Exch.	9:30 A.M.-4:15 P.M.	$500 × index	.05	$25.00	no limit	
VALUE LINE INDEX	Kansas City Board of Trade	9:30 A.M.-4:15 P.M.	$500 × index	.05	$25.00	no limit	
SUGAR world #11 domestic #12 & #14	Coffee, Sugar & Cocoa Exch.	10:00 A.M.-1:43 P.M.	112,000 lbs.	1/100¢	$11.20	no limit 1st 2 contracts others ½¢	¢/lb.
T-BILLS (13 weeks)	IMM–Chicago Merc. Exch.	8:20 A.M.-3:00 P.M.	$1,000,000	.01	$25.00	no limit	pts. of 100%
T-BONDS (Long Term)	Chicago Board of Trade	9:00 A.M.-3:00 P.M.	$100,000	1/32 pt.	$31.25	64/32	pts. 32nds of 100%
T-NOTES (10 Year)	Chicago Board of Trade	9:00 A.M.-3:00 P.M.	$100,000	1/32 pt.	$31.25	64/32	pts. 32nds of 100%
U.S. DOLLAR INDEX	New York Cotton Exch.	8:30 A.M.-2:40 P.M.	$500 × index	.01	$5.00	no limit	index

* Limits expand under certain conditions (except for C.M.E. livestock and K.C. wheat).

ity and to gain experience in the mechanics of futures trading without having to put money on the line. It also requires them to develop a trading strategy and test it in real time.

Students should be supplied with trading forms and a record sheet. They are to make decisions to sell or buy a futures position and record them on the forms. They record their position, commodity, number of contracts and REASON for their position after the market close on one day and turn it in before the market opens the next day. Their reason should be based on a fundamental or technical fact rather than "I think the market is too low (or high)." Why is it high or low? They are to maintain an additional record of profits, losses, and account balance over the period of time of the exercise. In the end, they should write a summary of their results and what they learned about both market mechanics and their speculative abilities. Grading should be upon their justification of trades rather than amount of trading profits.

FUTURES TRADING EXERCISE

Name _____ Date _____

New Position _____ Cancelling Out Old Pos. _____

Transaction	Kind of Grain	Quantity	Month	Price
_____	_____	_____	_____	_____

Explain why you made this decision and give a complete and detailed description including *dates* of the *sources* of information you used in making this decision.

Futures Trading Exercise: Record of Trading

The form on the next page can be used to record daily activity in the account. Each student should use a photocopy of this form for recording transactions. To illustrate the process, open an account with $20,000 of paper money. On the next day, suppose a student of the market decides corn price is too low and is likely to rise because dry weather threatens production. To initiate a position, he or she buys one December corn (5,000 bushels) by completing the order form and explaining the reason for the decision. On the accounting form the student records the position as **Buy 1 Dec. corn** under the "Transaction" column. Assume it is Tuesday afternoon and the closing price is $3.05. The transaction must be recorded before the opening at 9:30 a.m. CST on Wednesday. An initial speculative margin requirement of $1,200 per contract calls for a "Margin Deposit" of that amount, dropping the cash balance to $18,800.

Each day following initiation of the position, the account is settled by adding gains or deducting losses due to price change. The effect is recorded in the "Change in Value of Position" column. "Equity in Excess of Maintenance" refers to the balance in excess of minimum maintenance margin requirements. Minimum maintenance on a corn position is $600. Should "Equity" fall below minimum maintenance, additional deposits are required to satisfy "margin calls."

Suppose that Wednesday the market closes at $3.08. This figure entered in the "Price" column translates into a $150 gain as recorded in "Change in Value of Position." The "Equity" column notes the $150 of "profit." On Thursday price falls to $3.07. The "Change in Value of Position" drops $50 and "Equity" changes to $650. A limit down move on Friday is recorded in the "Change in Value of Position" as $−500 and drops "Equity" to $150. On Monday another limit move to $2.87 loses $500 and drops "Equity" below maintenance to $−350. A margin call for $400 is issued to bring the equity balance back above minimum; the cash account balance drops to $18,400. The student panics and closes out the trade with a "sell" order. He or she deducts the commission paid to the broker (assume $50) and absorbs the 18 cent per bushel ($900) loss. The student is left with $18,350 in the cash account, since $600 of initial maintenance remains with the broker.

Now it is the student's turn to enter a trade and continue the accounting throughout the term. Each student should be assigned to trade three or four commodities, accounting for each one on a separate page. In the summary term paper the student should reflect on the practices and emotions that affected his or her profitability.

Footnotes for Chapter 3:

(1) Personal conversation and correspondence with Keith Moens, Department of Educational Services, Chicago Board of Trade, 1983.

(2) Lloyd Besant, *Action in the Marketplace: Commodity Futures Trading* (Chicago: Chicago Board of Trade, 1983).

(3) Thomas A. Hieronymus, *Economics of Futures Trading* (New York: Commodity Research Bureau, 1971).

(4) Hieronymus, *Economics of Futures Trading* 14.

(5) Julius B. Baer and Olin G. Saxon, *Commodities Exchanges and Futures Trading* (New York: Harper & Row, 1948) Chap. VI.

(6) Frederick L. Thompson, *Agricultural Marketing* (New York: McGraw-Hill, 1951) 204.

(7) Besant, *Action in the Marketplace* 24.

(8) This discussion is based on copyrighted material from Chicago Board of Trade publications and is used with permission. Besant, *Action in the Marketplace*.

ACCOUNTING FOR TRADING EXERCISE

Transaction	Date	Price	Margins: Initial (I), Deposits (D), or Withdrawals (W)	Change in Value of Position	Equity in Excess of Maintenance	Trade Profit or Loss	Commission	Balance in Cash Account
Initial deposit	6/1							20,000
B1 Dec corn	2	3.05	1200 (I)		600			18,800
	3	3.08		150	750			
	4	3.07		-50	650			
	5	2.97		-500	150			
	8	2.87	margin call 400 (D)	-500	-350 / 50			
S1 Dec	9	2.87			650	(900)	-50	18,350

(Students to continue by making trades of their own.)

(9) Material for this section was gleaned from William J. Jiler, *Understanding the Futures Markets* (Jersey City, N.J.: Commodity Research Bureau, 1983) 23. Used by permission.

(10) Jiler, *Understanding the Futures Markets* 24.

(11) Commerce Clearing House, *Standard Federal Tax Reporter* (Chicago: Commerce Clearing House).

(12) USDA, *The Secret of Affluence* (Washington, D.C.: USDA, Office of Communication, June, 1976) ii.

(13) Hieronymus, *Economics of Futures Trading* 47.

(14) Commodity Exchange Authority, *Trading in Corn Futures* (Washington, D.C.: USDA, Aug., 1967).

(15) John W. Helmuth, *Grain Pricing,* Economic Bulletin No. 1 (Washington, D.C.: Commodity Futures Trading Commission, Sept., 1977).

(16) Blair Stewart, *An Analysis of Speculative Trading in Grain Futures,* Tech. Bulletin No. 1001 (Washington, D.C.: Commodity Futures Trading Commission, Oct., 1949) 4.

(17) Stewart, *An Analysis of Speculative Trading in Grain Futures.*

(18) Larry Williams, *How I Made $1,000,000 Last Year Trading Commodities,* 3rd ed. (Brightwaters, N.Y.: Windsor Books, 1979).

(19) Information for this section was gleaned from two sources: William M. Phelan, *How Commodity Futures Are Regulated* (Chicago: Chicago Mercantile Exchange, Aug., 1973), and Jiler, *Understanding the Futures Markets.* Used by permission.

(20) Jiler, *Understanding the Futures Markets* 31.

(21) For a discussion about speculating from a producer's perspective, refer to Merrill Oster, *How to Multiply Your Money — A Beginner's Guide to Commodity Speculation* (Cedar Falls, Iowa: Investor Publications). Materials used by permission.

(22) Richard I. Sandor, *Speculating in Futures* (Chicago: Chicago Board of Trade, 1977). Information comes from copyrighted material taken from Chicago Board of Trade publications and is used with permission.

Chapter 4

Price Determination: Market Fundamentals

THE MEANING OF DEMAND 102
    The Demand Principle 103
    Aggregate Demand 104
    Demand Elasticity 104
    Changes in Demand 105
        Population 106
        Income 106
        Habits and Preferences 106
        Prices of Substitute Products 106
    Understanding Demand 107
THE MEANING OF SUPPLY 108
    The Supply Principle 108
    Supply Elasticity 111
    Shift in Supply 113
THE EQUILIBRIUM PRICE 114
    Changing Price from Shifting Demand or Supply 116
    The Supply-Demand Balance 116
    Price Discovery 119
    Law of One Price 120
MARKET FORCES IN ACTION 121
APPLICATIONS OF DEMAND AND SUPPLY ANALYSIS 125
    Effects of Demand Elasticity on Price and Gross Receipts 125
    Conflict Between the Individual and All Farmers 129
    Incentives to Restrict Output 130
    Effects of Price Ceilings and Floors 131
    Relationship of Prices to Costs of Production in Agriculture 131
    Delay in Producer Response: The Cobweb Theorem 133
FUNDAMENTAL APPROACH TO PRICE LEVEL DETERMINATION 135
    Understanding Grain Prices 138
        Corn Price 138
        Soybean Price 140
    Understanding Livestock Prices 143
        Hog Prices 143
        Live Cattle Prices 145
SUMMARY 147

Price Determination: Market Fundamentals

Farmer and rancher marketing decisions and price forecasting are rooted in analysis of market fundamentals of demand, supply, and prices. The dual questions of "What determines value?" and "How are prices established?" stimulate interest in market analysis. In capitalistic economic conditions, answers are supplied by economic principles of demand and supply. But probably no more overused term is less understood than the "law of supply and demand." To form a foundation for understanding and evaluating commodity prices from the fundamental perspective of demand/supply, we need to sort out the meanings and implications of the terms.

THE MEANING OF DEMAND[1]

The economic definition of demand is **a schedule of the quantity of a commodity which buyers will purchase at different prices in a given market at a given time.** The numbers in Table 4-1 present a hypothetical demand schedule for corn to illustrate that relationship. In any production period with the price at $4.00 per bushel, buyers would be willing to take 6.1 billion bushels; if price were $3.00 per bushel, they would take 7.0 billion bushels; if $2.00, 8.5 billion bushels, and so on for all possible quantities and price combinations.

It is important to emphasize that the demand schedule does not indicate what the market price and quantity taken actually is; it indicates only the effect different prices would have on quantity purchased. A demand schedule expresses a price-quantity relationship from a buyer's point of view. The price that will exist in a market situation is not known; a demand relationship alone cannot establish price. What it does show is how quantity demanded will rise or fall as price decreases or increases.

Table 4-1: Hypothetical Demand Schedule for Corn

Price	Amount Purchased
($/bu.)	(billions of bushels)
4.00	6.1
3.75	6.3
3.50	6.5
3.25	6.75
3.00	7.0
2.75	7.3
2.50	7.6
2.25	8.0
2.00	8.5

The Demand Principle

Demand expresses the relationship between quantities purchased by consumers and the price of the commodity: The lower the price, the more that will be purchased; the higher the price, the less that will be purchased. The demand concept is formalized into the Demand Principle. It says that the quantity of a commodity that potential buyers are willing to take varies inversely with price. The principle may be stated another way: The greater the quantity of a commodity offered for sale in a given market at a given time, the lower must be the price per unit at which the entire amount can be sold.

The demand relationship can be illustrated as a demand curve as in Figure 4-1. Three reasons cause the characteristic downward, negative slope. The first

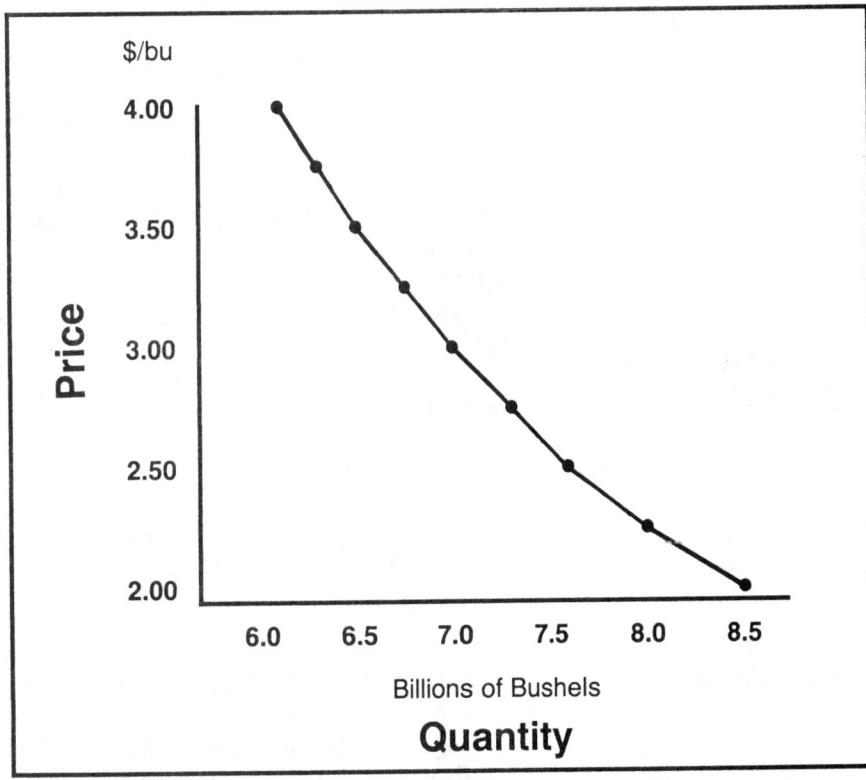

Figure 4-1: Hypothetical Demand Curve for Corn. Source: Kohls and Uhl 149.

derives from the diminishing marginal utility of a consumption good. The more consumers have of any given item during a given period of time, the less enjoyment they get from each successive unit of it. The first ice cream cone on a hot day is delightful, but the second and third produce successively less satisfaction. Because the enjoyment is less and less, a customer would be willing to purchase additional cones only if they were reduced in price. As stated by economists, diminishing marginal utility requires lower and lower price to induce people to consume more and more of a product during any one time period.

Second, people have different purchasing power. Wealthy people can afford to buy steaks at $7.00 per pound; those with lower income, having just as great a liking for steak, perhaps, could not afford to pay more than $2.50 per pound. Thus, the lower the price, the greater the number of potential customers who can afford to buy steak, and therefore, the greater will be the quantity purchased at lower prices.

Finally, people differ in their liking for a commodity and thereby in the price they are willing to pay for it. One person thoroughly enjoys lamb chops, possibly because they were a prestige dish during childhood. A second eats lamb upon occasion; and a third may not like it at all. Lamb would have to be sold at a moderately low price, relative to other meats, to get the second person to buy it, and at a very low price to induce the third. The lower the price, the greater the number of potential customers who will pay that price, and hence, the greater the sales.

Aggregate Demand

Demand principles explain how consumers pick and choose the products they buy as they allocate their income among the various items available for purchase. Consumers can often substitute one commodity for another, according to relative prices and the respective want-satisfying power. If pork is plentiful and cheap, consumers will buy more of it, relative to beef, than if it were scarce and expensive. Each individual buys more and more of a particular product until the satisfaction he or she gets from the last unit of each one is proportional to its price. If pork is $1.50 per pound and beef is $3.00 per pound, the consumer will buy enough of each one until the additional satisfaction he or she gets from the last pound of beef is twice the satisfaction he or she gets from the last pound of pork. The consumer allocates income among all the different goods and services bought until he or she gets the same additional satisfaction from the last nickel spent on each one. In so doing, each consumer establishes an individual demand equilibrium for the goods and services purchased. An individual consumer's willingness or reluctance to part with spendable income establishes demand for a product or service.

By adding together the demands of individual consumers, the aggregate demand for the market is established. It is the quantity of a product that all consumers in a market purchase at different prices.

Demand Elasticity

Elasticity of demand is the extent to which the quantity bought changes in response to a change in price. Demand elasticity is the relative, or percentage, change in the quantity bought resulting from a given percentage change in price. With some products, a change in price will bring about a proportionately large increase or decrease in the quantity that can be sold, while with others the quantity is only slightly responsive to change in price.

If the response in quantity taken is relatively greater than the change in price, demand is said to be **elastic.** In general, luxury or semi-luxury items have the most elastic demands. Consumers can do without the item or easily substitute another item for it. Leather coats provide an example of a luxury good with elastic demand.

Commodities for which consumption is relatively insensitive to price change have **inelastic** demand. The quantity of consumption of these products will hardly change as a result of higher or lower price. The classic example of a commodity with inelastic demand is salt. If salt price were doubled or tripled, people would consume about the same amount. If the price were halved, consumption would not increase materially. Salt has few substitutes and is a minor cost item. It fits the classification of necessities, all of which have inelastic demands.

The transition point between elastic and inelastic demand is called **unitary** elasticity. Here the proportional change in price is matched by the same proportional change in quantity.

Price elasticity of demand can be calculated mathematically by dividing the percentage change in quantity by the percentage change in price:

$$E = - \frac{\text{Percentage Change in Quantity}}{\text{Percentage Change in Price}}$$

The resulting value is always negative because demand curves slope downward and to the right. The size of the number calculated defines the degree of elasticity. Unitary elasticity has a value equal to -1; elastic demand has negative value from -1 to $-$infinity; and inelastic demand has values from -1 to 0. The importance of demand elasticity for farm commodities will be emphasized in later discussion.

Changes in Demand

A **change in demand** is a shift of the **entire** demand curve. An increase in demand shifts the curve to the right, a decrease to the left.

If the quantities that can be sold at each price are larger after a shift in demand, demand has increased. If the quantities that can be sold at each price are smaller, demand has decreased. Both changes are illustrated in Figure 4-2. The line $D_I D_I$ illustrates a demand increase; $D_D D_D$ a demand decrease relative to the original demand curve $D_O D_O$. With price fixed at P_O, quantity purchased is Q_I and Q_D respectively.

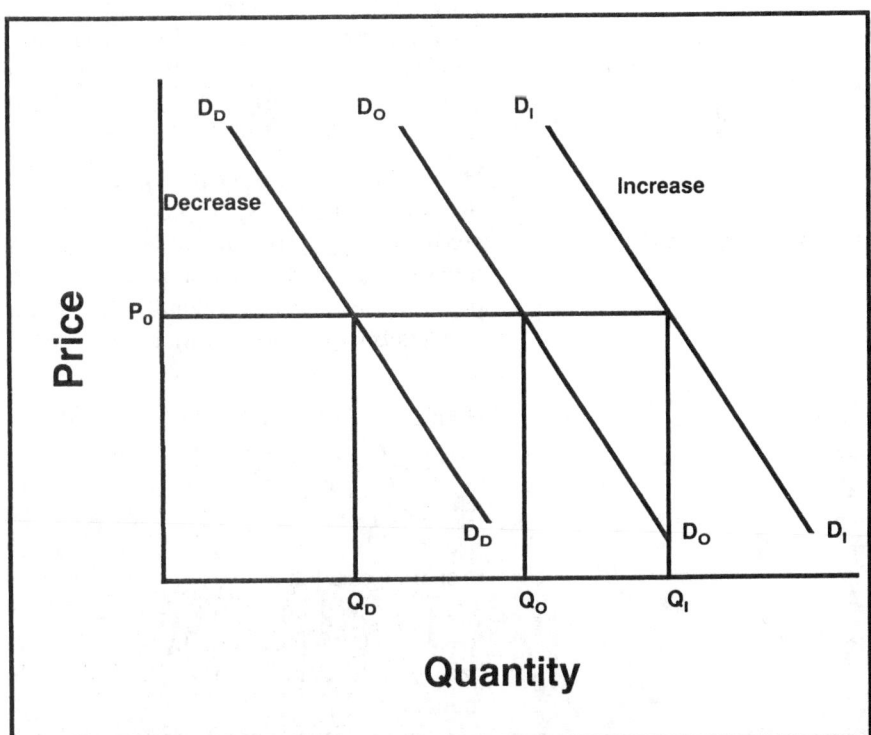

Figure 4-2: Changes in Demand.

Note that a change in demand is distinctively different from movements along a single demand curve. The changes in quantity demanded are price responses to **only** different supplies available on the market, while changes in demand are shifts of the entire curve.

Changes in demand for a commodity that moves the entire curve to the right (increase) or to the left (decrease) are responses to changes in:

1. Population, or the number of consumers in the market;
2. Income, or purchasing power of consumers;
3. Habits, tastes, and preferences of consumers; and
4. Price of substitute and complementary products.

Population. The effect of number of people on demand is obvious — more mouths to feed require more baskets of food. Over time, as population in the world increases, there is increasing demand for food. The opening of trade between two countries increases the demand in the exporting country as a larger population enters its market equation.

Income. The quantity of a commodity that is purchased is affected by level of income. Income elasticity is the term used to express relative change in quantity taken in response to relative change in income. In one case consumer purchases rise with increased income; and in the other, purchases decrease. The first is called a "superior" good and the second, "inferior." Meals away from home, highly processed items, and red meats qualify as superior goods; bread, rice, and potatoes are examples of the inferior goods. Income elasticity is high or positive for superior goods, indicating increased purchases in response to higher incomes. Income elasticity is low for inferior goods, indicating reduced purchase with higher income. Thus, the changing purchasing power of customers influences demand levels of agricultural products differently.

Habits and Preferences. Consumption habits and personal preferences affect demand by influencing the proportion of income that consumers will be willing to spend for a given product. Thanksgiving and Christmas holiday periods create different demand levels for turkey and cranberries than exist at other times of the year. Fads in consumption, nutrition education, or food scares can suddenly shift demand one way or the other.

Prices of Substitute Products. Virtually every commodity has substitute items as consumption alternatives. Grocery shoppers may select protein from vegetables or meat. In meats, they can choose among beef, pork, lamb, or poultry. Food processors can use corn sugar, cane sugar, or beet sugar in a canning process. They may substitute corn oil, soybean oil, cottonseed oil, or palm oil in preparing salad dressings. Cattle feeders can use corn, wheat, barley, or oats as alternative feed ingredients. Relative prices dictate a buyer's choice of which item to select among the alternatives. The effect that price change in one commodity has upon the demand for another is called price-induced cross-elasticity of demand.

Cross-elasticity of demand measures the nature of the effect and degree of response that a price change in one commodity has upon demand for another.

Two commodities are **substitutes** if a price increase in one causes increased consumption of the other. Beef and pork are examples of substitute meats. Rising beef prices encourage consumers to eat more pork. Some foods are **complementary** in that they are eaten together, such as bacon and eggs. An increase in the price of one causes a decrease in consumption of both.

Understanding Demand

There are three important distinctions that must be made in understanding demand. These distinctions are:

(1) demand versus consumption,
(2) need versus effective demand, and
(3) farm level versus retail level demand.

Demand and consumption are often confused. Consumption is the amount of a commodity that is actually used; demand is a schedule of the various quantities that will be consumed at various price levels. Demand is a barometer of consumer buying pressure; consumption is actual use of a commodity. Consumers will consume what farmers produce — **at some price level.** For example, pork consumption may increase significantly from one period to another. Does that mean pork demand has increased? Not necessarily! Consumption may have increased because hog slaughter increased. Because hogs must be marketed and moved into consumption at the time they reach proper weight, an increase in pork supply will cause prices to drop to encourage higher consumption. Thus, increased consumption implies large supplies, not necessarily strong demand.

On the other hand, a low rate of consumption may not mean slack demand; it may reflect reduced quantities available for consumption. Pork producers may reduce production in response to low prices. In time, consumption figures fall as a result. In this case, reduced consumption implies short supplies, not weak demand. Price will rise to limit consumption and prevent pork from disappearing from grocery cases as production decreases.

Demand indicates the differing quantities that will be purchased at differing prices, not the amounts **needed** by consumers. Distinctions must be made between need, want, and effective demand. **Need** is what the other person thinks is good for you. It is what we mean when we talk about millions of people being hungry in the world; they need an adequate diet. **Want** is what I would like to have if I had my choice. I may not be able to pay for it, but I strive to get it. **Effective demand** is what I choose to buy when I can afford it. For demand to be *effective,* consumers must have the income (purchasing power) to pay for it. In the 1960's and 1970's, farmers were encouraged to increase production to feed a hungry world because people needed it. The result was surplus production, depressed prices, and continued need for food for hungry people. It was a mistake to call that need "demand" or encourage farmers to produce to fill the need. There was inadequate purchasing power to back up the need to make it effective demand.

Farm level demand for agricultural commodities is different from retail level demand of grocery products. Consumers do not buy significant amounts of corn; rather they purchase corn products in the form of cornflakes or corn-fed beef. The demand for corn is therefore derived demand, derived from the multitude of products which require corn as a basic resource. Derived demand has important implications upon elasticity designations and determination of important demand factors, as we shall see later.

THE MEANING OF SUPPLY

Supply is a schedule of the varying quantities that will be offered for sale by producers at varying prices at a given time and place. Supply represents the price-quantity relationship from a producer-seller perspective.

The Supply Principle

The principle of supply states that more will be offered for sale at higher prices and less at lower prices. That is, the quantity of the commodity that will be produced and offered for sale varies directly with price.

The relationship can be illustrated in a hypothetical supply schedule of quantities and prices as in Table 4-2. It shows the amount of corn that would be produced and offered for sale at different prices in a production period. At $4.00 per bushel producers would supply 8.45 billion bushels; at $3.00, 7.65 billion bushels; at $2.50, 6.95 billion bushels, and various combinations in between. The figures can be plotted on price and quantity axes to show a supply curve, as in Figure 4-3. The supply curve slopes upward and to the right, since larger quantities are induced as a result of higher prices and smaller quantities result from lower prices.

Table 4-2: Hypothetical Supply Schedule for Corn

Price	Amount Offered
($/bu.)	(billions of bushels)
4.00	8.45
3.75	8.35
3.50	8.20
3.25	7.95
3.00	7.65
2.75	7.30
2.50	6.95
2.25	6.55
2.00	6.00

There are three factors that affect the extent of producers' increased supply in response to higher prices. They are (1) willingness to sell an existing supply from storage, (2) increased use of variable inputs to increase production from a fixed resource base, and (3) the shifting of resources between alternative uses.

With storable commodities, the total amount available is fixed by the production cycle. The supply cannot be changed until the next harvest. However, the day-to-day availability of a portion of the total supply, or the rate at which it filters into the market place, creates an important submarket for commodity producers. Higher prices entice grain out of storage and movement into marketing channels; lower prices encourage continued withholding of grain from the market. A farmer's assessment of prospects for higher price in the future and the cash flow requirements of the business affect the decision of whether to sell now or continue to hold for a higher price in the future. A loan that is due and must be paid now often forces poor decisions about when to sell.

Withholding storable commodities affects only short-run supplies of a commodity. Production was initiated by the farmer to generate income. It is simply a matter of time, usually within the production cycle, before the crop must be mar-

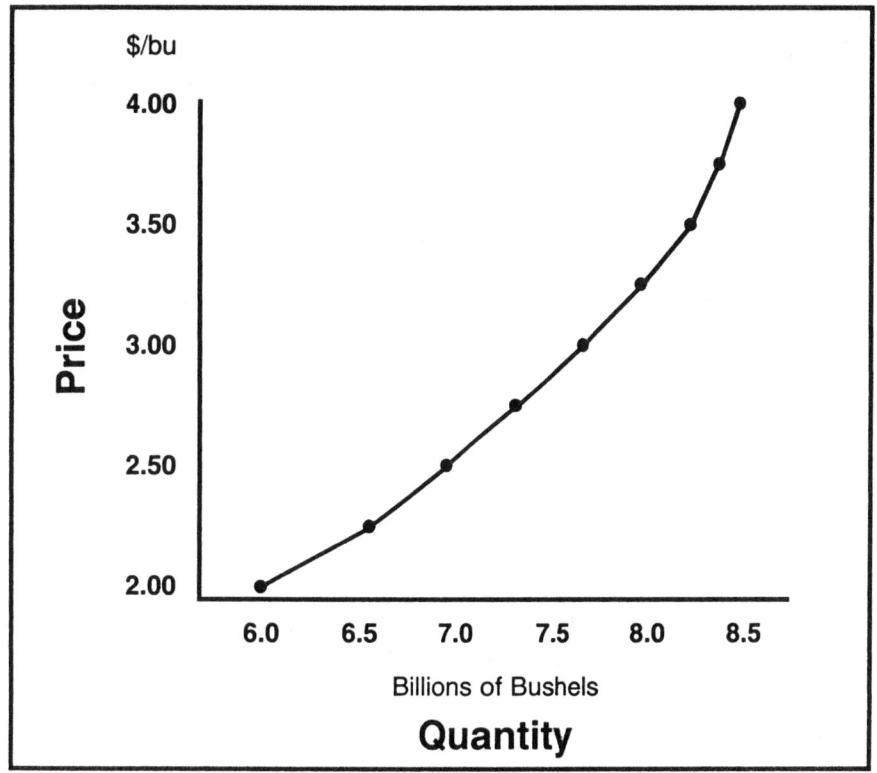

Figure 4-3: Hypothetical Supply Curve for Corn. Source: Kohls and Uhl 152.

keted. Few farmers can finance the accumulation of several years of production in hopes of higher and higher prices being offered by the market. Furthermore, holding commodities in inventory incurs significant costs for facilities and for interest on money tied up in inventory. Prospective price increases can be quickly offset by these holding costs.

For non-storable commodities, day-to-day market supplies are more rigidly fixed by the production process. Producers do not have the opportunity to significantly influence the time span when non-storable commodities flow through the marketing system. Hams, peaches, and tomatoes can be canned and stored, but the bulk of fresh commodities flood onto the market when they are market-ready. The system accepts all the beef cattle, hogs, eggs, fluid milk, peaches, tomatoes, lettuce, and strawberries offered and rapidly moves these quantities to consumers, regardless of how low price must fall to clear the market or how high price must rise to ration a limited supply. Non-storable commodities necessitate a different approach to market decisions than storable commodities. The differences will be examined later.

The second supply response to higher price arises from a producer's ability to increase output by increasing the amount of variable inputs committed to the production process. The classic example is nitrogen application on corn. Corn yield can be increased with increased nitrogen application, all other things remaining the same. The physical transformation of nitrogen into corn defines how much additional production arises from additional nitrogen application. As depicted in Figure 4-4, each additional unit of nitrogen produces a smaller and smaller increase in additional corn. This is referred to as the diminishing return from additional units of a variable resource. The ability to increase production is limited by physical relationships inherent in nature. Excessive quantities of nitrogen will actually decrease yield.

Figure 4-4: Illustration of Physical Production Response of Corn Yield to a Variable Input, Nitrogen.

If the price of corn increases and nitrogen price remains the same, the additional production is worth more. Thus, a farmer can afford to apply more nitrogen to increase production and increase the supply available. On the other hand, a lower corn price makes the additional bushel of corn produced worth less. Consequently, the producer is encouraged to reduce the nitrogen application and be satisfied with lower yield. The market consequently receives lower production. With changes in the profit-maximizing yield level, the market, therefore, receives increasing or decreasing supply in response to changing price of a commodity.

The production relationship described for nitrogen exists for all variable production inputs (e.g., feed, fertilizer, seed, chemicals, fuel, and extra labor). If the crop is worth more, there is more effort and expense applied to increase yield. Thus, total aggregate production increases as more bushels are produced on each acre in response to higher price. If crop value is decreased by lower price, the variable inputs are reduced and a smaller aggregate supply is available for sale.

The third supply response involves the relative market prices of competing enterprises. An acre of cropland can be utilized to produce cotton, soybeans, or rice in the South; corn, wheat, or soybeans in the Corn Belt; and wheat, barley, or grain sorghum in the Great Plains. Given time for adjustment, land use will be dictated by relative prices of the competing crops. To maximize profit, each producer will calculate the net return to land from the various alternative uses and plant the most profitable crop.

Relatively higher soybean price will encourage acreage shifts out of corn and cotton and into soybean production. Table 4-3 shows a simplified budget of comparative returns to corn and soybeans. The important parts of a budget are prices, yields, and production costs. Within any one production period, relative production costs of each crop do not change materially. Relative yields are also fairly constant. The only decision variable in determining what to plant is, therefore, relative price. If a soybean price 2.5 times as great as corn generates the same $100 per acre net return to land, a price ratio higher than 2.5 would encourage more soybean planting at the expense of corn acreage; less than 2.5 would encourage a shift to more corn. Most crop acres have several alternative uses and therefore have critical price ratios that will encourage shifting use from one crop to the other. Relative price dictates land use and the resulting supply of competing crops.

Table 4-3: Illustration of Net Return to Corn Compared to Soybean Production

Item	Corn	Soybeans
Price per Bushel	$2.50	$6.25
Yield	150 bu.	50
Gross Receipts	$375.00	$312.50
Production Costs	$275.00	$212.50
Net to Land	$100.00	$100.00

Relative Prices: $6.25/$2.50 = 2.5:1

Supply Elasticity

Supply elasticity is a measure of the extent of increase or decrease in quantity supplied in response to higher or lower market prices. Supply responses have the same elasticity designations as demand except that they are positive since increased price encourages increased production. Commodities that are highly responsive to price change have **elastic** supply curves. With them, a small change in price will encourage relatively larger increases in supply. The supply responses that react relatively little to price changes are **inelastic.** A large price change produces only a small supply change. In between elastic and inelastic is **unitary** response, where price change and supply change are proportionately the same.

Time designations are very important in supply elasticity considerations. Time is classified as short run, intermediate run, and long run. Corn, wheat, and cotton have annual production cycles and relatively short harvest periods. Once the planting period has passed, supply is fixed for the most part until after harvest of the next planting period. Weather conditions during the growing season and imports can modify the available supply within a country, but changes in supply during a production period are generally relatively small and are largely beyond control of producers.

Livestock supplies are also fixed by production cycles from breeding to market weight. Regardless of price change, little can be done to change the number of animals coming to the market during that period. It is only after a production cycle, when females can be diverted from the slaughter market into the breeding herd, that livestock supplies can be increased. In the short run, supplies are largely fixed by production cycles. Changes in supply are confined to changes in market weight and numbers of animals kept for breeding. Short-run supply is almost perfectly inelastic; it is a vertical line in Figure 4-5.

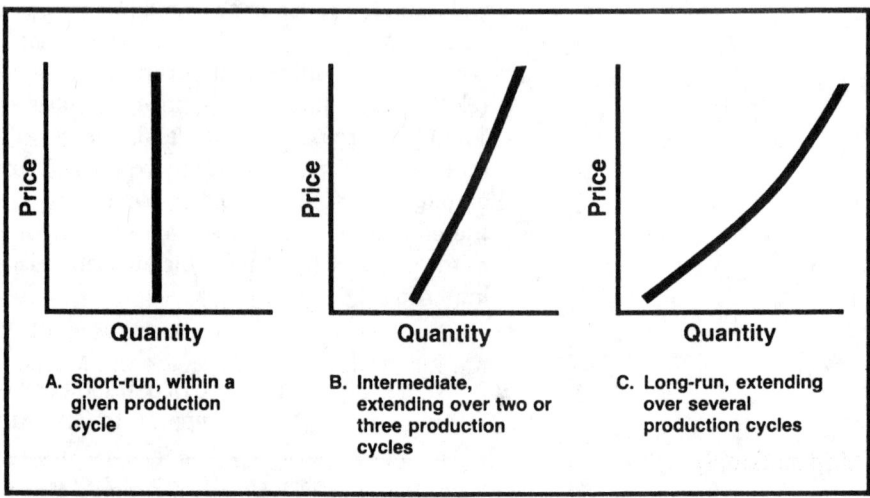

Figure 4-5: Hypothetical Supply Curves Illustrating the Effect of Time on Elasticity. Source: Kohls and Uhl 161.

In the intermediate period, production can be increased or decreased within the limits of existing facilities. Crop producers can apply more production inputs which increase yields; livestock producers can add females to the extent they have surplus grazing or building space. Supply increase due to changes in application of variable input occurs over two or three production cycles. It is more responsive to price change than short-run adjustments but is still relatively inelastic.

Once existing production facilities are used to capacity, it takes longer-run adjustments, over several production periods, to further increase supply in response to higher price. This consideration represents long-run supply adjustments in Figure 4-5. In this case supply increases depend on building more facilities, bringing more land into production, adding equipment, and/or applying technology.

Within the intermediate and long run adjustment periods, supply elasticity depends upon the ease with which supply adjustments can be made and the costs associated with changing production levels. If changes can be made easily and inexpensively, supply curves will be more elastic. If changes are difficult and expensive, supply responses will be more inelastic.

Supply reductions are fairly inelastic in response to price decline. It takes a relatively long time and continued low prices to force farmers to reduce production. This occurs because of the mix of fixed and variable costs making up total production cost. Land, labor, and machinery charges represent a large proportion of the total cost of producing agricultural products. They constitute the fixed costs of production and are incurred whether or not there is any production. The variable out-of-pocket production costs (e.g., seed, fertilizer, fuel, chemicals, and other inputs) are incurred only if production is initiated. Cutting production, by leaving resources idle, eliminates these variable costs. However, fixed costs go on. Thus, farmers will continue to produce if they can get more for the product than they spend for variable out-of-pocket costs. Even though they are not covering total cost of production, they minimize their losses by producing when they cover variable costs and have something left over to apply to a part of these fixed costs.

The time period of adjustment is important in supply response to lower price. In the short run, farmers may actually increase production in response to lower price in hopes of maintaining income. Since price per unit is lower, they reason that more units need to be produced in order to maintain a cash flow. In the intermediate run, production will decline as machinery wears out and facilities deteriorate. In the long run, producers quit agriculture, and land and facilities are taken out of production or revert to some other use. Supply eventually adjusts to equate total production cost with price, but the time period may be long and the human cost of adjustment may be extremely painful.

The almost totally inelastic supply of agricultural commodities within a production cycle and the highly inelastic responses in both intermediate and long run periods help explain much of the volatility of agricultural prices. Price changes are much more severe because of inelastic supply responses.

Shift in Supply

Up to now we have looked at the change in quantity supplied as it relates to price level. We looked at one supply curve and the relationship between changes in quantity and price. Figure 4-6 illustrates a shift in the entire supply curve. The original supply curve is S_O; the curve S_I shows an increase in supply; and the other curve, S_D, shows a decrease in supply. Notice that price, P, does not change. An increase in supply is defined by producer's willingness to provide a larger quantity to the market at the same price levels. A decrease in supply is defined as a reduction in quantity supplied at the same price levels.

In contrast to a change in quantity supplied, a shift in supply is a different economic concept. A shift in supply means that the entire supply curve moves to the right (increased supply) or to the left (decreased supply). To identify shifts in supply, the price level stays constant, but the quantity of supply available in the market changes.

Several factors lead to shifts in supply:

1. Technology: By definition, technology means more production from the same or similar resources. Thus, cost per unit of output is lowered as higher yield is produced. Producers are willing to sell more at each price level. Hybrid seed is often cited as the example of technological improvement that increased corn yield per acre and reduced per-unit production cost. Its adoption as a production practice led to increased supply. Farmers were willing and able to produce more at the same price because technology increased output and lowered average cost.

2. Innovation: This refers to the improvement in production practices that increase yield or reduce cost. Yield increases due to fall tillage of heavy soils are but one example. Plowing costs do not change by adopting the practice, but weather effects enhance the soil's productive capacity. Adoption of the innovation lowers unit cost of production and, thereby, shifts the supply curve to the right.

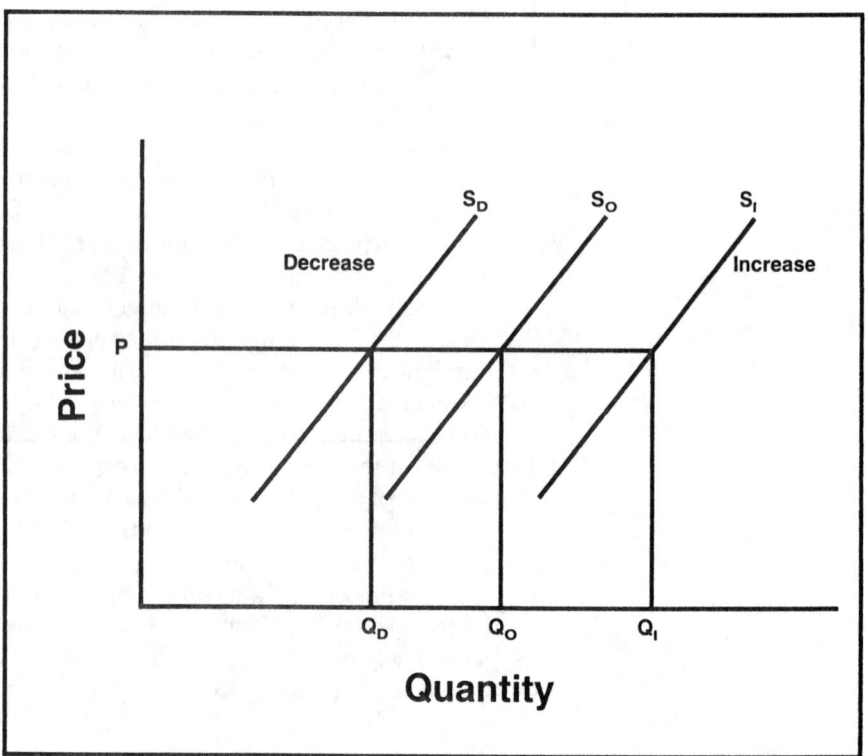

Figure 4-6: Changes in Supply.

3. Rules and regulations: Compliance with governmental regulations affects costs and, in turn, supply. Regulations specifying the handling of livestock wastes may require construction of facilities for handling effluent. Increased cost of operation shifts the supply curve to the left (reduce supply) as producers are unwilling to maintain production at the same price level in view of the higher costs mandated by new regulations.

4. Input cost: As resource costs change relative to each other, economic advantage shifts between crops. Higher nitrogen fertilizer price increases corn and cotton production costs to the advantage of soybean production.

5. Government programs: Incentives to improve drainage, develop irrigation, or grow conservation crops and idle cropland shift supply curves right or left and affect how much production is forthcoming at a given price level.

THE EQUILIBRIUM PRICE

The forces of demand and the forces of supply acting alone affect consumer or producer decisions separately and individually as they react to market price. Buyers want low prices; sellers want high prices. Somehow the two sets of forces must be brought together in a market arena to resolve that conflict. Table 4-4 shows both the demand schedule and supply schedule for corn which might exist in a market. At a price of $2.75 per bushel, buyers will take 7.3 billion bushels and sellers will produce and offer for sale 7.3 billion bushels. The amount supplied **equals** the amount demanded; this is called the equilibrium quantity. The equilibrium price is $2.75. It is the market clearing price level which will be established by demand and supply forces in a competitive free market economy. Figure 4-7 presents a picture of the relationship. Where the demand curve, D, crosses the supply curve, S, the two are equal. P_E is the equilibrium price, and Q_E is the equilibrium quantity.

Sellers would prefer a higher price, such as $3.50; buyers would prefer $2.25. Not all of the participants are satisfied with $2.75, but there is no other acceptable compromise. At a higher price, say $3.50, sellers would increase production to offer 8.2 billion bushels; but buyers would be willing to take only 6.5 bil-

Table 4-4: Hypothetical Demand and Supply Schedules for Corn

Demand	Price	Supply
(billions of bushels)	($/bu.)	(billions of bushels)
6.1	4.00	8.45
6.3	3.75	8.35
6.5	3.50	8.20
6.75	3.25	7.95
7.0	3.00	7.65
7.3	**2.75**	**7.30**
7.6	2.50	6.95
8.0	2.25	6.55
8.5	2.00	6.00

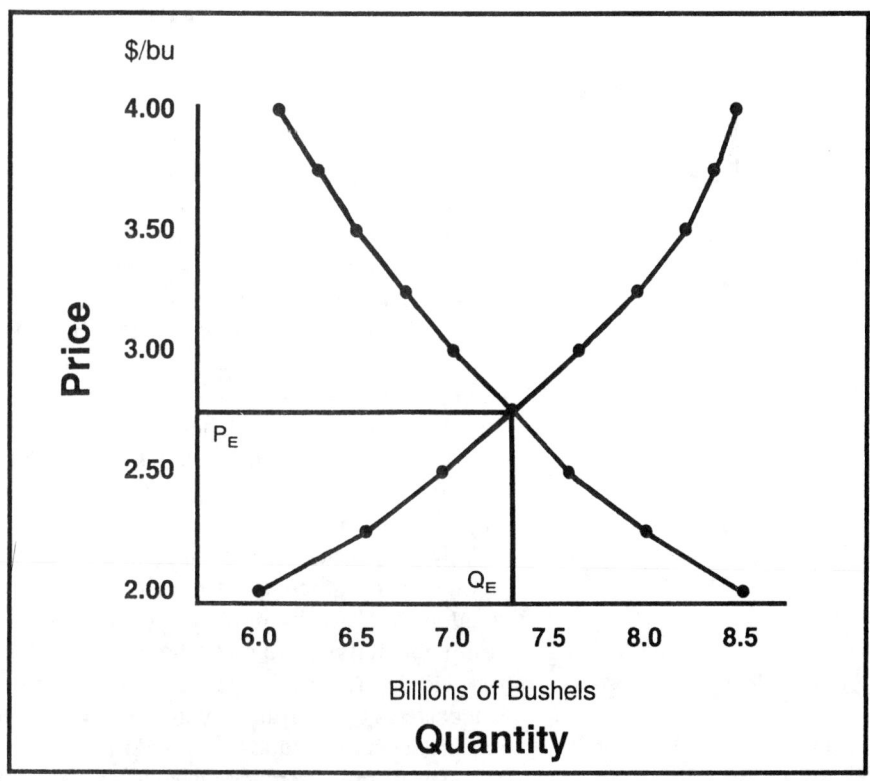

Figure 4-7: Equilibrium Quantity and Price.

lion bushels. A surplus of 1.7 billion bushels would hang over the market and begin to drive price down. In a market with price above the equilibrium price, it is a buyer's market in which more is offered than is demanded. Price cutting occurs, and the general price level declines to move the available supply.

Should the price be below equilibrium at a level of $2.25, sellers offer only 6.55 billion bushels. Buyers want 8.0 billion bushels at that price. A shortage of 1.45 billion bushels exists and pulls up the price level. It is a seller's market, and the price level rises to ration the limited supply.

At $2.75, buyers and sellers who enter the market are satisfied; 7.3 billion bushels change hands. An equilibrium price and quantity are established.

Changing Price from Shifting Demand or Supply

Various and opposing factors may be present in a market at any particular time to shift demand or supply curves, possibly both at the same time. Fortunately, the forces causing shifts in demand are distinctly different from those that change supply. It is, therefore, possible to trace the expected results of changes on equilibrium price and quantity by shifting one of the curves and moving along the other. Figure 4-8 illustrates whether rising or falling price can be expected due to demand or supply shifts. The solid lines represent the original situation, and the dotted lines represent the shift in either demand or supply, or both at the same time.

Either increased demand or reduced supply can be expected to increase equilibrium price. Likewise, decreased demand or increased supply will reduce price. If both curves are shifting at the same time, the price effects of one curve may be offset by the other, or they may magnify price effects of each other. Further on, we will discuss the important factors which shift demand curves or supply curves of farm commodities and try to assess the extent of price change to expect from those changes.

The Supply-Demand Balance

In the real world, a stable demand-supply equilibrium is seldom achieved. Supply and demand conditions are constantly shifting. On the supply side, variations in production decisions, government policy, technology, weather, disease, and unpredictable events affect acreage, yield, and total production. Furthermore, the time span required to adjust to changing price levels greatly affects the supply response. Demand changes occur from shifts in exports, population, consumer incomes, tastes and preferences, employment, business conditions, and political events.

The forces at work in achieving a supply-demand balance are illustrated in Figure 4-9. Equilibrium is established in relative terms by the balance between supply and demand. As the balance is tipped one way or the other by supply or demand factors, price simultaneously changes.

The price signals elicit opposing reactions from consumers and producers. Reduced supply increases the relative price and signals producers to increase production. At the same time, the price increase signals consumers to reduce consumption. In time, as a response to the increased supply and lower consumption, the price indicator swings back toward the right of the scale. The resulting lower price signals the consumer to increase consumption while at the same time signaling producers to reduce production.

The supply/demand balance illustrates both the communication role of price in marketing and the built-in stabilizing feature of price-coordinated mar-

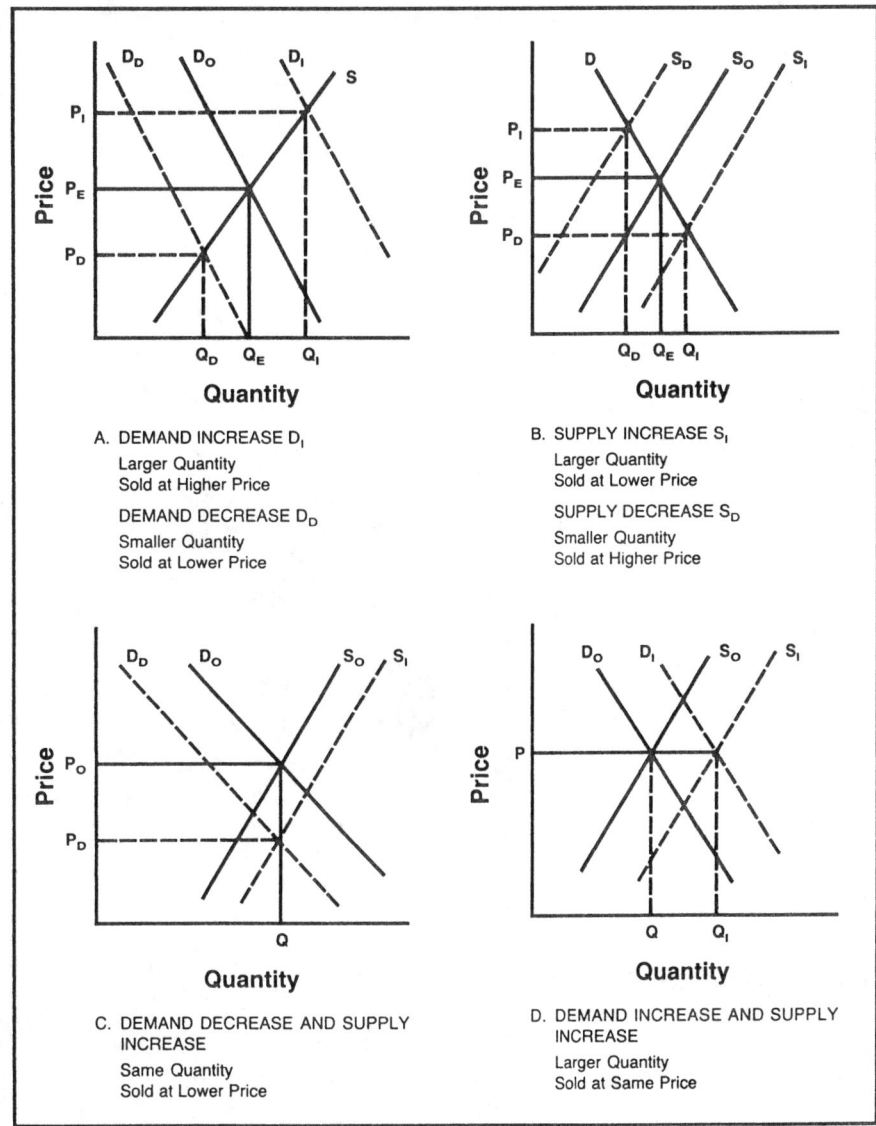

Figure 4-8: Illustration of Expected Price Change Due to Changes in Demand or Supply.

kets. Higher price simultaneously tells producers to increase production and tells consumers to use less. Lower price tells producers to reduce production and consumers to buy more. Adjustments to price changes maintain a supply/demand balance. But in the meantime, prices rise and fall in order to induce the adjustments.

The location of the fulcrum on the indicator needle of the supply/demand balance emphasizes the inelastic demand for agricultural commodities. Small changes in supply or in demand produce large changes in commodity prices.

In theory, equilibrium price and equilibrium quantity are specific and clearly defined. In reality, prices undershoot or overshoot the mark in search of the price which will clear the market. Prices are dynamic and change day to day, hour to hour, minute to minute. No one really knows exactly how much supply is available

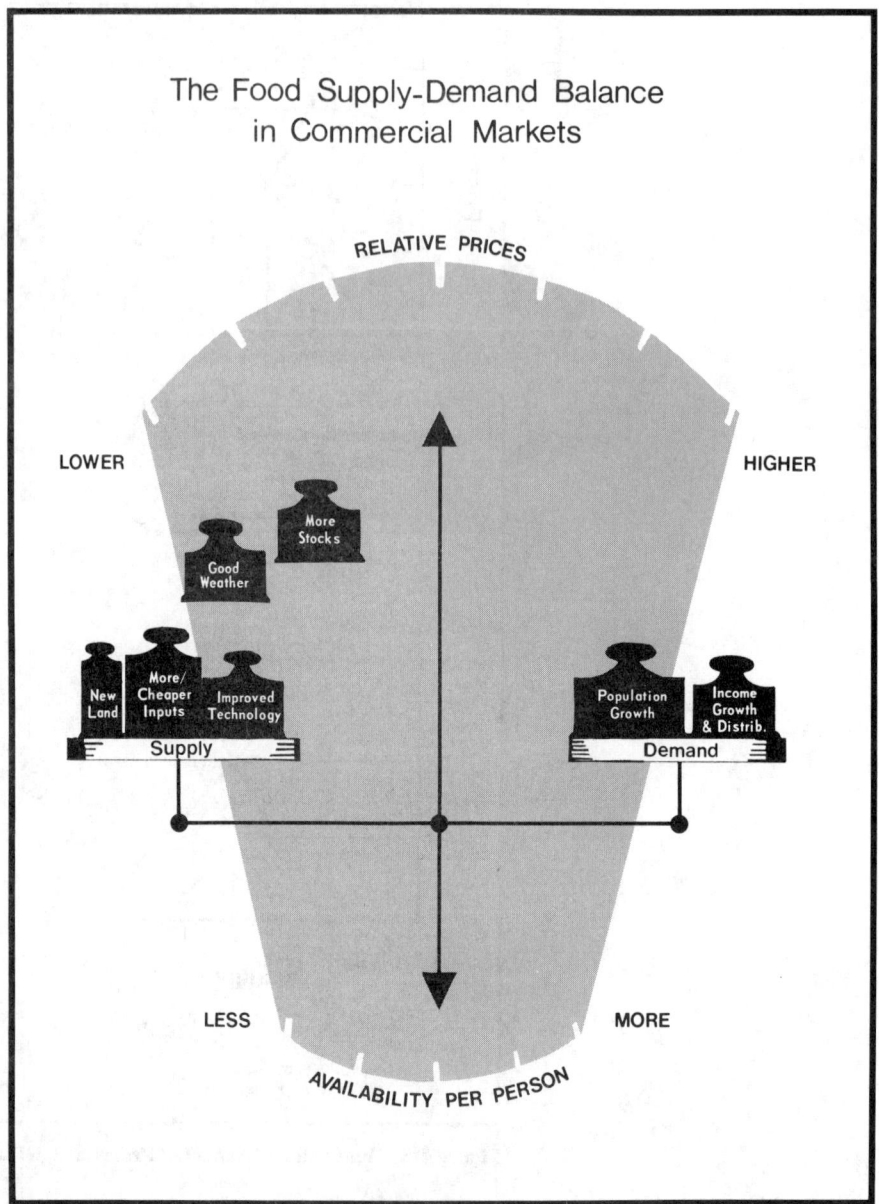

Figure 4-9: The Food Supply-Demand Balance. Source: James P. Houck and Wallace Barr, *Your Food: A Food Policy Basebook* (Columbus: Ohio State University Press, revised 1977) 6. Reprinted by permission.

or required. Furthermore, the time element greatly complicates the process. Large receipts on today's market may or may not depress price level. A large crop might be followed by gradually increasing price throughout the year, and a small crop may experience a falling price level following harvest. Because of imperfect knowledge, day to day observations may be the opposite of what theory suggests should be true. It is more nearly correct to say that equilibrium price is that price toward which price level tends to move but somehow never reaches. Understanding the principle is nonetheless important in trying to sort out price-making factors.

Price Discovery

At each point in the food chain from farm to fork, whenever title to a commodity changes hands, a transaction price must be determined. This price determination process needs to be related to the economic principles that explain equilibrium price. To begin, we must distinguish between price equilibrium and price discovery. **Price equilibrium** is the process by which the broad forces of demand and supply establish a general, market-clearing price for a commodity. **Price discovery** is the process by which buyers and sellers arrive at a specific price for a given lot of a commodity at a given location on a particular day. The overall demand-supply equilibrium price must be "discovered" and applied to each transaction in the market place.

The process of price discovery has two distinct phases:

1. Evaluating the conditions of demand and supply to determine the general level of price for the commodity. This is the price level around which prices for particular lots of the commodity, at different locations, of differing qualities, and in different time frames, will fluctuate.

2. Determining the value of a specific lot of the commodity being exchanged, relative to the general market level. This phase takes into consideration grade differences, location, buyer and seller services, and relative bargaining power of seller and buyer.

"Price discovery is a human process, beset by errors of fact and judgment."[2] There is no guarantee that buyers and sellers will always discover the equilibrium price, but the automatic stabilizing features tend to move in that direction. In real life, each buyer and seller operates in his or her own best interest according to the profit motive. The variations and imperfections in the price-discovery process make it profitable for both buyers and sellers to shop around among alternatives and to bargain on price and other terms of trade. This bargaining process leads to price discovery and price equilibrium.

Pricing errors occur at all stages of the price discovery process. But the errors can be minimized and pricing improved by high-quality market information and commodity grading programs so that all parties to the trade are acting on comparable information.

Two people in the food chain are most concerned about price level: the farmer on one end and the consumer on the other. Intermediaries usually operate on spreads, or price margins, between what they receive for the goods as they pass them along the food chain and what they must pay their suppliers. Intermediaries are often considered the "bad guys" who pay producers too little and charge consumers too much. Actually, intermediaries do not "determine" prices, since they do not determine consumer demand or market supplies. They merely "discover" prices based on their evaluation of the quantities of supply available and what customers are likely to pay for the amounts offered to them. Since intermediary profit margins are relatively stable, the **fluctuations** in price received by producers reflect, for the most part, changes in consumer demand and changes in producer supplies available on the market. The farmer must decide if the current price level makes it worth the effort to allocate resources to produce commodities for the market. The consumer decides if the products offered in the supermarket satisfy the family's needs and desires at an acceptable price. If both concur, the system will continue to function as it is; if either does not concur, there will be a change. This is how individual decisions are translated into consumer demand, market supply, and equilibrium price.

Law of One Price[3]

Despite the tendency for market-making forces to establish an equilibrium price, numerous different prices exist in various markets. The *law of one price* provides an explanation of these differences to confirm the existence of one equilibrium price for the market as a whole.

The Law of One Price (LOOP) states that "all prices in a market are uniform, after taking into account the costs of adding place, time, and form utility to products within the market."[4] The law defines the relationship among prices in terms of space, time, and form utilities created by the marketing process. Understanding the law of one price is important in understanding differences in price at different points in the country or world, at different times of the year, and in different product forms available in the market.

Figure 4-10 illustrates the law of one price in terms of location differences in the international corn market. The map shows local prices from three shipping points in the Corn Belt (Columbus, Champaign, Farnhamville), transportation

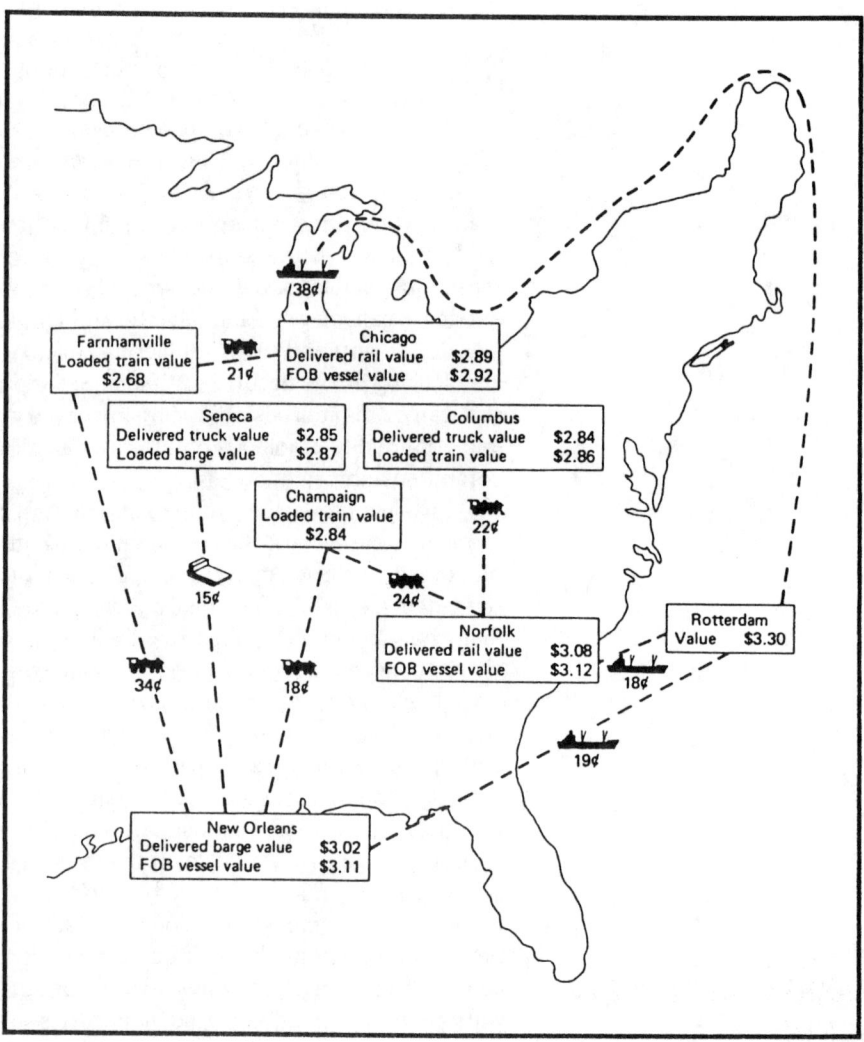

Figure 4-10: Relative Prices in the International Corn Market. Source: Kohls and Uhl 169.

costs to different markets, three export markets (Chicago, New Orleans, Norfolk), and a major port of entry in European trade (Rotterdam). Grain tends to flow from surplus production areas through the marketing channels to meet the needs of deficit production areas. Prices at the different locations are aligned according to the law of one price. They differ only in the cost of transportation, or the cost associated with providing location utility.

Under competitive market conditions, prices tend to maintain the same relative values over space, time, and form. If European customers are willing to pay an additional 50 cents per bushel, grain traders see the extra profits that could be made by increasing the grain trade from the U.S. to Europe. Assuming that the shipping charges are constant, traders willingly increase their bid price to entice a larger flow through the system. Competing traders make successively higher bids trying to increase their volume of supply going to the European market. After the adjustments are made, the U.S. – Rotterdam price difference will return to the cost of transportation, but both the Rotterdam and U.S. price levels will be higher, U.S. price pulled up by a higher Rotterdam price. Traders bidding against each other to capture the additional volume serve to reduce the extra profits promised to them. Price differences within the market return to levels consistent with the law of one price. In the meantime, the traders perform a useful function in shifting the corn from the lower-value market to a higher-value market.

LOOP is useful in explaining differences in price over time. Crop harvest occurs in relatively short periods of time, but consumption is fairly stable each day of the year. The harvest surplus must be carried forward to satisfy daily consumption requirements before the next crop is harvested. There are costs of building storage facilities to hold the crop, and there are interest costs incurred by owning the commodity in storage. The law of one price leads to the obvious conclusion that the difference between market price late in the crop year and the earlier harvest-time price will have to be large enough to pay the costs of holding the commodity in storage. If the difference is narrow, grain merchants will sell immediately rather than hold until later because there would be no profit to be realized from storage. The increased volume coming to the market at harvest tends to drive harvest-time prices lower. Processors and other users, seeing a potential shortage of supply later in the season, raise offers for later delivery. Thus, the price difference will tend to widen to cover the costs of storage and reestablish relative prices between harvest and later consistent with LOOP.

Finally, the law of one price explains the addition of form utility in the creation of different products from a common source commodity. A corn producer can sell cash grain or change the form of corn grain into milk, pork, beef, or poultry products. Producers will allocate their supply of the raw commodity to alternative products in such a way that the price they receive for the products, less their respective production costs, will equal the cash corn price. Whichever product will generate a larger return above cash costs and additional production costs will be increased in volume. When cash corn is worth $2 per bushel and hogs are worth $60 per cwt., farmers will convert corn into hogs and "walk" $5 per bushel corn to market. Again, LOOP explains the differences in price of different forms of a commodity.

MARKET FORCES IN ACTION

An interesting analysis for understanding basic price-making forces of demand and supply was presented by Robert L. Heilbroner in tracing market interactions of the Soviet Grain Sale.[5]

In midsummer of 1972, the Nixon administration startled the world by announcing that it had sold one-fourth of the American wheat crop to the Soviet government. It was the largest agricultural commodity sale in history, worth more than $1 billion. As a result, wheat prices began to soar from a level of $1.70 per bushel in mid-July before the announcement to an all-time high of $5.00 in late summer of 1973.*

The first point to understand is that the 1972 crop had been planted and, to a large degree, completely harvested when the announcement of the sale was made. Thus, the 1972 supply of wheat was fixed; the supply curve was totally inelastic. Production could not be increased until the next crop year. The additional demand from the U.S.S.R., therefore, made its effect wholly as an increase in price, as shown in Figure 4-11. Had the Soviets contracted to buy grain for 1973 delivery rather than 1972 delivery, the situation would have been quite different. More grain could have been planted in the next crop year to satisfy the additional demand. As a result, the equilibrium price, P_2, with sufficient forewarning to increase supply, would have been considerably less than the equilibrium price, P_1, given no advance warning.

The rise in wheat price was only the initial shock. The ripples spread to other commodities through the substitution effect. Since corn, oats, and rye are close substitutes for wheat in many uses, the cross-elasticity of demand is high. That is, an increase in the price of one feed crop will cause substantial increases in the quantities demanded of other feed crops. As a result, the rise in wheat prices

Figure 4-11: Illustration of Price Effect of Increased Demand from Russian Wheat Deal. Source: Heilbroner.

---

*Farmers called the sales "The Great Russian Grain Robbery" because price was so low. On the other hand, the Soviets paid for the grain with $35-an-ounce gold. Subsequently, gold price rose to about $850 per ounce. Farmers thought they were underpaid; probably the Soviets thought they overpaid.

caused a shift in the demand curves for other grains, as Figure 4-12 shows for soybeans. Although the increased demand was confined to wheat, the purchase quickly induced price increases among all the grains that could be substituted for wheat. Combined with other factors that changed demand, the effect was to boost the price of oats from $0.80 to $2.06 only 13 months later, rye from $1.01 to $3.86; and soybeans from $3.50 to a peak of $12.00.

Figure 4-12: Illustration of Increased Demand for Soybeans Due to Cross-Elasticity of Wheat Demand. Source: Heilbroner.

The Russian Wheat Deal, though important in the commodity price explosion, was not the only cause. Because of El Niño, the Humboldt ocean current off the west coast of South America shifted position in 1972. The change caused a significant drop in the anchovy crop — so significant, in fact, that the Peruvian government ordered a halt to all anchovy fishing lest anchovy breeding capacity be severely damaged. Anchovies are an important source of world protein supplies and a direct substitute for soybean meal. Once again, the substitution effect on prices was important. As fish-meal prices soared in response to reduced supply, there was another induced demand for soybeans and feed grains as substitutes for anchovies.

A final influence on commodity prices was devaluation of the U.S. dollar in 1972 and 1973. Devaluation made exported items less expensive to U.S. trading partners. Devaluation induced a shift in the dollar demand curve for all American crops. Japanese buyers may have been purchasing U.S. grain at an exchange rate of 375 yen to the dollar. For them, $4.00 soybeans cost 1,500 yen per bushel. Devaluation meant that the dollar was cheaper; it took fewer yen to exchange for $1.00. At 300 yen to the dollar, $4.00 soybeans cost only 1,200 yen. In response to the effectively lower price, the Japanese increased their soybean purchases.

Because the supply of soybeans was fixed by the crop year, large increases in price occurred due to the change in demand. The response is illustrated in Figure 4-13. The graph shows the rise in soybean prices from P_0 to P_1, as induced demands shifted the demand schedule to the right.

The effects on commodity prices did not stop with the grain sector of the economy. Soybean products are used for animal production as well as for human consumption. Higher soybean prices increased the cost of producing meat. Other things being equal, higher costs reduced profitability of beef production and caused feeders to decrease production. The supply of meat output was reduced, or in economic terms, the meat supply shifted to the left.

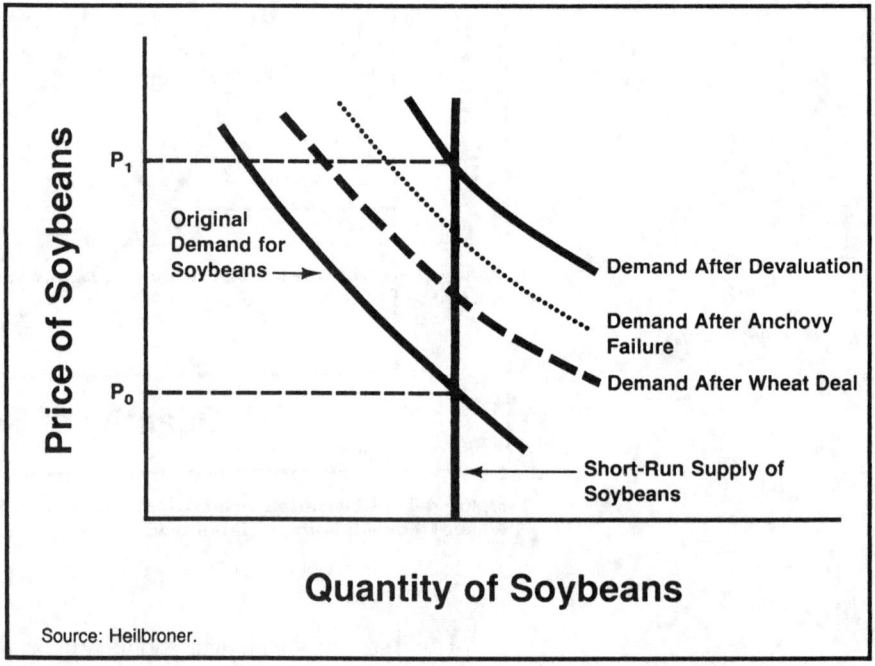

Figure 4-13: Illustration of Price Effect of Induced Shift of Demand Schedule for Soybeans. Source: Heilbroner.

The adjustment phase in livestock production created one of the peculiarities of agricultural markets. When manufacturers want to reduce production, they lay off workers and close the factory. To reduce production, the livestock breeder must reduce the size of the breeding herd. The only way to do that is to sell females that would otherwise be used for breeding. In other words, to get a long term decrease in the overall breeding herd, agriculture cannot avoid a short term increase in the supply of meat brought to the market. Thus, the price of beef or pork first goes down and rises only after breeding herds have been reduced. These effects are shown in Figure 4-14. They created a lagged response between the occurrence of events and the price effects of these events.

These interactions help us to understand what happened to beef prices. In midsummer 1972, Omaha steers were selling for $37.50 per hundredweight. A few months later, price had fallen to $33.00. Thereafter, prices rose to well over

Figure 4-14: Illustration of Price and Quantity Responses in Meat Production Due to Changing Profitability. Source: Heilbroner.

$50.00 as the supply curve for beef shifted to the left in response to reduced supplies.

In this review of effects of the Russian Wheat Deal, we have applied the economic terms of demand and demand schedules, supply and supply schedules, prices, substitution, elasticity, and lagged responses. Economic principles explain price-making events in the real world as the free markets act and react to changing conditions.

APPLICATIONS OF DEMAND AND SUPPLY ANALYSIS

A study of economic principles forms the foundation for understanding demand and supply influences upon price determination. However, definitions of economic terms and concepts are rather sterile until they are applied to real-world situations. We will develop some of those applications to improve our understanding of the principles and to provide a framework for analyzing effects of changing conditions on prices.

Effects of Demand Elasticity on Price and Gross Receipts

Commodity producers need to understand price elasticities of agricultural commodities if they wish to project the effects that changes in supply will have upon price and income.

Table 4-5 reports price elasticities of agricultural commodities at both the farm level and the retail level. There are several important implications that can be gleaned from the elasticity coefficients. First, all food commodities have inelastic demands at the farm level and almost all at the retail level. The elasticity coefficients show that a 1 percent price change is associated with less than a 1 percent change in quantity sold at the new price. Inelastic demand means that consumers are relatively insensitive to food price changes. It also means that increased sale volume can be attained only with relatively large price concessions. On the other side of the coin, inelastic food demand means that relatively small decreases in market supplies will produce proportionately larger price increases.

Second, farm-level price elasticity is always lower than the retail-level price elasticity. On graphs of the same scale, the more inelastic demand curves can be illustrated with steeper slopes, as shown in Figure 4-15. More elastic demand

Table 4-5: Retail and Farm Price Elasticities for Selected Foods

Commodity	Effect of a 1 Per Cent Change in Price on the Percentage Change in Quantity Demanded	
	Farm Level	Retail Level
Turkey	—	-1.56
Margarine	-.69	- .84
Chicken	-.60	- .78
Apples	-.68	- .72
Butter	-.46	- .65
Beef	-.42	- .64
Ice cream	—	- .52
Cheese	—	- .46
Pork	-.24	- .41
Fresh milk	-.32	- .34
Eggs	-.23	- .31
Potatoes	-.15	- .30
Bread, cereals	—	- .15

Source: P.S. George and G.A. King, *Consumer Demand for Food Commodities in the U.S. with Projections for 1980*, University of California, Giannini Foundation Monograph No. 26 (Berkeley: University of California Agricultural Experiment Station, March, 1971). Reprinted by permission.

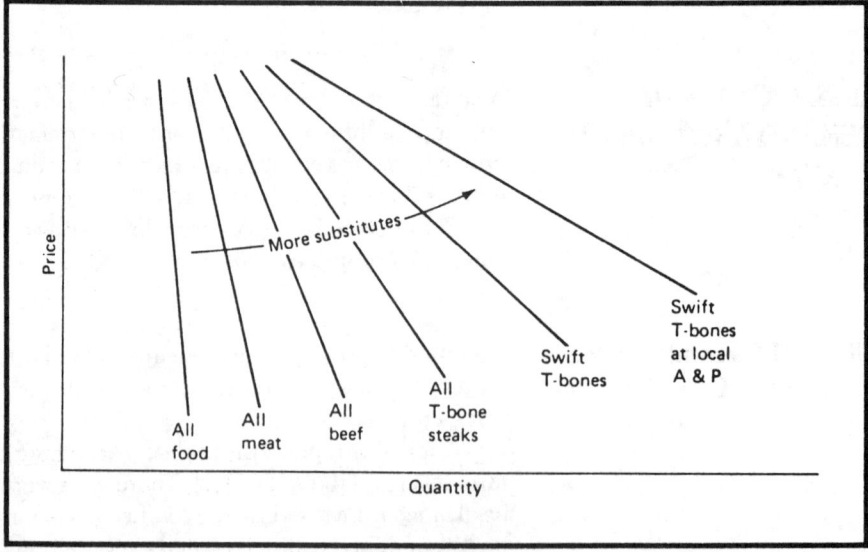

Figure 4-15: Demand Elasticity at Various Market Levels. Source: Kohls and Uhl 158.

curves at comparable prices are depicted with flatter slopes. Food products generally increase in price elasticity as they move closer to consumption. This reflects the greater substitution possibilities for branded items at a local store compared to raw commodities at the farm gate. Raw commodities at the farm gate have more inelastic demands. As a consequence, the price volatility for farm commodities will be much more severe than the price volatility at the grocery store.

The final note is the wide range in elasticities associated with different products. Consumption of basic foods such as bread and potatoes is almost totally insensitive to price change. Fruits, vegetables, and meats are somewhat more price-responsive. Thus, the kind of commodity that a farmer produces determines the price volatility he or she might experience as a consequence of demand elasticity.

From the producer's perspective, demand elasticities provide a measure of the price change that can be expected from supply changes. Inelastic demand coefficients indicate that relative price changes will be much greater than relative supply change. The price responses in Figure 4-16 bear out this point. Commodities with very low demand elasticities, such as potatoes and bread, are subjected to large price changes as the result of a small change in supply. A lesser inelastic demand for chicken produces less severe price changes.

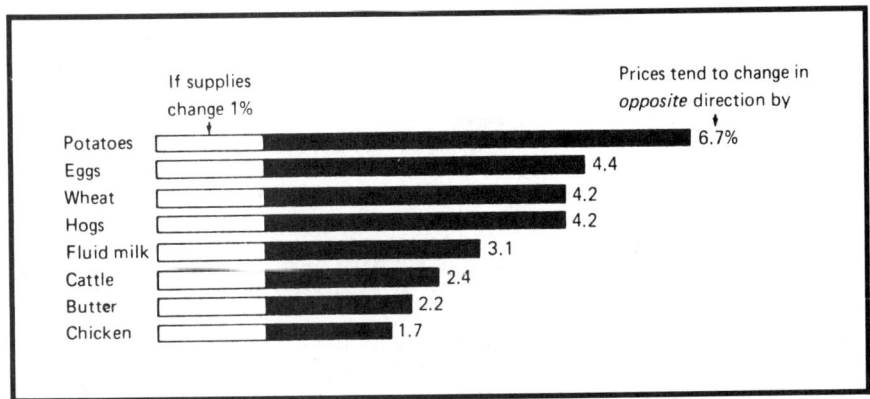

Figure 4-16: How Change in Supply Affects Price Received by Farmers. Source: Kohls and Uhl 198.

Are farmers better off selling a large quantity at a lower price or a smaller quantity at a higher price? Demand elasticity explains how receipts will be affected by price and quantity changes. Total receipts from a sale are computed by multiplying market price by quantity sold. For a commodity with unitary elasticity, a 10 percent decrease in price increases the amount purchased by 10 percent, and the gross receipts are exactly the **same** before and after the price cut. The increase in volume of sales exactly offsets the price decline and thus gross receipts are constant. For a commodity with elastic demand, the amount purchased increases by more than 10 percent and thus gross receipts are **greater** than before. A commodity with inelastic demand experiences less than a 10 percent increase in quantity purchased, so that gross receipts **decline.** These calculations are illustrated in Table 4-6.

The impact that inelastic demand for farm commodities has on farmers' gross receipts is especially important in commodity marketing. Demand elasticity is important to farmers in that it tells them what to expect as price and income change as supply increases or decreases. A farmer thinks that producing more product will increase his or her income. But the world does not work that way. More total aggregate production clears the market at a lower price. Inelastic demand means that a 10 percent increase in total supply produces as much as a 20 percent price decrease. An increase in supply and proportionately lower price reduces total revenue. Table 4-7 shows hypothetical corn consumption levels and price. The total value of the crop is determined by multiplying price times the quantity that would be consumed at that price. From the example, a low price of $2.00 per bushel and a large crop of 8.5 billion bushels has a gross value of $17.0 billion, and a high price of $4.00 per bushel on a small crop of 6.1 billion bushels has a gross value of $24.4 billion. Clearly, there is more income from the smaller crop. Furthermore, the larger crop probably came from more acres in production and/or more inputs per acre. Either way, greater costs were incurred to produce more. Farmers' income suffers from both lower price and higher costs. Farmers work harder to produce more and end up earning less. Why would they do so?

Table 4-6: Illustration of Change in Gross Receipts Due to Demand Elasticity

Elasticity	Price per Unit	Quantity Sold	Gross Receipts (Price x Quantity)	Change in Gross Receipts
Elastic	$1.00	90	$90.00	Increase
	.90	110	$99.00	
Unitary	1.00	90	$90.00	Same
	.90	100	$90.00	
Inelastic	1.00	90	$90.00	Decrease
	.90	95	$85.50	

Table 4-7: Hypothetical Corn Production, Market Price, and Gross Value of Corn Crop, Illustrating Inelastic Demand

Production	Price	Gross Value
(billions of bushels)	($/bu.)	($ billions)
6.1	4.00	24.4
6.3	3.75	23.6
6.5	3.50	21.1
6.75	3.25	21.9
7.0	3.00	21.0
7.3	2.75	20.1
7.6	2.50	19.0
8.0	2.25	18.0
8.5	2.00	17.0

Conflict Between the Individual and All Farmers

The aggregate demand for corn is highly inelastic. Relatively small increases in total supply must be sold at large price concessions to encourage additional consumption. An individual farmer, however, faces a totally elastic demand curve, as illustrated in Figure 4-17. One farmer can sell the entire production at the market

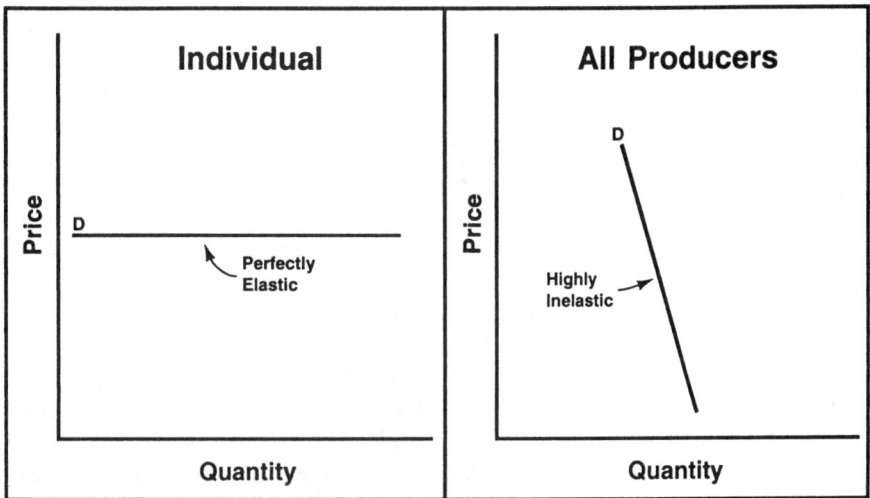

Figure 4-17: Demand Curves for Corn for the Individual Producer and for All Producers.

price without the necessity of cutting price to encourage buyers to take a few more bushels. By the same token, an individual producer cannot raise the price level by withholding corn from the market. There are so many corn producers and each one produces such a small portion of the total crop, no one of them acting alone appreciably affects the supply. While 200,000 bushels of corn represents a fairly large farming operation, 200,000 bushels is miniscule in an 8 billion–bushel crop. When corn production is profitable, it behooves the individual farmer to produce as much as possible. Total profit is increased by producing more bushels. Furthermore, since many costs, such as labor and machinery, are fixed, the farmer can lower total costs per unit by increasing production. Thus, a farmer has double incentive to produce more. But when all producers increase production, the larger total supply lowers farm price. A farmer, as an individual, makes rational, profitable decisions to produce more. However, more total output is not necessarily in the best interest of farmers as a whole. The collective impact of all those individual decisions to produce more serves to drive down the market price, and consequently total income and profit.

The conflict in objectives arises because agricultural producers come the closest of any industry to operating under economic conditions of pure and perfect competition. Those conditions are:

1. A great number of buyers and sellers so that no one is sufficiently large to influence price through individual actions, and therefore farmers are price takers;
2. Both buyers and sellers have complete and perfect knowledge of all factors that affect market conditions. Furthermore, each one will utilize the information in a rational manner to maximize individual profit;

3. There is no collusion or operating restrictions among either buyers or sellers to limit the supply of product coming to the market (except when the government attempts to intervene with production controls);
4. The product is homogeneous in that one lot is exactly like another lot within quality specifications. There is no product differentiation;
5. The factors of production are perfectly mobile and free to enter or leave an industry in order to achieve greater returns.

These competitive conditions significantly impact the way farmers operate. Because any one individual cannot influence the market, farmers are price takers; they take the price dictated by aggregate demand and supply conditions. Furthermore, individuals feel that they can produce and sell as much as they like without influencing the price. Thus, a farmer's attention is focused on production efficiency to improve profit. There is a strong incentive to lower operating costs in hope of improving profitability. That explains why agriculture is continually subject to the price-cost squeeze. Anytime that profits increase, additional production is stimulated. More production, in turn, reduces market price back to "normal" profit levels.

The conditions of pure and perfect competition are modified to a degree by real-world situations. Some farmers and marketing organizations are large enough to influence market price; information is never 100 percent perfect; there can be product differentiation; and factors of production are not perfectly mobile. These imperfections provide farmers with opportunity to increase profits through improved marketing decisions.

Incentives to Restrict Output

Inelastic demand provides producers, acting collectively, an opportunity to raise price by restricting output through production quotas or acreage controls and shifts the supply curve leftward, as shown in Figure 4-18. Net income is boosted by the combination of higher price and lower production costs.

Some farm groups promote the idea of output restriction by their members. But they are not successful in reducing supply enough to raise total return. It is against a farmer's nature to leave land and other resources idle. As an individual, a farmer has the economic incentive to produce more in response to higher prices. Thus, there are strong incentives not to cooperate in voluntary group action to reduce production or to cheat against those who do. A less-than-altruistic farmer might hope that neighbors or some other region might cut acreage or suffer poor yield to reduce the overall crop, but the individual producer prefers bumper crop yields. A large crop sold at a high price is the best of all possible worlds for the individual.

Successful agricultural supply control programs usually require governmental authority to determine output restrictions and to enforce compliance. Any kind of market order or production quota, in essence, restricts supply. Both "carrot and stick" approaches are used. Land retirement programs, in the name of conservation, are encouraged by payment schemes to farmers to hold land out of production. Commodity groups with specialized products and limited production areas form producer marketing cooperatives and subscribe to production quotas. Consumers, of course, pay higher food costs as a result of successful supply control programs. Furthermore, many of the benefits are subsequently capitalized in the value of the scarce resource. Tobacco and peanut allotments, for ex-

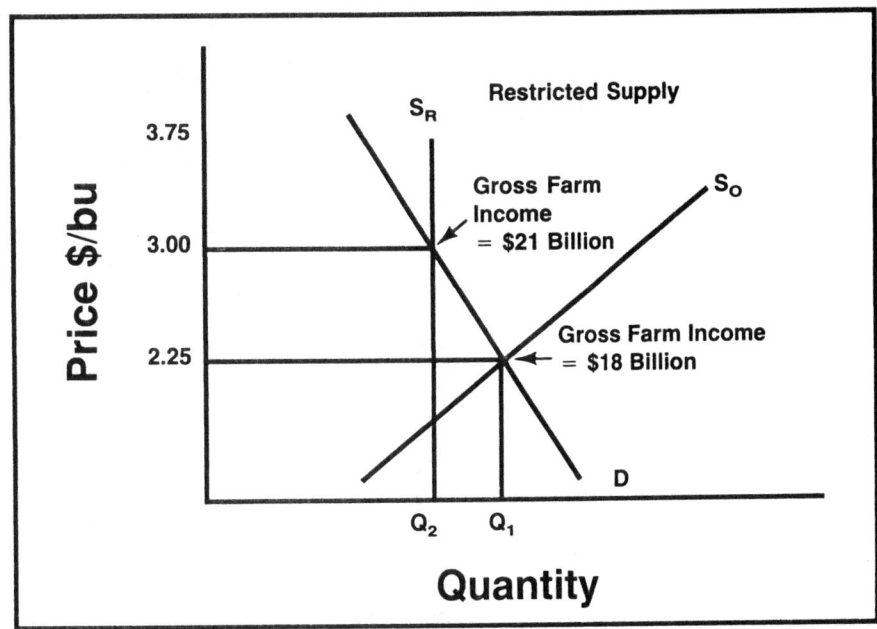

Figure 4-18: Effects of Supply Restriction on Gross Receipts.

ample, have taken on values all their own, separate from the land to grow the crop.

Effects of Price Ceilings and Floors

Price ceilings or floors are government-imposed interference in the market place to limit price rise or prevent price decline. Price ceilings to hold prices low are set below equilibrium price, and price floors are set above equilibrium price. Price ceilings are imposed to control inflation and hold down consumer costs, although it is debatable how effective they are in the long run. Price floors are used to support farm prices and bolster farm income.

Price floors and ceilings prevent buyers and sellers from reaching equilibrium price and quantity. Surpluses or shortages are created by market interference. As shown in Figure 4-19, a price floor set above equilibrium at P_f signals buyers to consume Q_m, but it signals producers to produce Q_g. The difference is surplus production which must be controlled by marketing quotas, stored, dumped, plowed under, or sold in non-competing markets.

A price ceiling, P_c, set below equilibrium price creates a shortage. The low price encourages buyers to consume Q_m but encourages farmers to produce only Q_g. A rationing procedure has to be devised to allocate the restricted supply. Black markets tend to develop to circumvent the rules and regulations and raise the price to P_b.

Relationship of Prices to Costs of Production in Agriculture

The question is often asked, "Is there a relationship between commodity price and cost of producing the commodity?" Economic principles state that equilibrium price is established where demand and supply are equal for short run periods as well as long run. In a free market economy, with strong demand, price rises

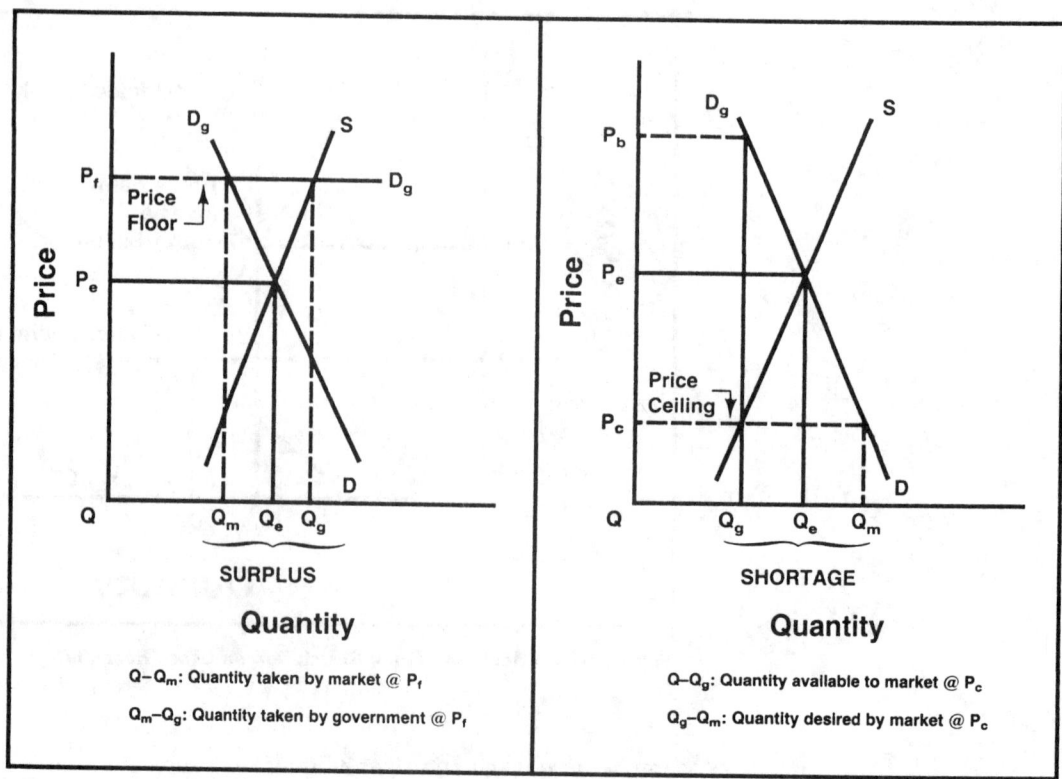

Figure 4-19: Effects of Price Floor (Left) and Ceiling (Right).

high enough to ration the commodity to highest-value users. Price will rise as long as buyers continue bidding for the existing short supply. But as price rises, more and more potential buyers are priced out of the market. High price will thus ration a limited supply to those who valued it the most or who have the means (income) to acquire it. At the other end of the scale, a low level of consumption relative to available supply means that the commodity must be sold at a low price to dispose of it. Low price encourages consumption and discourages future production.

The time frame is important in its effect upon the level of the equilibrium price. Once a crop such as corn is harvested, the operator can either sell it at the prevailing market price or store it to hold it off the market temporarily. The farmer grew it to sell, and eventually grain will move into the market. Because there are so many sellers, one producer cannot influence price by either selling his or her product or withholding it from the market. At harvest, the demand situation may allow the producer to enjoy large profits from high price or substantial loss from low price, but he or she cannot alter the previously made decision to produce. Therefore, in the short-run period of a production cycle, no relationship exists between cost of production and market price. Perishable production and competitive economic environment force producers to be price takers. **The market does not care what it costs to produce the product.** Its only concern is the price necessary to balance available supply with effective demand. With a short supply, price rises to ration it, and with a surplus supply, price falls to move it.

Price Determination: Market Fundamentals

In the intermediate time period before planting occurs, the producer can make certain adjustments in response to price level. He or she may plant only the best land and reduce fertilizer rates and chemical applications if prices are low. A producer will restrict inputs and consequently reduce production somewhat. But as long as he or she covers the variable, or direct out-of-pocket, costs of production, the producer will continue to produce even though he or she is not covering the fixed portion of costs. Thus, in the intermediate time frame, production will continue as long as a producer covers variable production costs, that is, seed, fertilizer, chemicals, and fuel, even though fixed costs of capital, labor, land, and management are not paid. At the other extreme, if price is high, farmers exert every effort to produce as much as possible by applying more inputs. They perform those production practices that return more than cost in the manager's effort to balance added returns against added costs. Thus in the intermediate period, high or low price encourages some modest supply adjustment within the fixed farm plant.

In the long run, larger changes in production are possible. Low-quality land will be taken out of production and either lie idle or be put to some other use. Some farmers will not replace facilities or equipment. If product price is persistently low, high-cost producers will be forced to quit. The supply of the commodity will decline and prices will rise until remaining producers share the demand at a price where there no longer are losses. On the other hand, if price is high, operators will maintain land in high productivity and expand the operating base. Producers will clear forests, drain swamps, and irrigate new land, expand facilities, and adopt new technologies to share in the profits. The supply curve will be shifted to the right, causing price to be under pressure from the increased production. Therefore, in the long run, production is significantly affected by price level. Supply will adjust so that average price will tend to equal average total cost of production. All production costs will be covered, and returns to capital, land, labor, and management will be no more, or no less, than they could receive if they were used in other activities.

Again, the market does not care what it costs a producer to supply the commodity. The "invisible hand" of the market concerns itself with moving the available supply at whatever price is determined by the interaction of supply and demand.

Delay in Producer Response: The Cobweb Theorem

The principles of supply and demand explain how price reaches an equilibrium point to allocate supply between buyers and sellers. The cobweb theorem explains why price may not settle at an equilibrium point but fluctuate about it and constantly rise and fall to create commodity price cycles. If the quantity of a commodity could be adjusted instantaneously to price change, price cycles would not occur. That is, however, not the case. The biological nature of agricultural production gives rise to production cycles, and production cycles lead to price cycles.

Price cycles run counter to production cycles. When supplies increase, prices fall; when supplies decrease, prices rise. Price changes are much greater in magnitude than production changes because of the inelastic nature of demand and supply of farm products.

Periodic expansion and contraction of supply of the product causes price cycles. Price cycles are, therefore, supply-based; they reflect producer output decisions. Suppose, for example, hog production was low at Q_1, and prices were high at P_1 in period I, as shown in Figure 4-20. Hog producers look at the favorable

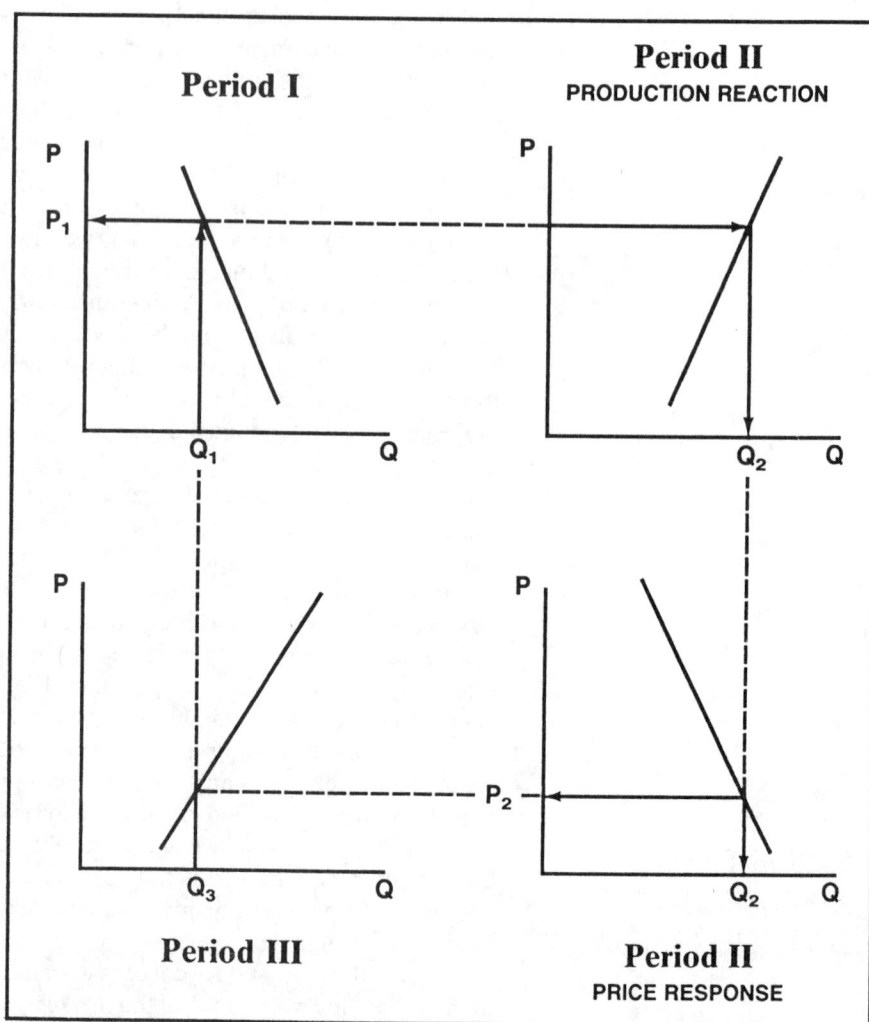

Figure 4-20: Illustration of Supply Response to Market Price in Explaining Price Cycles.

price and base their future production plans upon it. High price encourages the retention of breeding stock to expand production. New producers are encouraged to enter the business. Breeding, gestation, farrowing, and finishing the market animals require about nine months before production decisions result in increased output. As a result, producers bring Q_2 to the market in period II. Since they must move animals when they are ready for market, the increased output is literally dumped on the market. In response to the larger supply, price falls to P_2. Producers now reevaluate the situation in light of depressed prices and decide to raise fewer hogs. Many quit production altogether. Consequently, production falls to Q_3 in period III. This is the same production level as period I, where the cycle began. Price rises as high as P_1, and the stage is set for the cycle to repeat itself.

Agricultural price cycles are caused by the tendency of farmers to base future production plans on current price and profit rather than on future price expectations. They act as though present price will continue indefinitely into the future. Production response is often off target as a result.

Four conditions are necessary for price cycles to occur. First, there must be a time lapse between a change in price and producers' production response to the change. Second, producers must base future production plans on current prices rather than future price expectations. Third, producers must have reasonably good control over output as they do in livestock. There are no well defined price cycles in crops because producer decisions are only partially responsible for crop size (with nature being the major force). Fourth, the product is not storable and is therefore dumped on the market. Finally, price cycles develop in competitive industries because each producer believes that his or her own production decision will not affect the overall supply; and he or she will not, therefore, affect price.

Length of agricultural price cycles depends upon both biological and psychological time lags. The biological lag in livestock is the time period from expansion of the breeding herd to market maturity of the resulting offspring. The cycle is longer for cattle than chickens because of biological differences. Agricultural price cycles are usually longer, however, than the biological time lag. Though the biological production cycle in hogs is 9 to 10 months, the hog price cycle usually runs 3 or 4 years. Cattle production cycles are about 4 years, but the cattle price cycle runs 8 to 12 years. Psychological factors account for the remaining time lag. They explain the time required for current market conditions to convince producers that current prices are truly here to stay. The combination of biological and psychological lags explains the length of price cycles. However, the variability in psychological factors makes the length of the price cycle somewhat unpredictable.

FUNDAMENTAL APPROACH TO PRICE LEVEL DETERMINATION

Economic principles provide essential understanding of demand and supply relationships in establishing price. But they fall short of providing workable information for decision making. Thousands of supply and demand forces tug and push at equilibrium price; some buoying it up, others pulling it down. Analysis of market price from a fundamental approach attempts to sort out the important price-making forces, while discarding unimportant ones. Price analysis further measures the degree of influence exerted by such forces.

Econometricians combine economic theory and mathematics into analysis of price-determining forces. Agricultural price analysts investigate changes in supply and demand and the resulting changes in commodity price. It is not enough to simply show that agricultural prices decline when the supply increases or demand decreases. Price analysis provides a quantitative estimate of **how much** the price changes in response. Price analysis begins with determining which supply and demand forces are most important in setting price, then it attempts to measure the degree by which changes in demand and supply affect prices.

Once relationships are defined and respective influences measured, price forecasting becomes possible. Statistics showing increased crop acreage, increased yield, and increased carryover of supplies can then be translated into projected price change as the result of increased supply.

The underlying principles of supply-demand relationships are common to all commodities. However, application of these principles to individual commodities varies greatly because each commodity has its particular set of price-making forces. Furthermore, all commodities interrelate to each other and together fit into the much larger macroeconomic picture.

Darrel Good, Thomas Hieronymus, and Royce Hinton developed mathematical models to measure the impact of individual forces on price.[6] Their master model of Midwest agriculture is presented in Figure 4-21. It is a systematic description of the factors that affect prices and price relationships. Their model is an interrelated system in that a change in one part of the system (such as a new corn production estimate) changes every other value in the system in due course. A goal of price analysis is to quantify the relationships and measure the values so that each new piece of information can be fed into the calculations and its impact on the other factors of the system instantly estimated. Despite the fact that "the current state of the art of price analysis and forecasting falls far short of providing all the data needed to measure all of the relationships in the system, we must be aware of the total system and know where each segment fits in the system if we expect to understand price determination in the market process."[7]

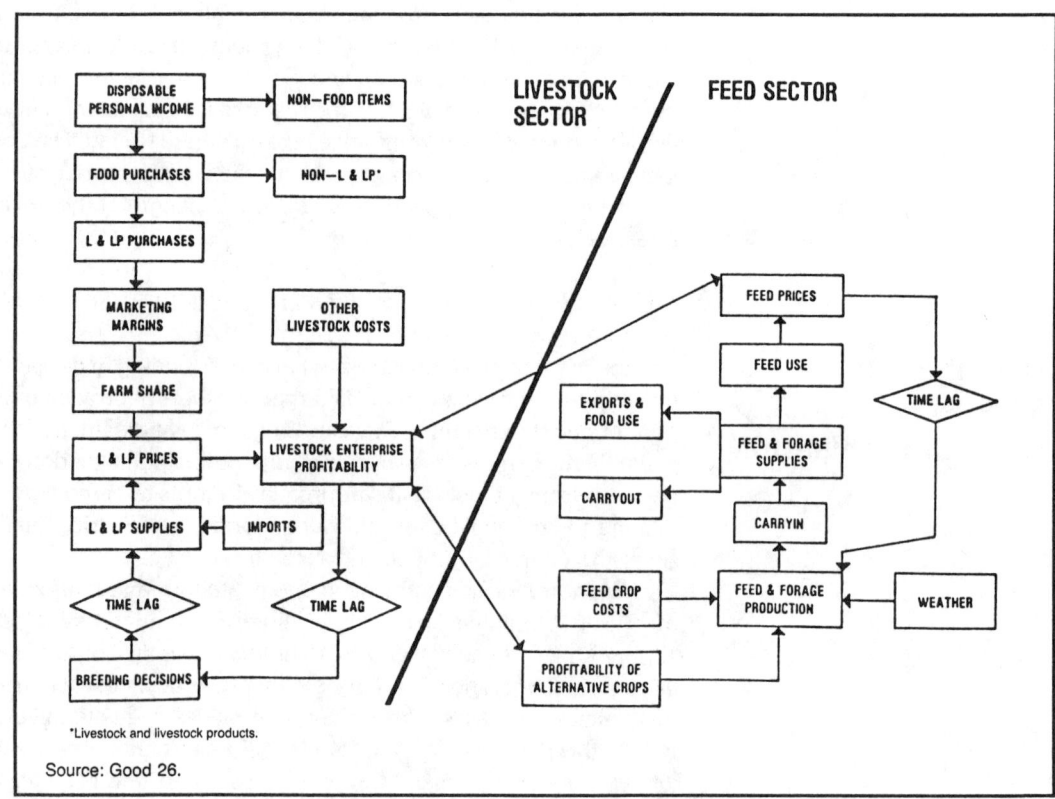

Figure 4-21: Master Model of Midwest Agriculture. Source: Good 26.

The two major segments of the Illinois model are the livestock sector and the crop sector. Corn Belt farmers are primarily concerned with crop production, but most of their grain is marketed as livestock or livestock products. The relationship of livestock enterprise profitability and feed costs (crop prices) links the two major sectors together.

The dominating factor that in large measure determines values in the entire system is disposable personal income (DPI). As a measure of consumer purchasing power, it is income remaining after taxes are paid. The amount of disposable personal income that consumers spend for the end products of agricultural production ultimately determines the level of farm income. Out of the total livestock product portion of food expenditures, marketing margins paid to intermediaries are subtracted to determine farm share of consumer dollars. These dollars constitute the demand side of the livestock and livestock product price equation. On the supply side of the equation is the quantity of livestock and livestock products available for consumption. Domestic production plus imports less exports make up the total supply available to consumers. At this point in the model, net consumer expenditure, expressing demand, interacts with livestock product supplies from farms to determine prices.

The Illinois regression analysis explained 79 percent of the variation in consumer meat expenditures by the two variables of consumer income and meat supplies.* The thousands of other variables accounted for only the remaining 21 percent of expenditures.

Relative to the first variable, meat consumption changes directly with income change. As incomes increase, meat consumption also increases. When incomes decrease, meat consumption decreases.

In the supply relationship, a larger supply sells for less total money and a smaller supply sells for more money due to price inelasticities. Consumers are reluctant to change consumption patterns and are encouraged to do so by relatively large price changes. As an example, a 1 percent reduction in supply may result in a 2 percent price increase. As the quantity of meat supply decreases by 1 percent, total expenditures rise by 2 percent. A smaller quantity multiplied by a higher price produces an increase in total revenue.

In the next stage of the model, livestock production decisions are based upon enterprise profitability. Each producer's assessment of livestock prices, compared to feed costs and other production costs, determines his or her expected profitability. Livestock profitability thus establishes the demand side of the crop price equation. Losses discourage production; high profits encourage livestock expansion. Livestock numbers and feeding rates therefore translate into demand for feed grains.

The crop sector represents the supply side of the crop price equation. Acreage planted multiplied by yield produces annual production figures. Supply carried over from the previous year is added to production to make up total supply available. Total supply then moves in four alternative demand directions: food, industrial use, export, and feed. Carryover is the amount left over that was not sold at a reasonable price for feed. In the later stages of a crop year, next year's expected production is a factor in determining how much of old crop is used immediately and how much is saved. If prospects point toward a large crop, price declines to encourage immediate feeding. If prospects suggest a small crop, price level rises and more of the old crop is saved for consumption the next year.

It is out of these aggregate market interactions that prices of individual commodities are formed. We shall follow the analysis further to identify important price-making forces in particular commodities. By understanding those forces and assessing their importance, farmers are in a better position to evaluate market information. They can better sort out real influences from market noise.

---

*Meat supplies of beef, pork, veal, broilers, and turkeys.

Understanding Grain Prices

Grain markets must establish a price level at the beginning of the season that will encourage consumption up to an appropriate level of carryover at the end of the year. If price is too high, usage will be curtailed and carryover will be too large. If the price is too low, the rate of use will be too rapid and pose a threat of running out before the next harvest. The solution is to establish a market clearing price at the beginning of the year, but the problem is to anticipate the "correct" market clearing price. The market as a whole looks at all factors and forms a judgment about the appropriate price. "It is a collective speculative judgement; it is almost always in error"[8], but it continually redresses the errors as prices fluctuate up and down day to day and week to week. The collective judgment changes continually as market forces and their interpretation change. Daily price changes are a series of adjustments to changing supply-demand factors.

Corn Price. A balance sheet method was used by Good to estimate availability of corn for feed use, as in Table 4-8. Estimates for food, industrial use, and exports were subtracted from total supply, leaving the remainder available for feed and carryover. The estimates for the first three uses were made without regard to price. In the case of food and industrial use, the products were high in value and had no close substitutes. The amounts of corn used in these production processes were consumed largely without regard to price.

Foreign demand was highly inelastic. In the past, an important destination of U.S. corn exports was the European Economic Community. Prices there were higher than world price and were maintained at high levels by a variable levy that could be changed daily. When the price outside the EEC Common Market went down, the levy was increased and vice versa. Changes in U.S. prices had little effect on the amount used in EEC. Japan, as the largest importer, did allow prices to fluctuate somewhat but not enough to affect the quantity imported. The U.S.S.R. was another important destination of U.S. corn. Soviets appeared to buy on the basis of need and political goals rather than price. Other countries that imported relatively small amounts of corn also responded primarily to need rather than price. Only during years of short U.S. crops, with prospects for abundant supplies after the following harvest, did it appear that importers modified their demand.

Since the other demands were relatively fixed, the focal point in forecasting corn price was livestock feed demand. Corn was priced low enough to make its use attractive to livestock producers but high enough to prevent feeders from attempting to use the entire supply. Corn price was consequently demand-determined, given the available supply. The key to corn price forecasting was the calculation of the supply available for feeding. When corn was scarce relative to livestock numbers and level of feed use, its price was high in relation to livestock prices. When corn was abundant, the price was low in relation to livestock prices.

The Illinois study, therefore, tied value of corn to three factors: its feeding value, increase or decrease in carryover, and speculative error.

The comparison of the feed value of corn and its market price for 10 years is given in Table 4-9. Though feed value and crop price were not the same year to year, corn price tended to return to feeding value as its central value.

Another factor in explaining corn price is projected change in carryover. When the carryover supply is expected to increase, we should expect corn price to be below feed value to encourage consumption. When carryover is expected to decrease, corn price should be higher than feed value to discourage usage. Com-

Table 4-8: Corn Balance Sheet: Production, Use, and Carryover, 1959-1978 (Millions of Bushels)

Crop Year	1959	1960	1961	1962	1963	1964	1965	1966	1967	1968
SUPPLY										
Acres harvested (mil.)	72.1	71.4	57.6	55.7	59.2	55.4	55.4	57.0	60.7	56.0
Yield (bushels/acre)	53.1	54.7	62.4	64.7	67.9	62.9	74.1	73.1	80.1	79.5
Carryin	1,524	1,787	2,008	1,653	1,365	1,537	1,147	842	826	1,169
Production	3,825	3,908	3,625	3,606	4,019	3,484	4,103	4,167	4,860	4,450
Total Supply[1]	5,350	5,696	5,635	5,260	5,385	5,022	5,251	5,010	5,687	5,620
DISAPPEARANCE										
Food, Industry, Seed	283	286	307	323	340	349	360	364	362	359
Export[2]	212	276	415	416	500	570	687	487	633	536
Feed & Residual	3,068	3,126	3,260	3,156	3,008	2,956	3,362	3,333	3,523	3,607
Total Disappearance	3,563	3,688	3,982	3,895	3,848	3,875	4,409	4,184	4,518	4,502
Carryout	1,787	2,008	1,653	1,365	1,537	1,147	842	826	1,169	1,118
Change in carryout	−263	+221	−355	−288	+172	−390	−305	−16	+343	−51
PRICE, E. Cent. Ill.										
Current $	$1.08	1.02½	1.04	1.15	1.16	1.23	1.24½	1.28	1.05½	1.17
1972 $	1.58	1.48	1.48	1.61	1.60	1.66½	1.64	1.63	1.29	1.36½

Crop Year	1969	1970	1971	1972	1973	1974	1975	1976	1977	1978
SUPPLY										
Acres harvested (mil.)	54.6	57.4	64.0	57.4	61.9	65.4	71.3	70.0	70.0	70.3
Yield (bushels/acre)	85.9	72.4	88.1	97.1	91.2	71.4	86.2	87.9	90.8	100.8
Carryin	1,118	1,005	667	1,126	709	483	359	399	884	1,104
Production	4,687	4,152	5,641	5,573	5,647	4,664	5,797	6,266	6,226	7,087
Total Supply[1]	5,806	5,161	6,309	6,700	6,357	5,149	6,158	6,668	7,312	8,192
DISAPPEARANCE										
Food, Industry, Seed	365	385	390	429	448	450	491	513	550	575
Export[2]	612	517	796	1,258	1,243	1,149	1,711	1,684	1,948	2,133
Feed & Residual	3,824	3,592	3,997	4,304	4,183	3,191	3,558	3,587	3,710	4,260
Total Disappearance	4,801	4,494	5,183	5,991	5,874	4,790	5,760	5,784	6,208	6,906
Carryout	1,005	667	1,126	709	483	359	399	884	1,104	1,286
Change in carryout	−113	−338	+459	−417	−226	−124	+40	+485	+220	+182
PRICE, E. Cent. Ill.										
Current $	$1.24	1.40½	1.14½	1.81	2.82	3.01½	2.64	2.19½	2.17½	2.46½
1972 $	1.37½	1.48	1.15½	1.74	2.49	2.40½	2.00	1.57	1.46	1.52

Source: Good 70.
[1] Includes imports.
[2] Does not include exports of corn products.

parison of the difference in corn price versus feed value and change in inventory for 10 years is also presented in Table 4-9. Again, the relationship was not perfect, but there was a tendency for corn price to be lower than feed value when carryover increased and to be higher than feed value when carryover decreased.

The third factor in the analysis is speculative error. It refers to the market's inability to accurately assess the forces at work, and the occurrence of unanticipated events such as drought, corn blight, early frosts, changes in exports, and political events. Speculative error represents the unexplained portion of price variability.

In summary, the Illinois authors noted that government loans and inventory programs dominated the general price level of corn for many years prior to 1972. During the period 1972-1982, the market over- and under-estimated the price impacts of both shortage and surplus. The market's biggest problem was identifying value in a climate of rapid inflation. An equilibrium value of $2.50 in one year was the same as $2.75 the next if inflation was running at a rate of 10 percent per year. However, when other forces were accounted for, the price foundation of corn depended on its value as livestock feed in the United States, plus or minus changes in carryover stocks.

Establishing a central value based on livestock prices is therefore a useful point of departure in forecasting average price of corn for a given year. Feed value is determined by the price livestock feeders can afford to pay for corn as a production input.

In another work, Scott Irwin related carryover corn stocks to the price:loan ratio.[9] His work illustrated the effects of both the carryover supply and the price-supporting effect of a commodity loan on price, as in Figure 4-22. When stocks exceeded roughly 2 billion bushels, the loan rate became the market price. At the other extreme, with the small carryover of 400 million bushels in 1974, corn price was more than 2 times higher than loan. Farmers can use Irwin's analysis to project from expected year-end corn supplies to price level direction. Shrinking carryover supplies point toward higher prices; increasing carryover, lower prices.

Soybean Price. Good, Hieronymus, and Hinton observed that soybeans are the most complex and difficult of all commodities to forecast. The soybean is not one commodity but two, soybean meal and soybean oil. One cannot be produced without the other. Yet these products have totally different markets and uses. Soybean oil is a high-grade vegetable oil used primarily in shortening, margarine, and salad and cooking oils. Meal is a protein supplement for livestock feeding. Both commodities move extensively into world trade; both have numerous close substitutes in use. This makes supplies of competing products important in price determination. Price elasticities of demand for both oil and meal are low, making prices

Table 4-9: Feeding Value and Average Price of Corn, and Change in Carryover, 1969-1978 Crop Years

Years	Feed Value	Average Price	Corn Price/ Feed Value Difference	Change in Carryover
	($/bu.)	($/bu.)		(mil. bu.)
1969-70	1.36	1.24	- .12	- 113
1970-71	1.27	1.41	+ .14	- 338
1971-72	1.36	1.15	- .21	+459
1972-73	1.58	1.81	+ .23	- 417
1973-74	1.99	2.82	+ .83	- 226
1974-75	2.18	3.01	+ .83	- 124
1975-76	2.20	2.64	+ .44	+ 40
1976-77	1.84	2.19	+ .35	+485
1977-78	2.18	2.17	- .01	+220
1978-79	2.49	2.46	- .03	+ .68
Mean	1.85	2.09		

Source: Good 80.

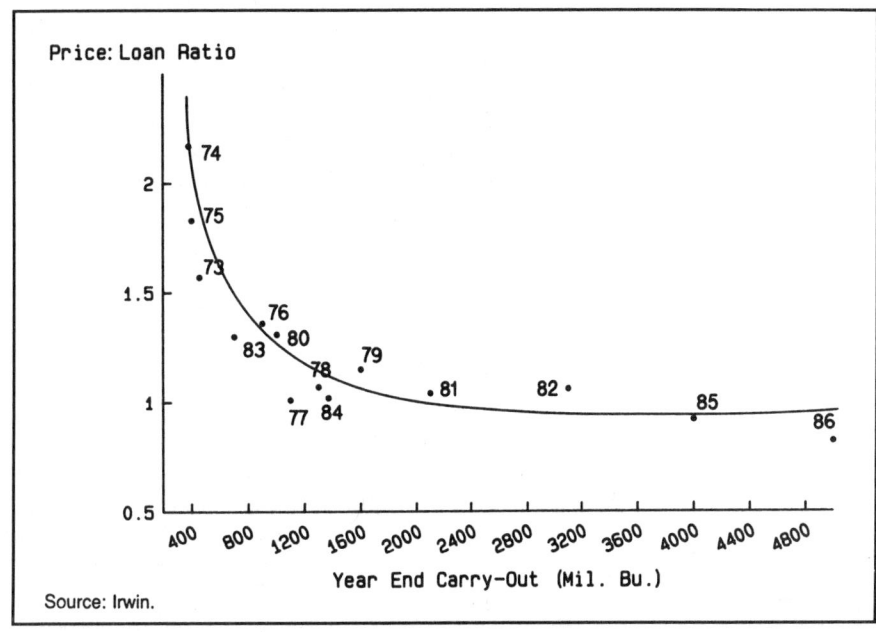

Figure 4-22: Corn Prices Relative to Carryover Stocks, 1973-1986. Source: Irwin.

highly volatile. Market demand for both products expanded rapidly since 1940, but the rate of demand expansion is uncertain in the future.

Since soybean products have different price-making forces, they must be considered separately and their values added together to obtain a value of soybeans. Each 60-pound bushel of soybeans contains about 11 pounds of oil and 47.5 pounds of 44 percent protein meal, with the remainder being waste or loss. The value of a bushel of soybeans can be calculated from the value of the products. The formula is:

$$\begin{array}{l}\text{Soyoil Price} \quad \underline{\qquad} \times 11 \quad \underline{\qquad} \\ \text{plus SBOM Price/lb.} \underline{\qquad} \times 47.5 = \underline{\qquad} \\ \text{equals Product Value, per Bushel} \quad \underline{\qquad}\end{array}$$

The difference between the product value and soybean price is the crusher's (processor's) margin.

In the 1970's the "normal" processing margin averaged about 40 cents per bushel. Deviations from normal were related to farmers holding soybean supplies from the market. In years when prices were high and farmers were ready sellers, margins tended to be high. In years of low price levels and tight farmers holding, margins tended to be small.

Soybean price is, therefore, determined by the world market for its products. Soybean oil prices are a function of world price of edible fats and oils, since exports form a major part of the market. Furthermore, the domestic demand for oil is highly inelastic. Price of oil is relatively cheap and U.S. consumption is constant regardless of price change. Price in the U.S. must remain close to world price. On the other hand, soybean meal prices are made in the U.S. Demand outside the U.S. is highly inelastic, so the amount exported is not materially affected by price.

Quantities imported by other countries are usually tied to perceived needs for the commodity and are not price-sensitive. When the supply situation dictates an adjustment in the use of soymeal, the adjustment is made in the U.S. Thus, factors outside the U.S. tend to set oil price, and factors inside the U.S. set meal price.

The soybean balance sheet becomes a basic reference point in forecasting soybean price. The projected increase or decrease in carryover of supply suggests price direction. Table 4-10 is a long-term soybean balance sheet. The outstanding features are the rapid and persistent growth in acreage, almost tripling, and the rather consistent yield. Production consequently more than tripled during the period.

The first step in calculating a balance sheet is to determine acreage, yield, and production. The first planting intentions report for the year is released in Jan-

Table 4-10: Soybean Balance Sheet: Production, Use, and Carryover, 1959-1978 (Millions of Bushels)

Oct. 1 Year	1959	1960	1961	1962	1963	1964	1965	1966	1967	1968
SUPPLY										
Acres harv. (mil.)	22.6	23.7	27.0	27.6	28.6	30.8	34.4	36.5	39.8	41.4
Yield (bu./ac.)	23.5	23.5	25.1	24.3	24.4	22.8	24.5	25.4	24.5	26.7
Carryin	88	52	27	78	46	67	30	36	90	166
Production	533	555	679	669	699	701	845	928	976	1,103
Total Supply	621	607	706	748	745	768	875	964	1,066	1,269
DISAPPEARANCE										
Crush	394	406	431	473	437	479	537	559	576	606
Export	140	135	149	181	187	212	251	262	267	287
Seed, Feed, Res.	35	39	47	48	54	47	52	53	57	52
Total Disapp.	569	580	627	702	678	738	840	874	900	945
Carryout	52	27	78	46	67	30	36	90	166	325
Charge in carryout	-36	-25	+51	-32	+21	-37	+6	+54	+76	+159
ECI Price—Curr. $	$2.13	2.55	2.41	2.53	2.61	2.83	2.98	2.83	2.61½	2.57½
1972 $	3.11	3.69	3.43	3.54½	3.60	3.83	3.92	3.61	3.20½	3.01

Oct. 1 Year	1969	1970	1971	1972	1973	1974	1975	1976	1977	1978
SUPPLY										
Acres harv. (mil.)	41.3	42.2	42.7	45.7	55.8	52.4	53.8	49.4	57.9	63.0
Yield (bu./ac.)	27.4	26.7	27.5	27.8	27.7	23.2	28.8	25.6	29.6	29.2
Carryin	325	230	99	72	60	171	185	245	103	161
Production	1,126	1,127	1,176	1,283	1,547	1,215	1,546	1,265	1,762	1,870
Total Supply	1,451	1,357	1,275	1,355	1,607	1,386	1,731	1,510	1,865	2,031
DISAPPEARANCE										
Crush	738	760	721	722	821	700	867	787	927	1,018
Export	429	434	417	480	539	421	555	565	700	753
Seed, Feed, Res.	54	64	65	93	76	80	65	55	77	87
Total Disapp.	1,221	1,255	1,203	1,295	1,436	1,201	1,487	1,407	1,704	1,585
Carryout	230	99	72	60	171	185	245	103	161	173
Change in carryout	-95	-131	-27	-12	+111	+14	+60	-142	+58	+12
ECI Price—Curr. $	$2.58½	3.04½	3.27	6.49½	6.11	6.12	5.33	7.23	6.25	7.15½
1972 $	2.86½	3.21	3.30	6.24½	5.39	4.88½	4.03	5.18	4.19	4.41

Source: Good 87.

uary; the second in April; and a report on planted acres at the end of June. Typically, 98 percent of planted acres are harvested, so planted acreage is multiplied by 0.98 before being entered on the balance sheet. The first yield report is released in August. Until then, yield estimate is based on trend in historical yield and modified by weather conditions to date. Production estimates are based on conditions on the first of the month and are released about the tenth of each month, August through November. The final estimate of production is made in late January.

Current estimates of carry-in stocks are added to production estimate to obtain total supply for the coming year. After harvest, when the consumption year starts, the balance sheet gradually becomes an estimate of carryover, based upon disappearance to date. Disappearance occurs as exports or domestic use. Exports are reported weekly and fairly quickly set a trend for the year. They must be adjusted, however, by estimated production in South America. Brazil and Argentina became important producers in the 1970's. Their crop is planted in October-December; January is a critical weather month; and harvest is in the March-May period. Crop estimates for South America are made by the Foreign Agricultural Service of the USDA and announced periodically throughout the growing season. Calculation of domestic processing in the U.S. is available from the National Soybean Processors Association on a weekly schedule and the Bureau of the Census on a monthly schedule. The rate of processing or crushing to date becomes increasingly reliable and important as the crop year progresses. Carry-out is what is left of supply at the year end.

The carry-out figure gauges whether soybeans are relatively scarce or abundant. For budgeting purposes, 180 million bushels was selected as a reasonably comfortable carryover. Below 150 million bushels, supplies are light and prices stronger. Above 200 million bushels, large supplies weigh heavily on price.

Changes in carryover supplies on the soybean balance sheet, thus, become a basic reference point in forecasting soybean price. The soybean balance sheet is an indication of the demand situation in the product markets of soybean oil and meal as demand interacts with available supply to set price direction.

Irwin calculated the price level of soybeans relative to loan as compared to stocks on hand at the end of the crop year for the period from 1976 to 1986.[10] The carryover supplies were calculated as a percent of annual usage. When stocks exceeded 30 percent of annual use, the soybean loan rate became the effective market price (Figure 4-23). With stocks about 10 percent or less of use, price level was 1.5 to 2.6 times higher than the loan rate.

Similar data for wheat price and annual carryover stocks are presented in Figure 4-24. Stocks above 1.6 billion bushels caused the loan rate to set the floor on price level. Carryover stocks below 1 billion bushels produced prices 1.5 to 3 times higher than loan rate.

Understanding Livestock Prices

Hog Prices. The nature of pork production makes hog price prediction relatively straightforward. Hogs come to market at fairly uniform weights on a biologically consistent schedule. Since there is very little inventory buildup or liquidation, there is hardly any inventory speculation on the part of producers. Pork is consumed shortly after it is produced and is not held in storage. Since no major trend elements in either production or consumption exist, price of hogs "lends itself to systematic quantification and to derivation of functional relationships."[11]

Figure 4-23: Soybean Prices Relative to Carryover Stocks, 1976-1986. Source: Irwin.

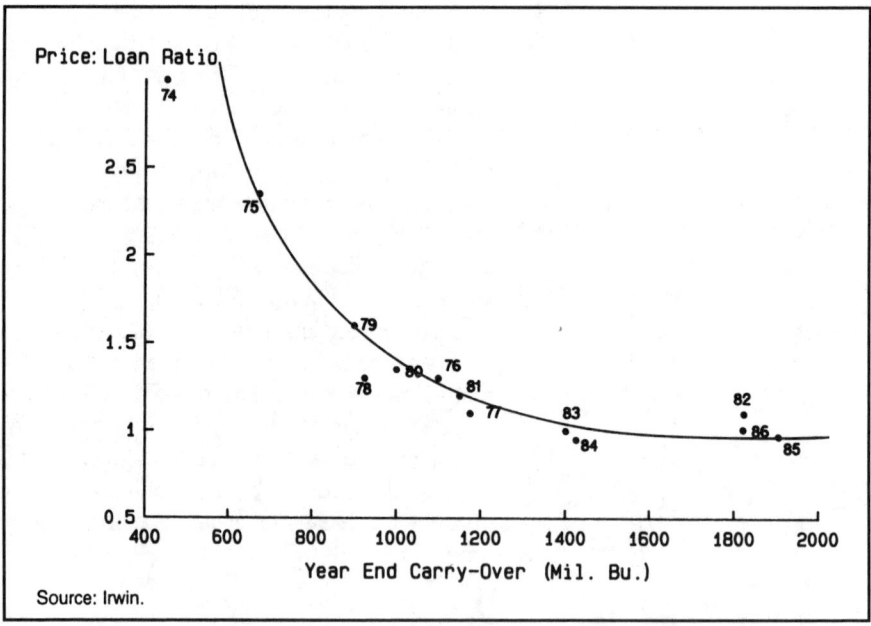

Figure 4-24: Wheat Prices Relative to Carryover Stocks, 1974-1986. Source: Irwin.

The factors which were found to explain retail pork price include consumer demand for meat in general, supplies of pork, supplies of competing meats, and consumer preferences. A rather consistent marketing margin was subtracted from the retail price to get back to farm level hog price. The Illinois demand formula accounted for about three-fourths of the variability in quarterly average price of live hogs. It showed that the price of hogs depended upon the amount of pork production, beef production, broiler production, and real disposable income. A 1.0 percent change in pork production was associated with a 1.64 percent change in hog prices in the opposite direction; for each 1.0 percent change in beef production, there was a 0.72 percent change in hog prices in the opposite direction; for each change of 1.0 percent in broiler production, there was a 0.77 percent change in hog prices in the opposite direction; and for each 1.0 percent increase in disposable personal income, there was a 1.75 percent increase in price of hogs.

A mathematical forecast, based upon historical price relationships, can be treated as an indicator of value.[12] The Illinois procedure for forecasting hog price is to measure anticipated pork, beef, and poultry production for the current and next three quarters of the year, adjust for seasonality price changes, account for income effects, and adjust for inflation. Results of their work are shown in Table 4-11. Though the projected price seldom matched the actual market price, the forecasts did project periods to expect significant price change due to changes in the factors that affect price.

Live Cattle Prices. Accounting for factors to explain live cattle prices is more difficult than live hog prices. The production cycle is longer and there is a wider range of sources of supply from dairy, breeding stock, and fed and non-fed steers and heifers. The mixture of kinds and quality continually changes as feeding profitability changes.

Table 4-11: Illinois Demand Model for Hogs

Period[1]	Hog Price (1972 $)	Production			DPI (1972 $)	GNP Deflator	Adjustment Factor	Hog Price, Current $		
		Pork	Broilers	Beef				Computed	Actual	Error
	($/cwt.)	......(mil. lb.)......			(bil. $)	(1972 = 1.0)		......($/cwt.)......		
1976 I	37.17	2,860	2,035	6,228	875.8	1.3055	.939	46.31	48.53	-2.22
II	36.29	3,044	2,260	6,300	883.3	1.3180	1.012	43.86	47.83	-3.97
III	35.78	2,827	2,397	6,504	889.4	1.3329	.956	43.35	47.69	-4.34
IV	25.70	3,605	2,268	6,572	895.5	1.3498	1.063	35.57	34.70	.87
1977 I	28.65	3,188	2,090	6,253	905.7	1.3701	.939	44.10	39.25	4.85
II	27.84	3,420	2,338	6,169	915.3	1.3920	1.012	41.06	38.76	2.30
III	31.57	2,965	2,460	6,379	926.5	1.4148	.956	46.14	44.67	1.47
IV	28.27	3,522	2,304	6,295	940.9	1.4333	1.063	44.46	40.52	3.94
1978 I	31.79	3,172	2,250	6,076	950.6	1.4556	.939	49.38	46.27	3.11
II	32.07	3,397	2,502	6,049	959.6	1.4831	1.012	46.24	47.57	-1.33
III	31.61	3,109	2,608	5,909	964.7	1.5170	.956	50.08	47.95	2.13
IV	32.49	3,507	2,509	6,113	976.5	1.5450	1.063	49.22	50.20	-.98
1979 I	32.97	3,276	2,413	5,669	985.9	1.5786	.939	54.15	52.04	2.11
II	28.54	3,797	2,809	5,129	986.2	1.6142	1.012	43.85	46.07	-2.22
III	23.71	3,786	2,932	5,325	986.6	1.6484	.956	39.27	39.08	.19
IV	21.65	4,230	2,803	5,336	990.3	1.6833	1.063	36.27	36.44	-.17

Source: Good 47.

[1]Quarter I = Dec-Feb, II = Mar-May, III = June-Aug, IV = Sept-Nov.

Regression Equation:
Price of Hogs (1972 $) = 70.6357 - .0141 Pork Prod. - .0111 Broiler Prod. - .00375 Beef Prod. + .0610 DPI
(1972 $) "t" values   (6.04)   (- 9.86)   (- 3.16)   (- 2.68)   (3.91)
price flexibilities   Pork: - 1.64   Broiler: - .77   Beef: - .72   DPI: + 1.75
plus 100 mil. lb.   -$1.41   -$1.11   -$ .38

r^2 = .768 r^2 after adjustment = .937

Several factors were identified as important in affecting cattle prices. Disposable personal income and consumer willingness to spend money for meat made up the most important demand variable in the equation. A second factor was beef supply, which was a function of production. The third factor was the supply of competing meats: lamb, broilers, veal, and turkey. Other substitutes such as eggs, dairy products, or vegetables were not found to be important competitors.

The mathematical model explained about 78 percent of the variation in cattle prices by using the four variables of beef, pork, and broiler production, and disposable personal income. The results showed that a 1 percent change in beef production was associated with a 1.9 percent change in price of cattle in the opposite direction; each 1 percent change in pork production brought a change of 0.39 percent in cattle prices in the opposite direction; each 1 percent change in broiler production brought a change of 0.2 percent in cattle prices in the opposite direction; and each change of 1 percent in real disposable income brought a change of 1.13 percent in the price of cattle in the same direction.

On balance, the formula provided satisfactory ballpark estimates of cattle prices. Results are shown in Table 4-12. In particular years, the forecast differed significantly, but the method established a central value. When prices departed

Table 4-12: Illinois Demand Model for Cattle and Wholesale Beef

Period	Cattle Price (1972 $)	Whlsl. Price (1972 $)	Production			DPI Annual Rate	GNP Defl.	Cattle Price (Curr. $)			Whl. Beef Price (Curr. $)		
			Beef	Pork	Broil.			Computed	Actual	Error	Computed	Actual	Error
	($/cwt.)	(¢/lb.)	.....(mil. lb.).....			(bil. $)	(1972=1.0)	..($/cwt.)..		(%)	...(¢/lb.)...		(%)
75 I	28.76	48.93	5,842	3,146	1,833	825.5	1.2421	41.98	35.72	+14.9	68.64	60.77	+11.5
II	38.22	63.10	5,593	2,991	2,061	867.1	1.2596	48.34	48.14	+ .4	77.00	79.48	- 3.2
III	37.94	63.41	5,942	2,557	2,080	858.3	1.2820	45.57	48.64	- 6.7	73.27	81.29	-10.9
IV	35.38	58.19	6,295	2,901	1,992	864.9	1.3017	40.02	46.05	-15.1	65.77	75.74	-15.2
76 I	29.48	48.68	6,490	2,957	2,116	877.5	1.3130	37.29	38.71	- 3.8	61.55	63.92	- 3.9
II	31.19	49.77	6,144	2,848	2,314	881.2	1.3279	42.87	41.42	+ 3.4	68.89	66.09	+ 4.1
III	27.77	44.41	6,617	3,014	2,371	888.4	1.3430	35.30	37.30	- 5.7	58.03	59.64	- 2.8
IV	28.61	46.03	6,410	3,668	2,185	896.7	1.3634	37.07	39.00	- 5.2	61.18	62.75	- 2.6
77 I	27.38	44.04	6,329	3,276	2,156	903.6	1.3834	41.62	37.88	+ 9.9	67.89	60.92	+10.3
II	28.93	45.90	6,162	3,186	2,399	912.5	1.4093	44.95	40.77	+ 9.3	71.91	64.68	+10.1
III	28.38	45.65	6,320	3,074	2,424	928.0	1.4259	44.38	40.47	+ 8.8	71.11	65.09	+ 8.5
IV	29.29	47.41	6,218	3,499	2,249	939.9	1.4482	46.11	42.42	+ 8.0	74.21	68.66	+ 7.5
78 I	31.13	49.98	6,104	3,243	2,327	948.7	1.4705	50.41	45.77	+ 9.2	80.17	73.50	+ 8.3
II	36.51	57.98	5,936	3,264	2,547	953.0	1.5082	53.60	55.06	- 2.7	84.07	87.45	- 4.0
III	35.03	54.88	5,922	3,158	2,567	962.2	1.5345	55.90	53.75	+ 4.0	87.48	84.12	+ 3.7
IV	34.95	53.86	6,040	3,540	2,442	973.2	1.5668	54.18	54.71	- 1.1	85.52	84.39	+ 1.3
79 I	40.83	62.98	5,546	3,399	2,551	981.3	1.6022	65.40	65.42	0.0	101.18	100.90	+ .3
II	44.26	66.85	5,076	3,760	2,845	977.8	1.6381	71.45	72.51	- 1.5	108.66	109.51	- .8
III	39.40	60.41	5,219	3,779	2,854	980.9	1.6720	70.22	65.88	+ 6.2	107.02	101.00	+ 5.6
IV	39.16	61.39	5,416	4,346	2,665	986.7	1.7058	65.61	66.86	- 1.9	101.73	102.36	- .6

Source: Good 63.

Regression Equations:
    Cattle Price (72 $) = 82.1626 - .0116(Beef Prod) - .0039(Pork Prod) - .0037(Broil. Prod) + .0463(DPI-72 $)
        "t" values   -11.5   -3.8   -1.3   +3.6
        Price flexibilities   1.895   .368   .216   1.125
        r^2 = .779
    Wholesale Beef Price (72 $) = 129.887 - .0164(Beef Prod) - .0056(Pork Prod) - .0077(Broil. Prod) + .0641(DPI-72 $)
        "t" values   -10.4   -3.5   -1.7   +3.2
        Price flexibilities   1.666   .347   .347   .974
        r^2 = .736

substantially from central value, they subsequently returned and usually overcorrected.

SUMMARY

Fundamental supply and demand principles explain equilibrium price and value of an agricultural commodity and commodities in general. However, discovering the market-clearing equilibrium price is not an easy matter in a dynamic, changing real world. Shifts in supply and demand conditions result in new prices being established; inelastic demands and supply responses cause exaggerated price changes. As a consequence, farm prices and incomes tend to be highly variable. But, if farmers and ranchers understand price-making principles and their application to various situations, they can improve their skill in making price decisions for the commodities they produce.

Our objective in this chapter was to develop the notion of a general PRICE LEVEL established by FUNDAMENTAL demand and supply considerations. We furthermore sought to identify important price-making forces and evaluate their impact in causing price change.

By sorting out important price-making variables from the unimportant ones, we identify the fundamentals to watch in making price projections. By using the calculated elasticities of the important variables, we can project expected price change based upon change in the fundamental factors.

By knowing what is important and how important individual variables are, farmers are in a better position to project prices from fundamental information.

Questions for Study and Discussion:

1. Define and illustrate the demand principle.
2. What does elasticity of demand show?
3. What is meant by a change in quantity demanded? What factors shift the demand curve?
4. What is effective demand?
5. Define and illustrate the supply principle.
6. What is supply elasticity, and what factors affect it?
7. What is meant by changes in quantity supplied? What factors lead to shifts in supply?
8. What is the meaning of equilibrium price and quantity?
9. Illustrate the effects on price that arise from changes in (a) demand and (b) supply.
10. How does the supply-demand balance operate to achieve equilibrium price?
11. Explain price discovery as distinct from, but a part of, equilibrium price.
12. How is the Law of One Price used to explain price differences?
13. What is the effect on total receipts from increased supply of a commodity with inelastic demand?
14. Farmers work harder to produce more but wind up earning less. Why?
15. What are created by price ceilings and price floors? Why?
16. How does the cobweb theorem explain price cycles?
17. What are the objectives of price analysis in a fundamental approach to price determination?

18. What fundamental factors are important in determining grain prices?
19. What fundamental factors are important in determining hog and cattle prices?

Exercise: Maintaining a History of Price-Making Events

Each day the commodities page of *The Wall Street Journal* reports on traders' comments about market forces in play for the previous day. Make a copy of those comments and maintain them over a period of time. A periodic review of a month or two of comments is enlightening in evaluating published reports of what were thought to be important market-making forces. Time helps to sort out whether the events were important or not. After a while, you may determine that newspaper reports of current events are really records of history and comments on events were really "noise," not true price-making factors.

One handy way for maintaining this record is to glue the newspaper clippings in a spiral notebook. Headline news stories of demand and supply events from other sources can be attached to the record also.

Students can hold a group discussion or write a paper as a review of their records of market actions after three months. These activities will help students understand the forces at work and their relative importance in price determination.

Footnotes for Chapter 4:

(1) Demand, supply, and price sections of this chapter are adapted with permission of Macmillan Publishing Co. from Richard L. Kohls and Joseph N. Uhl, *Marketing of Agricultural Products,* 6th ed. (New York: Copyright 1985 by Macmillan Publishing Co.).

(2) Kohls and Uhl, *Marketing of Agricultural Products* 170.

(3) Kohls and Uhl, *Marketing of Agricultural Products* 167.

(4) Kohls and Uhl, *Marketing of Agricultural Products* 168.

(5) Robert L. Heilbroner, "The Economic Problem Newsletter," Vol. V, No. 2 (Englewood Cliffs, N.J.: Prentice-Hall, Inc., Winter, 1973).

(6) Darrel L. Good, Thomas A. Hieronymus, and Royce Hinton, *Price Forecasting and Sales Management* (Champaign: Cooperative Extension Service, College of Agriculture, University of Illinois, 1980). Copyright 1980 by the Board of Trustees of the University of Illinois. Used by permission.

(7) Good, Hieronymus, and Hinton, *Price Forecasting and Sales Management* 27.

(8) Good, Hieronymus, and Hinton, *Price Forecasting and Sales Management* 67.

(9) Scott Irwin, unpublished data, The Ohio State University, Columbus.

(10) Irwin.

(11) Good, Hieronymus, and Hinton, *Price Forecasting and Sales Management* 35.

(12) The Illinois analysis of projected price is available on a subscription basis from Darrel Good, "Outlook Update," Department of Agricultural Economics, University of Illinois, Urbana, 61801.

Chapter 5

Technical Price Analysis

CHARTING	152
Bar Chart Basics	152
Reading Charts	153
Psychology of a Trend	155
Identifying a Trend	156
VOLUME AND OPEN INTEREST	157
CHART FORMATIONS	158
Psychology of Support and Resistance	158
Defining Price Objectives	159
Head and Shoulders Formation	159
Flags and Pennants	160
Triangles	161
Gaps	161
Retracement	164
Identifying Market Tops and Bottoms	164
Reversals	164
Saucer or Bowl	167
Double Tops or Bottoms	168
Bar Chart Rules	168
Confirmation	171
Reliability	171
MOVING AVERAGES	172
SUMMARY	177

Technical Price Analysis[1]

Farmers, grain elevator operators, speculators, and other businesspersons involved in commodity markets are keenly aware that commodity prices are influenced by thousands of variables. These may include number of animals on feed, size of crop, domestic use, exports, or range conditions. Governmental policies to support prices, restrict imports, or establish embargoes also influence prices. International tensions and economic progress in other countries impact on the value of commodities. To complicate matters further, prices are affected by rainfall or drought anywhere in the world, transportation costs, and currency exchange rates.

Forecasting prices based upon these supply and demand factors is called fundamental analysis. It is based upon the economic principle that the value of any commodity derives from a balancing of supply and demand forces registered as market price. But fundamental analysis has inherent problems which limit its usefulness in forecasting day-to-day or month-to-month price action. Nobody can know all the facts about what a price should be, despite being given all the fundamental information about real value of a commodity. First, factual information is subject to error. Data samples used as estimates of demand and supply statistics are based upon an acceptable statistical probability of being wrong. Second, facts change because fundamental conditions change. Today's valid price projection of expectations six months hence will change in the interim because of drought or changes in breeding intentions, for example. Third, the facts may, by design, not be presented correctly. What poker player will allow the face to show what cards he or she really holds? Reliability of information supplied on Brazilian production or intended Soviet purchases is often subject to question. Finally, the numerous fundamental facts available and the extent of influence each one exerts simply cannot be sorted out in one brain.

Even if all facts were known, true price discovery remains a problem because the constant flow of fundamental information into the market does not form a price until it is interpreted by people. The extent of knowledge and interpretation of what it means differ markedly. Since people are involved in pricing decisions of a commodity, it is impossible to rule out human emotions from the decision-making process. When a market is going up, people expect it to continue going higher; when it is going down, they expect it to continue moving lower. Before supply and demand facts become price, they are subjected to interpretation and human judgment. It is not necessarily the facts, but what people believe the facts to be, that is important in establishing price.

Because of the problems of fundamental analysis, many market participants pay little heed to fundamentals. They are technicians who rely upon technical analysis to judge price direction and objective. Technical analysis can be defined as **a study of market action with the purpose of determining probable**

price change. The working tool of technicians is a graph or chart which shows price movement over a period of time. Hence, technicians have become known as chartists. Technicians concede that bearish technical information which drives market price down or bullish information which drives it up sets the tone of the market, but it is the prevailing consensus of all market participants which determines market direction and extent of movement.

One purpose of chart reading is to measure the relative strength of buying and selling pressure. If chart patterns demonstrate that buying pressure at the prevailing price level is more powerful than selling pressure, it is logical to expect that prices will rise. On the other hand, if selling pressures overpower buying pressure, the expectation is that prices will fall. Technical analysis attempts to measure the collective judgment of buyers, sellers, and speculators. Charts graphically illustrate the collective opinion of the market participants.

Chartists arrive at their conclusions about price by diagnosing formations or patterns that appear on the charts. Formations suggest where other technicians are likely to place buy orders or sell orders — they define price objectives. Furthermore, charts can actually influence the market because thousands of traders make their buying and selling decisions from them. So many technicians are watching the same patterns that their collective analysis causes self-fulfilling prophecies. Charts work because chartists make them work.

Farmers are sometimes suspicious of "crystal ball gazing," seemingly divorced from reality. They are by nature fundamentalists. As practicing economists, they understand that drought-reduced crop supplies will fetch a higher price, or over-expansion of the beef cow herd will eventually produce too much beef and drive the price down. They fret and complain when the "gunslinger speculators" create distorted low prices on the various commodity exchanges. Their agitation is further compounded when they miss short-lived high prices. But technicians can work to the farmers' advantage. Because so many traders watch chart patterns, there is a tendency to over-buy a market in a bullish trend and over-sell it in a bearish trend. Instead of lobbying to close the exchanges when excessive selling pressure creates "over-sold" low markets, or enthusiastic buying creates "over-bought" high markets, producers should learn to use chart patterns to help them reap the fruits of the speculators' efforts. Producers can be assured that if price levels move excessively high or excessively low, fundamental factors will ultimately correct the distortion by setting a price where demand and supply are in equilibrium. Charts present a road map of how the market price arrives at its final destination, offering high and low market pricing opportunities along the way.

Charts are not a utopia in price forecasting, but they can be an important tool in a farmer's marketing toolbox. First, they can be helpful in developing good market judgment as a "check" on the accuracy of conclusions that are derived from studying fundamentals. For example, farmers who sold soybeans in the bull markets of 1980, 1983, and 1988 at prices of $9.00 to $10.00 per bushel had several chart confirmations that market objectives were in the $9.50 to $10.00 range. Fundamentalists were still bullish, expecting prices to reach $12.00 or $15.00 per bushel. Obviously, they were misinterpreting supply and demand information. Within a few weeks, the market slumped to the $7.50 price area.

Charts force the user to follow market activity on a daily schedule. This attention makes one a better student of the market. By providing a visual picture of past price activity, a historical record in charts gives one a broader perspective of market activity. A chart reader is not so elated by daily limit up moves, nor as depressed by limit down moves. Finally, following charts and learning their rules can help the producer predict price trends of crops and livestock to establish price objectives where profitable sales and low-cost purchases can be made. With charting, farmers should be better able to know what is a high price or low price and make improved marketing decisions. A cold, hard picture of price action tempers emotions when marketing decisions are being made.

CHARTING

Technicians base their analysis on three factors: (1) price movement, (2) volume, and (3) open interest. By far, the most important item is price movement.

Various charting systems show price movements. They may consist of lines, bars, step diagrams, or different symbols plotted on arithmetic, logarithmic, or square root scales. Bar charts illustrate one method used to define price objectives.

Bar Chart Basics

Each day's futures trading establishes three benchmark prices on bar charts: high, low, and settlement. A line connecting the high-low points corresponds to the price range of the day. It is a gauge of intra-day activity. The top of the line marks the highest transaction price of the day and the bottom, the lowest price. A horizontal "tic" mark indicates the closing (settlement) price as shown in Figure 5-1.

The settlement price is particularly important. Millions of dollars change

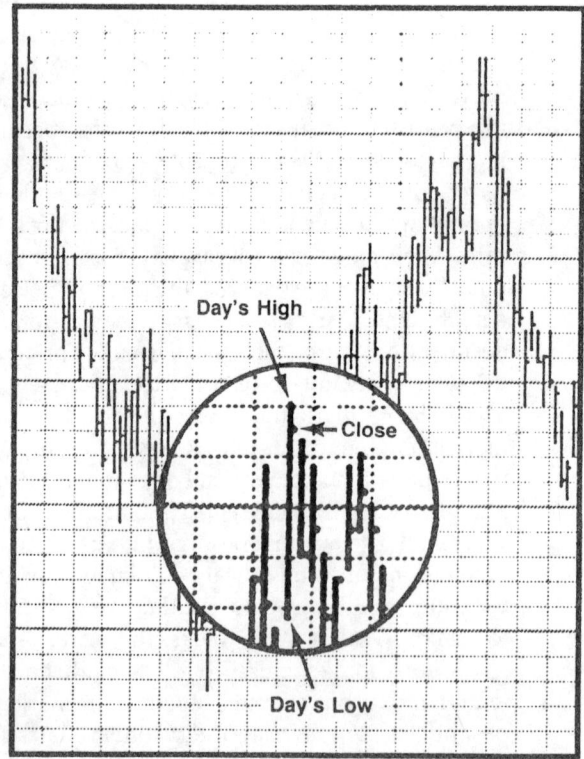

Figure 5-1: Bar Chart of Price Action.

hands between holders of respective long positions or short positions since accounts are balanced on the close each day, relative to the previous day's close. A higher close than the previous day means that holders of short positions transfer money to longs and vice versa if the close is lower. Real money changes hands based upon the settlement price.

In a broader sense, the close defines the winner in the struggle between bulls and bears. During the trading day, the shifting forces of demand and supply float the market upward or downward, but account balances change only with the trade at the bell. The close represents the sum total of everyone's opinion of value that day and an expectation of what will happen in the interim between the close and the opening the next trading day.

Bar chart lines are plotted on arithmetically ruled paper with the vertical scale referring to price and the horizontal scale referring to time. Commercially available paper has a heavy line for Monday and lighter lines for the remaining days of the week. Most paper has 52 weeks of space. Gradations on the vertical scale should be sufficiently wide to allow easy and accurate plotting, but sufficiently close such that price fluctuations during the year can be contained on the page. Uniformity and accuracy in plotting are extremely important. Inaccurate charting can give false impressions. Because charting can be a time-consuming and tedious process, many traders subscribe to one of several chart services available. These provide current and historic charts and chart analysis on a weekly or monthly schedule. There is no substitute for maintaining one's own charts, however. Without them, the advantage of daily attention is lost. Some publications provide a convenient method for individuals to keep daily records on the nearby options, and monthly reports from the publication company provide a permanent record of all trading months.

Weekly and monthly charts are drawn in the same way as daily charts, with each bar corresponding to price action covering time of either the week or the month, as the case may be. They are used for the study of long-term trends. Examples of daily, weekly, and monthly charts are shown in Figure 5-2.

Reading Charts

The art of forecasting from bar charts depends upon recognizing and interpreting formations which have historically been associated with price movement in a particular direction. However, there are sufficient examples of unexpected developments and unusual formations which rule out charting as an infallible forecasting device. Some regard chart analysis as "reading tea leaves."

The prevailing tendency, illustrated on charts, is for prices to move in a particular direction, or trend. A market is in an up trend when it develops a series of higher highs and higher lows over successive days. A down trend is just the opposite, lower highs and lower lows. An up trend line is drawn by connecting the lower points of the overall movement. A down trend line is drawn by connecting the higher points. High and low points on a sideways trend conform to bounds set by horizontal lines.

A key charting rule is that commodity prices tend to trade in channels. These are the boundaries through which prices pass as they generate a trend line. As long as price activity continues within the trend channels, the channels are useful for suggesting buying and selling levels.

The trend lines are also critically important for indicating important trend changes. The top line of channels is called "resistance," and the bottom is called a

Daily Charts

Weekly Charts

Monthly Charts

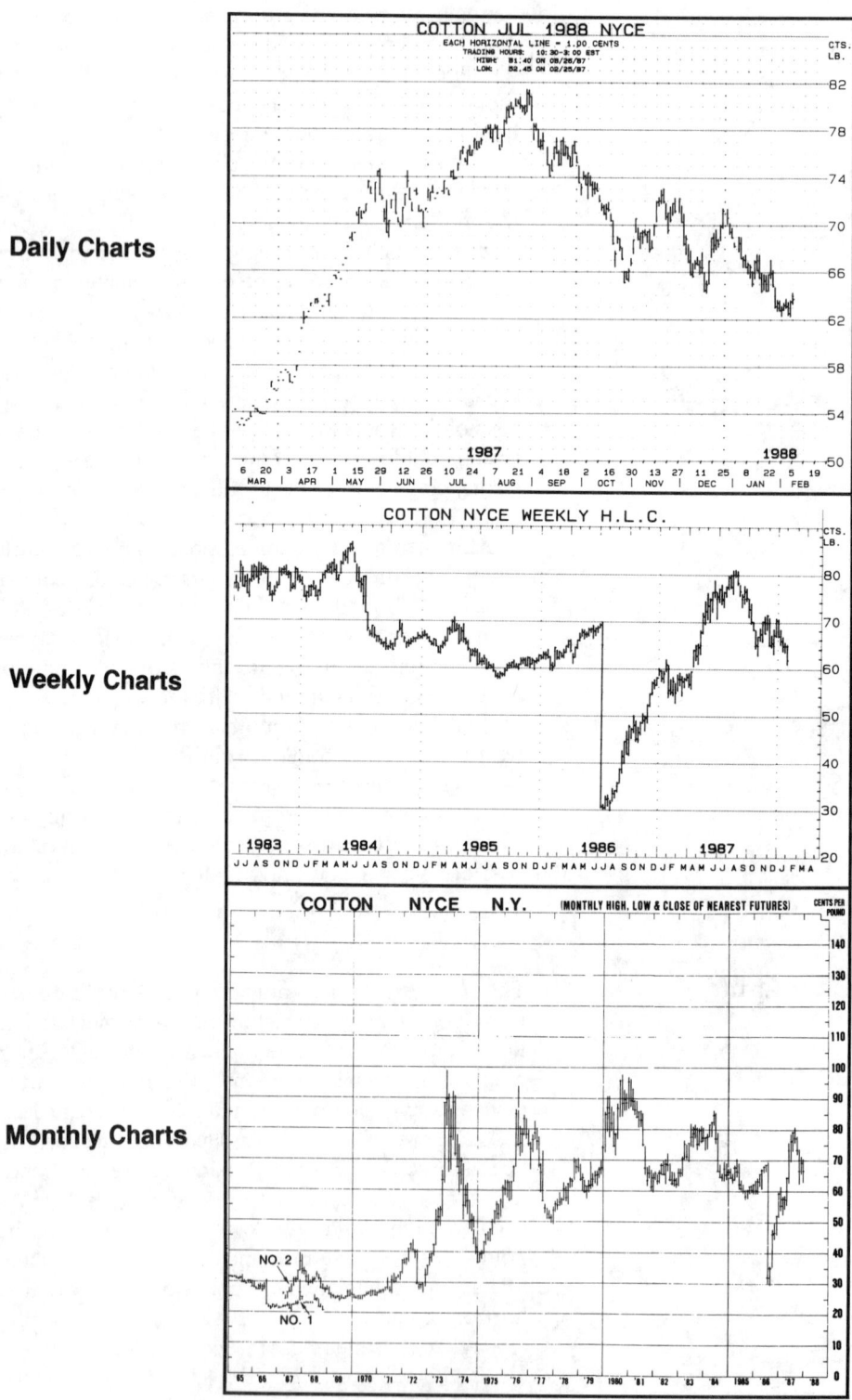

Figure 5-2: Daily, Weekly, and Monthly Charts of Futures Contracts.

"support" line. As shown in Figure 5-3, each time a price move approached support or resistance, it turned the other direction. The reason for this is that thousands of chart traders are using basically the same information. They buy on "dips" and sell "rallies." Thus, there is buying activity when the market dips to support and selling pressure when it rises toward resistance, until there is a breakout.

Figure 5-3: Illustration of Up, Down, and Sideways Trends, and Support and Resistance.

Psychology of a Trend[2]

Since no one is absolutely sure about the true value of the commodity in question, the relative buying or selling strength of the bulls or bears causes market price fluctuation. The fact that new buyers are willing to pay more for a commodity than the previous buyer in an up trend is based upon the expectation that price will keep going up. At some point, however, not enough buyers are willing to pay a higher price.

At the top of a market, hedging (by selling short) makes a larger number of contracts available. When farmer-producers begin selling against their production, or elevator operators begin hedging their long cash positions, the market can turn around in response to their action. Speculators on the sidelines can be drawn into the action and begin making sales. Prices drop. Previous buyers holding a long position have a natural inclination to hold their positions, expecting price to go back up. Some follow an old charting rule that states there should be two successive closes outside the trend line before expecting a change in trend. Even as prices erode further, many longs remain convinced that the dip is only temporary. They hang on, possibly buying more contracts offered by those who are less courageous. Their confidence may stem the tide, possibly causing a recovery. But if buying is not strong and more contracts are offered for sale, the buyers' confidence is shaken. They begin to liquidate their positions to protect what profits remain or limit their losses. Their selling pressure adds to the momentum, and a down trend is set into motion.

Eventually, prices decline to such a low point that both longs and shorts become convinced that prices cannot go lower. Price may be below cost of production. At that time, traders with short positions begin covering their positions to take profits. They must do so by buying. Their "buy" actions stem the down trend and attract new buyers who expect a turnaround in price. At low price levels, farmer-producers are willing to assume price risks themselves rather than hedge at a loss. They do not offer short contracts. Farmer-buyers who want to lock in low-cost input prices for protein supplements, feed grains, or feeder livestock enter the market to hedge-buy their needs. Speculators seeing the rise beginning rush to get in the early action. The market moves up and an up trend begins. Often, new fundamental information feeds into the market showing reduced supplies or increased demand. The psychology turns into a bull market and creates an up trend.

Identifying a Trend

It is not easy to identify a trend until it is fairly well established. Many trend lines, both major and minor, may have to be drawn before a true trend can be ascertained. An up trend is confirmed when the market closes above the top of the "short" covering rally (those holding short positions who buy in order to cancel their short position). Additional new buying reinforces the up trend. Many bull markets are accompanied by a rapid and large buildup of open interest, which shows new buyer willingness to get in on the action at a higher price.

A change from up trend to down trend is confirmed in a similar manner. At some point either a lack of buyers or an influx of new sellers forces the market to close below the up trend line. After a sell-off the market sometimes rallies at least to the downside breakout point. A close below the bottom of the first decline through the up trend line confirms the down trend. A decline in open interest further confirms the change in market direction as traders liquidate positions.

Commodity prices do not spend all their time in upward or downward trends. Often they move in sideways channels with no indication of which way they are headed. Typically a "shock" from fundamental information is required to start a new trend direction. Sideways markets are usually at the bottom of markets, seldom near the top. Up trend markets usually take considerable time in building to their ultimate objectives. Once a market tops out, though, it can drop suddenly. Good selling opportunities can pop up on charts and be gone before farmers have had a chance to react, unless they are closely attuned to the market.

Farmer hedge-sellers can use changes in up trends as a signal to implement their sell decision. This is particularly important if topping action given by other formations such as gaps and reversals (defined later) have also appeared in the charts. Likewise, farmer hedge-buyers looking for market bottoms to lock in input costs should watch for chart confirmations which signal price bottoms. The same chart rules of change in trend apply in both up and down markets.

Brock and Martin listed 10 trend-following rules which give purpose to chart watching:[3]

1. Commodities spend a large part of their time in sideways trends.
2. A down trend is valid until the market closes above a major down trend.
3. An up trend is valid until the market closes below the major upward trend.

4. An up trend is confirmed when the market closes above the top of the first rally through the down trend line.

5. A down trend is confirmed when the market closes below the bottom of the first decline through the up trend line.

6. A sideways (sometimes called trading) market exists until the market closes above the resistance line, signaling an up trend, as in Rule #4, or below the support area, signaling a down trend, as in Rule #5.

7. Sideways trends are generally at bottoms of a market — not tops.

8. Market bottoms often have narrow trading ranges and are dominated by overall bearish sentiment.

9. Down trends generally go into a sideways trend, rather than immediately into an up trend.

10. An up trend often turns suddenly into a down trend market.

VOLUME AND OPEN INTEREST

Volume and open interest are important tools used in conjunction with bar charts. They provide measures of the extent of activity coming into market or leaving it. Volume and open interest data are provided as part of many chart service reports, as shown in Figure 5-4.

Figure 5-4: Example of Volume and Open Interest Report.

Volume represents the number of trades for a particular commodity on a given day. As a barometer of trading activity, it measures the number of buyers and sellers participating in the market. Change in volume is often associated with price change in the same direction. For example, a gradual increase in volume during an up trend suggests that more buyers are willing to come into the market expecting a continued price rise. However, a rapidly accelerating volume following a substantial upward price movement often signals a major top and possible price reversal as traders liquidate positions.

Open interest is the number of open positions established, but not yet liquidated by delivery or an offsetting trade. Open interest is not related to volume; there can be a high-volume day without a change in open interest. Open interest increases only when new purchases are matched by new sales. For example, if a buyer makes a new purchase of 5,000 bushels of July soybeans and the trade is taken by another person who initiates a new short position of 5,000 bushels of July soybeans, the net result is an increase in open interest of 5,000 bushels. If a new purchase is offset by the sale of a previously purchased contract, then there would be no change in open interest. Decreases in open interest occur when previous purchases are sold and matched by liquidation of previously sold positions. This means that both buyers and sellers are closing out previous commitments. Open interest figures, therefore, show whether new positions are being established or old ones are being canceled out.

Analysis of volume and open interest information can often detect market conditions which are not readily apparent as changes in supply and demand conditions of the actual commodity. For example, new export business may be reflected in futures markets before it is publicly announced. Increased open interest is generated by commercial exporters buying futures contracts of the commodity they plan to export to protect themselves against possible price increases.

It is important, however, to adjust changes in volume and open interest figures to account for seasonal activity before applying general rules to making marketing decisions. Open interest in grains normally increases during harvest seasons as elevators establish routine hedges to offset cash purchases. Later in the year a decline in open interest can signal that country elevators are lifting hedges by buying back previously established positions once cash sales have been concluded. Their buying can result in temporary higher prices. Furthermore, sharp declines in open interest may occur in soybeans after the first of January because cash sellers waiting for a new tax year are selling and lifting hedge positions. Seasonal patterns do not vary much between years. Therefore, changes due to seasonality can be anticipated fairly closely. It is the changes in volume and open interest outside seasonal influences which support analysis of price action in detecting future moves, as summarized in Table 5-1.

CHART FORMATIONS

Psychology of Support and Resistance

Support and resistance psychology applied in trend analysis has broader application in the real world market. Any point in a chart that has been established by a price objective as the extreme of a price move, either up or down, creates an important psychological barrier to trading beyond the extreme. Resistance has several strong psychological barriers to continued market strength. First, buyers are reluctant to enter new higher price territory where others have feared to tread. There is a shortage of buyers at resistance points. Second, many holders of long positions see rallies to old resistance levels as an opportunity to exit their posi-

Table 5-1: Volume and Open Interest Signals
Associated with Price Action

When Prices Are Up	When Prices Are Down
Volume and Open Interest Up —Means STRONG MARKETS	Volume and Open Interest Up —Means WEAK MARKETS
Volume and Open Interest Down —Means WEAK MARKETS	Volume and Open Interest Down —Means STRONG MARKETS

tions. Those who bought before the rally have profits in their positions that they are willing to take. Those who bought near the top of a previous rally may want to get off the roller coaster ride. When the market went down after a period of strength, many longs refused to sell. They watched their positions decline in value and losses mount. Reluctance to realize actual dollar losses kept these traders from liquidating. When the market turns upward again, these longs are ready to sell out when contracts reach the price they originally paid. This tendency to let losses run in hopes of breaking even on a rebound is characteristic of neophyte traders who fail to establish stop losses.

Resistance levels entice sidelined traders to initiate short positions. In the absence of a change in fundamentals to send price higher, they think that a short position has less risk than a long one. New sellers enter the market. As price approaches an old high, there is an avalanche of sell offers from varied sources. Often, the market falls under the cumulative selling pressure, and another correction occurs.

The psychology operates in opposite fashion in down trending markets. Support refers to a level on the chart where a price decline may be expected to stop. A support level could be stopping points of earlier down moves and can be expected to halt subsequent declines.

All past price levels at which a significant reversal in direction has occurred are prime targets for future resistance and support areas. Several chart formations develop from this phenomenon.[4]

Defining Price Objectives

Head and Shoulders Formation. The oldest, most popular, and possibly most reliable trend reversal pattern is called the Head and Shoulders Formation. It can signal a down trend if it occurs at the top of the market, or an inverse head and shoulders can signal an upturn from a market bottom. An example is given in Figure 5-5.

Descriptively, the topping formation begins with a rally and a decline within an up trend to form the left shoulder. A second rally, carrying beyond the first rally but falling back to the starting point, outlines the head. The right shoulder is formed by a third rally and a decline that extends to the two earlier stopping points. The two stopping points form a neckline. The head and shoulders formation is completed when the neckline is broken; the break confirms the change in trend.

Figure 5-5: Example of Head and Shoulders Top.

A common measuring rule applied to head and shoulders formations is that once the neckline is broken, the subsequent move will at least equal the distance from the top of the head to the neckline. In the February hog chart the measure from $56.00 to $47.50 gave a downside objective of $39.00. The measuring rule applies conversely for up trend signals generated by an inverted head and shoulders.

Flags and Pennants. Appropriately named, flags and pennants fly from a mast to symbolically foretell and lead major advances and declines. They have decorated many rapid and extensive commodity price moves.

Flags and pennants have a quick vertical price move that forms a flagpole to the body of the formation. A period of consolidation follows the initial move as traders await new information. Flags usually extend in the opposite direction from the major move, and when tightly constructed they are especially reliable indicators of direction and price objective. Loosely formed flags that point in the direction of the move may be unreliable. Normally, if flags and pennants are not completely formed in a month, they are likely to be part of another formation and should be viewed with suspicion.

In measuring the move to be expected from flags or pennants, one guide is to use half the length of the flagpole. The distance from the initial breakout to the center of the flag would be half the distance of the next expected move out of the formation. In other situations the length of the flagpole, transposed on the lowest trading price prior to the breakout, establishes the price objective. As illustrated in Figure 5-6, price moved through resistance strongly upward from a level of $7.00 to form a pennant. Its center point of $8.05 suggested the next move would extend $1.05. The breakout occurred at $8.25, giving an objective of $9.30; the next up move actually reached $9.20. A bull flagpole was set at $8.25. Measuring from the breakout to the center of the flag at $8.75 suggested another objective of $9.25. The market eventually traded to almost $9.90 before declining significantly. The subsequent bear flag in the down move (from the $9.20 breakout to the $8.00 midpoint) projected a downside objective of $6.80. Flags and pennants can be used to measure objectives in either down moves or up moves.

Triangles. Triangle formations have been classified as ascending, descending, or symmetrical (Figure 5-7). The formation is also referred to as a coil. As trading activity approaches the apex, the balance between buying and selling gets narrower; the coil spring is wound tighter and tighter. In an ascending triangle market bottoms edge higher and higher, but overhead resistance holds; this action points to an upward breakout. In a descending triangle, market tops move lower and lower, but support holds; this action points to a downward breakout. In a symmetrical triangle, tops become lower and bottoms higher. They generally give no warning as to direction of breakout. Buying and selling become so closely balanced that a minimum of bullish or bearish fundamentals will upset the balance and spring a move in either direction. Triangles figure prominently as signs of impending activity, but they are the most untrustworthy forms as far as their ability to predict trends. It is extremely difficult to anticipate the direction of price movement out of a triangle. Frequently the breakout proves to be false and the trend is quickly reversed.

The measuring rule for triangles is that price will move a minimum of the vertical side of the triangle once the breakout occurs. Furthermore, the breakout generally occurs about two-thirds of the distance to the apex. Trading volume typically increases after the breakout. In the example of July cotton (Figure 5-8) the vertical side, measured from 33 cents to 52.5 cents, pointed to a move to 71 cents. Eventually, the price exceeded that level.

Gaps. Gaps are open spaces in charts where no trading occurs across a vertical distance. At times changes in market fundamentals occur overnight and set off a significant move in market direction. Such changes often take place after consolidation in one of the previously mentioned formations. An up-gap develops when buying enthusiasm floods the market and causes the lowest price of a particular day to be higher than the highest price of the preceding day. A down-gap results from heavy selling pressure, causing the highest price on one day's trading to be lower than the lowest price of the preceding day.

Four types of gaps occur: common gap, breakaway gap, runaway gap or measuring gap, and exhaustion gap, as shown in Figure 5-9. A common gap exists temporarily. It has no special significance because it is typically filled a short

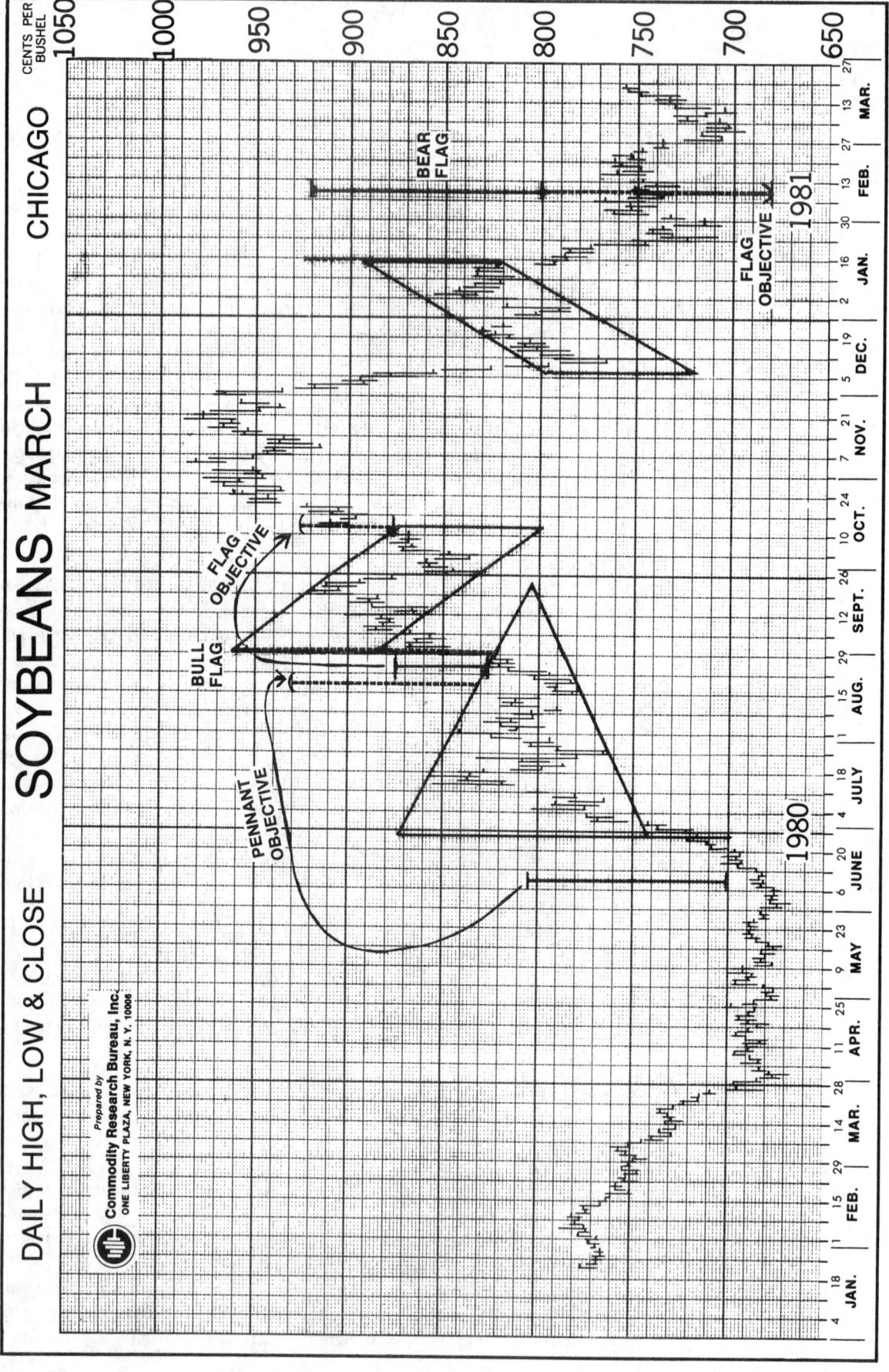

Figure 5-6: Examples of Pennant and Flag Formations.

Technical Price Analysis 163

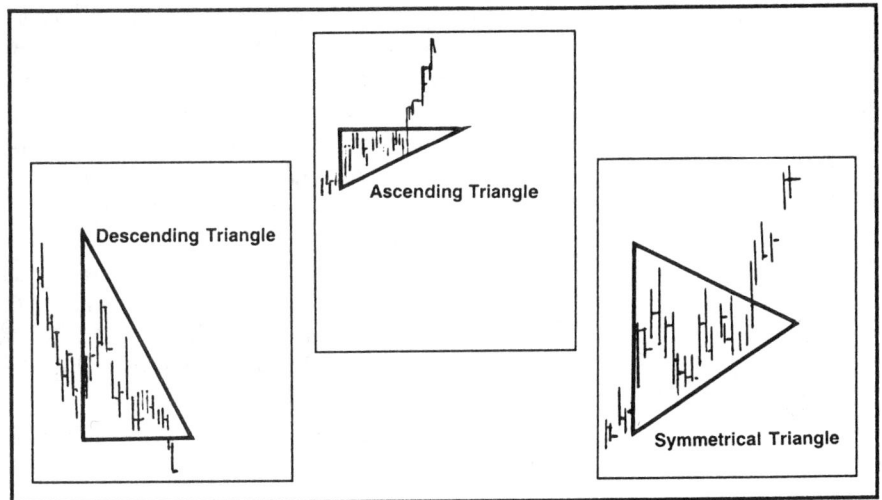

Figure 5-7: Illustrations of Three Types of Triangle Formations.

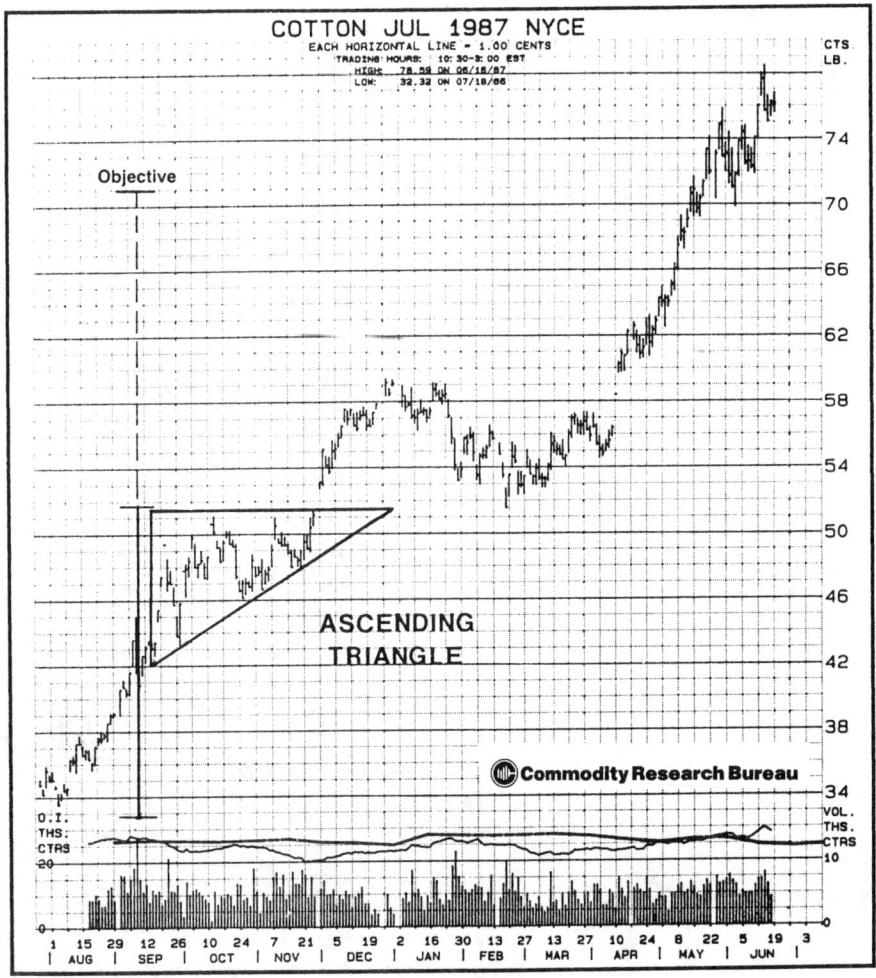

Figure 5-8: Example of Ascending Triangle Formation.

time later as the trading range extends to overlap the opening. It seems that the market wants to tidy up unfinished business by moving back to fill previous gaps. Gaps become price objectives, since traders believe that all gaps are eventually filled. Their trading takes prices back to fill common gaps.

Breakaway gaps signal the completion of consolidation formations where the market has traded for some time without direction and are indicators of a dynamic move to follow. Often a fundamental shock stimulates buying enthusiasm by drawing high volume and open interest into the market as traders sense a strong up trend. Trading pressure does not move the price back to cover the gap. Longs become bolder and add to their positions; shorts buy back their positions and go long. A little later, a flurry of buying snaps open the up trend and forms the runaway gap. With astonishing reliability, this also becomes a measuring gap of half the distance of the remaining up trend, calculated from the consolidation base which produced the breakaway gap originally. Toward the top of the trend, the market is flooded with bullish news. Conventional wisdom convinces everyone that strong demand will outstrip limited supplies and the market is going to record highs. Two groups of buyers are attracted into the market: small speculators off the street and belabored short sellers. Small traders waited two or three weeks to be convinced the market was going up, but once convinced, they buy aggressively. The belabored short sellers buy in panic to cancel their positions as margin calls mount. This wave of buying creates an exhaustion gap, or "sucker surge." The commodity is priced well above what any user can translate into value. At this point, the market is overbought and the stage is set for correction. Often downside gaps occur in the correction to set downside objectives in the same fashion as in the up trend.

In Figure 5-9, the breakaway gap occurred at $7.44 in early July. A measuring gap at $8.26 in early September signaled an objective of $9.08. The market moved to $9.20 later that month. In the down trend after the top, breakaway and measuring gaps signaled a downside objective of $6.91.

Retracement. Markets often react to sudden and extreme changes in trend by bouncing back in the direction from where the move started. On its way to the top of a move, the market fades and selling increases. As the mood shifts, often a 50 percent retracement, or rebound, occurs, as in Figure 5-10. At other times, sentiment pulls the price back a little further to a 66 percent retracement. Following a sell-off in an up trend, many buyers see the lower price as a bargain relative to recent market highs. Bullish sentiment, still hovering over the market, attracts new buyers who missed the previous upsurge. Their collective action bids price back toward the old high. Eventually, bearishness prevails and the change in trend does occur to send the market lower. Likewise, there can be a 50 percent or 66 percent retracement back toward a market low after the market has bounced off a bottom.

In the wheat example, a 50 percent retracement of its 92-cent move set a $2.70 objective, and a 66 percent retracement set an objective of $2.54. The July contract traded to $2.52 before going off the board.

Identifying Market Tops and Bottoms

Reversals. An important chart signal is a reversal day. Its use is to identify a market extreme by spotting the day of a market turnaround. When a reversal day occurs after a relatively large move, and prices start to retreat on subsequent days, the reversal may signal an important move in the opposite direction.

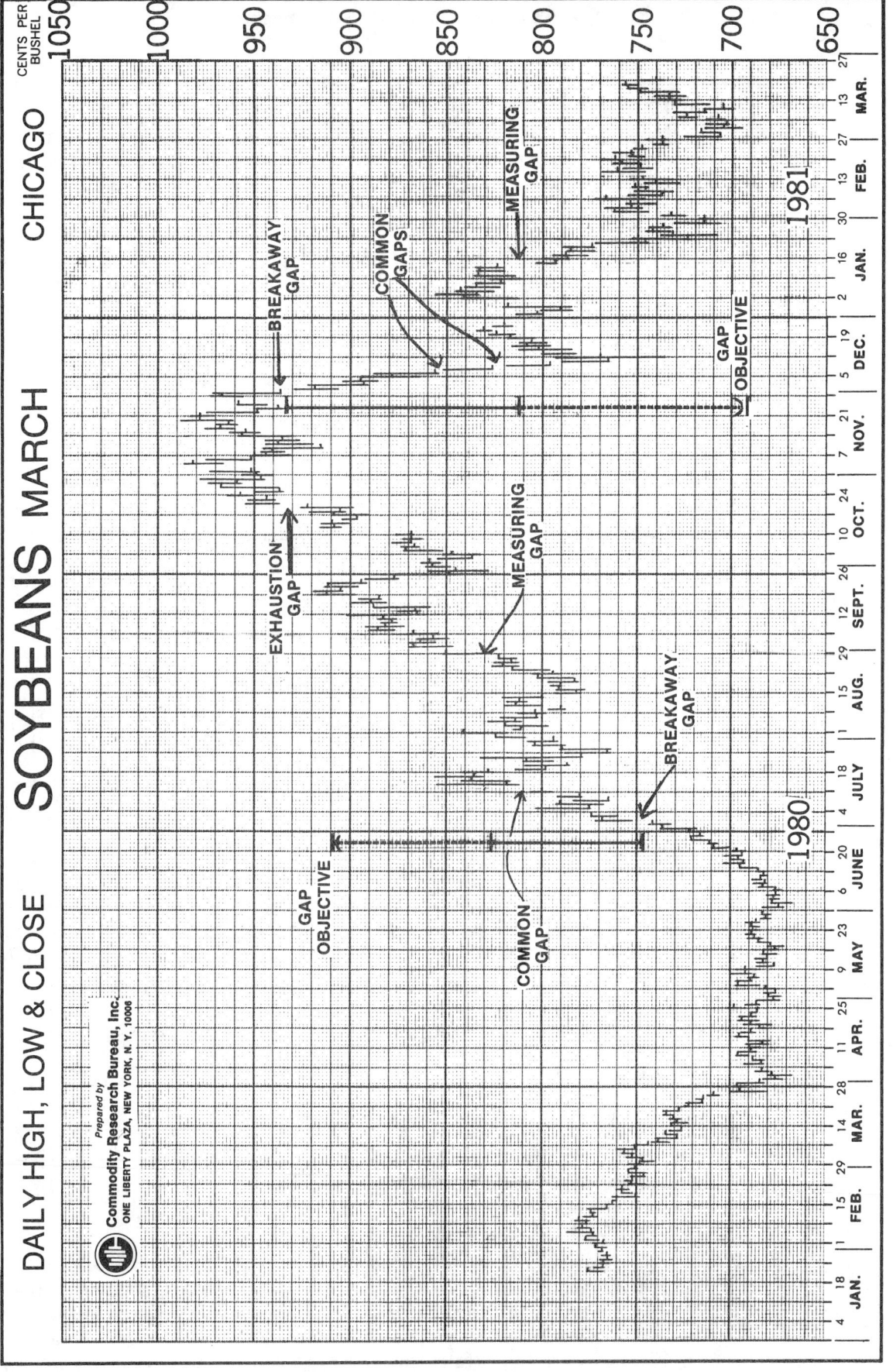

Figure 5-9: Examples of Gap Formations.

Three kinds of reversals are key reversals, hook reversals, and island reversals, as shown in Figure 5-11. They occur at both market tops and bottoms.

A bearish key reversal at the top of a market has the following characteristics:

> A high higher than the previous day.
> A low lower than the previous day.
> A close that is lower than the previous day's low.

A bullish key reversal at market bottom has:

> A high higher than the previous day.
> A low lower than the previous day.
> A close that is higher than the previous day's high.

Figure 5-10: Example of Retracements.

A key reversal provides a single day's picture of the battle between bulls and bears. Volume is unusually high, indicating lots of trading activity. In an up market, strong bullish buying pushes the high above the previous day's high. At the same time, the bears become heavy sellers to pull the low below the previous day. The close tells who wins the tug of war. If it is below the previous low, the bears

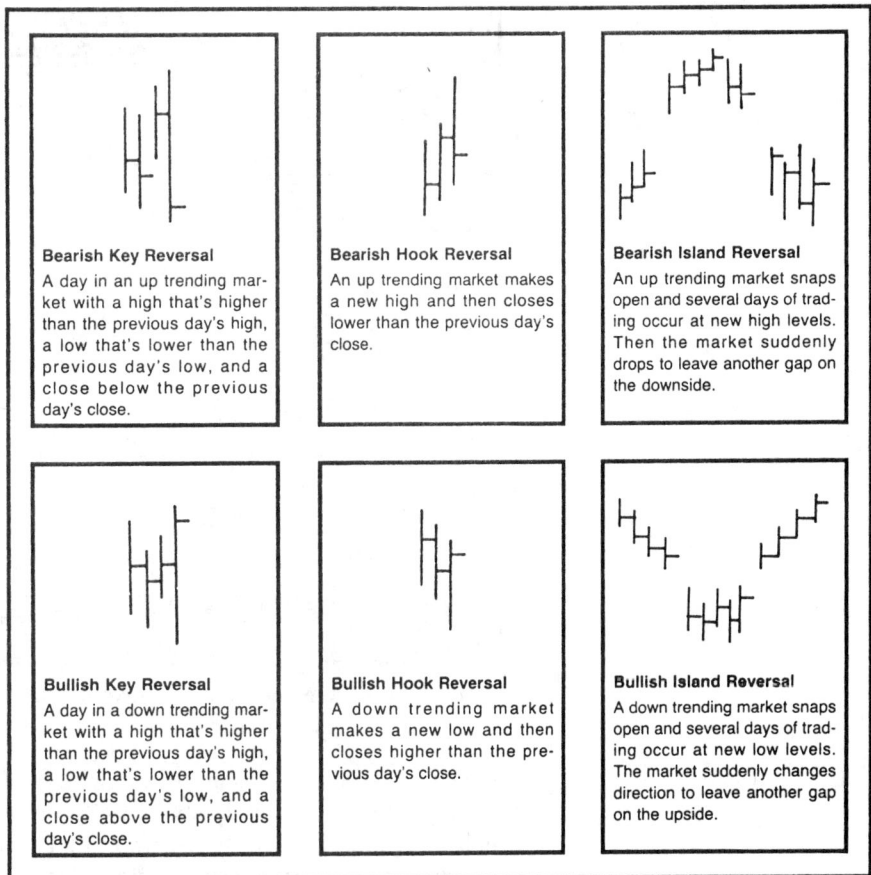

Figure 5-11: Examples of Reversals in Market Direction.

have won and the market is likely to turn down due to heavy selling pressure. At the bottom of a down trend market, a close higher than the previous day's high means the bulls have won and the market is turning up.

Hook reversals occur more often but are less reliable as indicators of change in market direction. Either the market high in an up market or a market low in a down market is exceeded, but the other end of the range is not exceeded. The key is the direction of the close. A lower close in an up market means market weakness, and a higher close in a down trend market means market strength.

An island reversal consists of a few days of market action set off from the main body of prices by gaps on both the up and down side. These reversals should be considered as warnings of impending potentially large price moves.

Reversals are not perfect predictions of market bottoms or tops, though they are more reliable for tops. Their importance is enhanced if they occur concurrently with other chart signals such as exhaustion gaps, high volume, or price objectives defined by other formations.

Saucer or Bowl. The name saucer or bowl describes a rounded bottom or top formation, depending upon the depth and slope of the picture formed (Figure 5-12). Over a period of time, prices gradually curve up or down, depending upon

168   Commodity Marketing from a Producer's Perspective

Figure 5-12: Example of Saucer Top Formation.

whether they are establishing a bottom or top formation. They indicate the probable direction of the subsequent move. Round tops and bottoms are viewed as reliable trend indicators. They also imply that a major move is shaping up. The width of the saucer, determined by the time element taken to form it, roughly indicates the extent of the move to follow. When a saucer formation portends a lengthy move, a resting plateau, or platform, may develop between the saucer and the main part of the move.

Double Tops or Bottoms. Double bottoms occur at areas of major support at market lows, and double tops occur where major resistance halts price advances (Figure 5-13). The formation develops in an up trend when price moves to the same level within a three- to five-week period but there is not sufficient bullish sentiment to carry it to higher levels once a topping action has formed. Similarly, a second effort to drive prices lower in a down market that fails to penetrate the support established by the previous move causes a double bottom.

The outstanding implication of a double top or bottom is that a large subsequent price move in the opposite direction will follow. Real double top and bottom formations are rare. More often they become parts of other patterns.

Bar Chart Rules

Farmers can use a number of rules with bar charts to form guidelines for marketing decisions. Several of these rules provided by Oster follow:[5]

1. Sell breakouts of up trends. Watch for breakout of an up trend as a selling opportunity. Initiate a short position after the second consecutive day the market closes below the trend line.

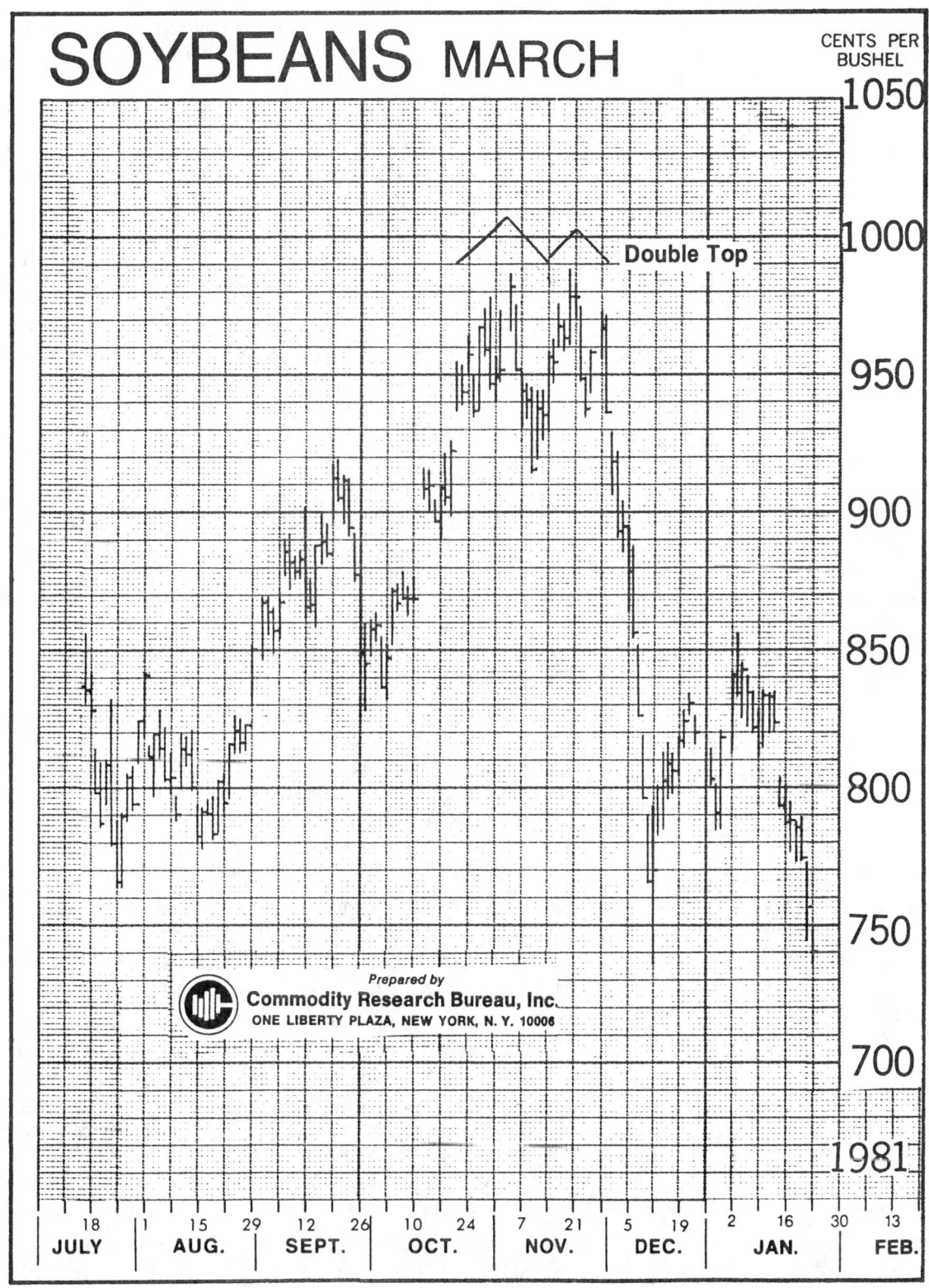

Figure 5-13: Example of Double Top Formation.

2. Buy breakouts of down trends. Initiate a long position after the second consecutive day the market closes above the up trend line.

3. Sell double or triple tops. Whenever a contract hits the same price area for the second or third time, it is a good time to sell. If price action cannot penetrate the same price after several tries, chances are that reduced demand for the moment will lead to a major price decline.

4. Buy double or triple bottoms. Commodity buyers looking for a buying opportunity can use a double or triple bottom as a signal that sellers trying to run prices lower cannot break support levels. This means that a price rise is probably imminent.

5. Sell when prices hit resistance areas. Areas of price congestion are future support or resistance levels. As prices move along in channels, the congestion areas become a hard level to penetrate. Many advisory letters suggest selling on rallies because reaction to resistance brings prices back up to support levels. If fundamentals dictate that the sideways channel is the right price level, the rallies are about the only sales opportunity to gain a few cents in price.

6. Buy when prices dip to support areas. Support areas set the floor to which prices fall. Unless there is a change in fundamentals, such as a larger supply that must be moved or a drop in demand levels, the traders will have a hard time forcing prices any lower. When the bottom is hit, an upward rebound is likely.

7. Sell a 50 percent retracement of a down trend. If a commodity has dropped severely from a top formation and begins to rise again, it is likely that prices will move to 50 percent of what they were when the the down trend began. Producers who missed a selling opportunity at the high of the market often get another opportunity when the market has retraced 50 percent. If prices continue to rise, the next peak can be a 66 percent retracement. A scale-up sales technique works well under these conditions. Initiate the first sale at a 50 percent retracement level and double the amount sold at the 66 percent level.

8. Sell head and shoulder tops. This is not so easy, but when a head and shoulders occurs, confidence of a down trend increases once the price breaks below the neckline after the second shoulder is is formed.

9. Use price gaps as objectives of both corrections and extremes. After an upgap occurs, wait a few days for the price to return to fill the gap. If the gap does not fill, prices often move strongly away from the gap. Learn to read breakaway, measuring, and exhaustion gaps. They indicate how high prices are likely to move or the extent of a fall. Gaps are a major tool in determining price objectives. Price gaps are usually filled by later trading. It may take weeks or months to return to price levels left open, but it is nearly certain that a gap will not be left open eternally. Old gaps become price objectives eventually.

10. Use flags and pennants as measures for price objectives. Their midpoint is often half the anticipated move. Though questionably reliable, their use in conjunction with other measures to establish market objectives may point to market extremes both high and low.

Confirmation

Confirming a chart formation enhances confidence in its reliability. A combination of gaps, flags, and reversals on a single chart often confirms a similar price level. Several formations pointing to the same price objective can give more confidence that the objective will be reached. Further confirmation is determined by consulting different trading months of a particular commodity. For example, when attempting to determine whether a head and shoulders is forming in July corn, every other corn contract being traded should be studied for confirmation that the formation is occurring. Weekly and monthly continuation charts should also be consulted to confirm longer-range trends, support, and resistance levels. Objectives defined by formations on the daily charts may also appear in the longer-range charts. As a rule, formations should be regarded with suspicion when a majority of charts fail to confirm the formation, unless there are fundamental factors which suggest that divergence should occur.

Reliability

Technical analysis of chart patterns is more art than science, and allowance must be made for errors. Unexpected occurrences can change price trends suddenly and without warning. Rain or hail, overnight political decision, wars, and any number of unpredictable events can invalidate forecasts made by interpretation of chart patterns.

Another factor limiting their use is the fact that chart formations are often very indecisive. Patterns are quite obvious after they have been formed; they are not as obvious during the formative stage. The example in Figure 5-14 shows

No Sign of a Top in Corn

"Corn prices have climbed steadily since establishing lows at harvest. Solid uptrends have formed and only mild downward corrections have occurred. The latest correction started on Apr. 11 at $3.25 and ended Apr. 20 at $3.09. Since no bearish reversals or other signs of a top have occurred, this could prove to be a bull flag formation and the midpoint of a rally to $3.46 in the July futures.

The $3.46 objective is calculated by taking the 37-cent distance from the Mar. 14 low at $2.88 and the Apr. 11 high of $3.25 and adding it on to the Apr. 20 low of $3.09. Rallies to this swing objective will be indicated by a close above $3.21, a new rally high close.

The failure of prices to rally through the previous rally high at $3.25 would be bearish. If this occurs, a potential double top could form, which would offer tough resistance to futures price advancement attempts.

Should prices sell off from current levels and close below $3.10, a downtrend will form and declines to $2.95 in the July contract should be expected."

Quote from Top Farmer Intelligence. Used by permission.

Figure 5-14: Example of Indecisive Analysis in Chart Reading.

three up trend lines and projects a $3.46 objective price. Six days after this analysis, prices dropped below $3.00. Subsequently, the market rallied to $3.41. Chartists usually maintain a neutral attitude until patterns become clearly discernible.

Robert J. Taylor reported a Technical Reliability Index to evaluate the price-forecasting dependability of chart patterns.[6] The Index measures the percentage of time that a completed formation resulted in the expected price movement for that formation. Since technical formations occur with greater frequency in some commodities than in others, the evaluation was further refined to indicate frequency of occurrence of each formation in each commodity. Three symbols, used in connection with the Technical Reliability Index, were F = frequent, A = average, and I = infrequent. The Index and frequency are reported in Table 5-2. It shows, for example, that breakouts from sideways trends occur frequently for cattle prices and are more reliable (75%) than ascending triangles (47%) in detecting market bottoms. Likewise, head and shoulder tops occur frequently in soybeans with a high reliability (82%) that they are the true tops.

The dual problems of reliability and accurate detection of the formation in the development stages limit chart formations in their use as a marketing panacea. A topping or bottoming formation may occur once in a year. The farmer is left with what to do in the interim until another objective is reached, especially if he or she missed the extreme of a move. Bar charts, therefore, become means to complement other methods of analysis in establishing price objectives at which marketing decisions should be executed.

MOVING AVERAGES

Whereas bar charts are used to identify price objectives, another approach to making pricing decisions is based upon trend analysis. This method of analysis does not attempt to predict price movement but simply follows price trends to take advantage of the major moves a market might make during the course of the marketing year. It relies on signals from the market analysis to define price direction. While market prices may be moving up one day and down the next, there is often an underlying direction to the market. Trend-following analysis attempts to identify the direction of major trends while filtering out minor corrections.

A moving average is an arithmetic average of prices over time. The time span is constantly moving forward to reflect only the most recent prices. For example, Table 5-3 reports closing July wheat prices during the month of April and shows the 3-day and 10-day moving averages. To calculate the 3-day moving average for April 5, the closing prices for April 1, 4, and 5 are added and divided by 3 to produce the average of $369.08. The 10-day moving average is calculated in similar fashion, but it includes 10 days of prices rather than 3, and the sum is divided by 10. The respective calculations for the next day are made by deleting the first day's price from the calculation and including the fourth or eleventh day's price. By so doing, each average always includes 3 or 10 days of most recent closing prices.

A variation on the technique of calculating moving averages is a front-weighted procedure. It is based on the assumption that most recent market activity is more important in detecting market action than earlier price data. More weight is given to the most recent day's price as compared to past prices. Table 5-4 shows how to calculate a front-weighted three-day moving average of the wheat prices from the previous example. April 1 is the farthest day away and consequently receives a weight of 1. The next day, April 4, receives a weight of 2. The

Table 5-2: Technical Reliability Index

Commodity	BOTTOM FORMATIONS					
	Head and Shoulders	Double Top	Ascending Triangle	Island Bottom	Breakout from Sideways	Saucer Bottom
Cattle	A/71%	A/67%	A/47%	I	F/75%	F/73%
Cocoa	A/72	A/69	A/57	A/67%	F/71	F/73
Copper	A/73	A/68	A/67	A/71	F/77	A/69
Corn	A/70	F/73	A/46	I	F/68	F/75
Cotton	A/68	A/67	A/52	I	F/81	A/71
Eggs	A/70	F/74	A/59	A/71	A/73	I
Hogs	A/72	A/68	A/60	I	A/74	A/69
Lumber	A/76	A/71	A/68	I	A/73	A/74
Orange Juice	A/74	A/71	A/69	I	F/77	A/76
Plywood	A/81	A/72	A/68	I	A/70	A/73
Pork Bellies	F/85	F/72	A/41	A/74	A/57	I
Potatoes	A/70	A/69	A/62	I	A/73	A/71
Silver	A/73	A/70	A/61	I	A/76	I
Soybeans	F/70	F/76	F/38	I	A/66	A/62
Soybean Meal	A/70	A/74	F/71	I	F/77	A/72
Soybean Oil	A/68	A/71	F/75	I	F/77	A/70
Sugar	I	A/60	A/62	A/62	A/71	A/69
Wheat	A/75	A/74	F/41	I	F/71	F/64

Commodity	TOP FORMATIONS				CONSOLIDATIONS	
	Head and Shoulders	Double Top	Descending Triangle	Island Top	Pennant	Flag
Cattle	A/69%	A/62%	A/61%	I	A/72%	A/67%
Cocoa	A/78	A/65	A/62	A/72%	A/71	A/73
Copper	A/77	A/69	A/66	A/81	A/60	A/57
Corn	A/81	A/73	A/54	I	F/73	F/71
Cotton	A/70	A/72	A/52	I	F/76	A/70
Eggs	A/68	F/74	A/51	A/76	F/79	F/73
Hogs	A/72	A/71	A/55	I	F/72	F/70
Lumber	A/71	F/70	A/65	I	A/69	A/66
Orange Juice	A/70	A/66	A/67	I	A/72	A/69
Plywood	A/80	A/72	A/66	I	A/69	A/67
Pork Bellies	F/87	F/75	A/40	A/77	F/65	A/60
Potatoes	A/80	A/72	A/70	A/78	A/65	A/61
Silver	A/77	A/71	A/70	I	A/73	A/69
Soybeans	F/82	F/73	F/55	I	F/68	A/61
Soybean Meal	A/72	A/65	F/71	I	A/64	A/60
Soybean Oil	A/65	A/72	F/73	I	A/67	A/62
Sugar	A/60	A/61	F/61	A/72	A/55	A/52
Wheat	F/81	F/77	F/51	I	F/64	F/58

F = Frequently   A = Average   I = Infrequently

Source: Robert J. Taylor, "Technical Personalities of Major Commodities." Used by permission.

Table 5-3: Calculation of 3- and 10-Day
Moving Price Averages, July Contract

Date	Closing Price	Moving Average	
		3-Day	10-Day
April 1	368.4		
4	368.4		
5	370.2	369.08	
6	370.6	369.83	
7	369.2	370.08	
8	372.4	370.83	
11	374.6	372.17	
12	366.0	371.08	
13	361.6	367.58	
14	361.4	363.17	368.38
15	362.0	361.83	367.73
18	359.4	361.17	366.83
19	356.2	361.08	365.43
20	352.2	359.41	363.58
21	352.6	357.16	361.93
22	357.0	357.41	360.38
25	359.2	359.83	358.83
26	362.0	362.90	358.43
27	363.4	365.08	358.6
28	361.4	365.83	358.6
29	361.2	365.58	358.57

Table 5-4: Calculation of Front Weighted 3-Day Average

Date	Price Closing	Weight	Weighted Value	Front Weighted 3-Day Moving Average
April 1	368.4	1	368.4	
4	368.4	2	737.0	
5	370.2	3	1110.75	369.38

most recent closing price of $3.702 is given a weighting index of 3. Multiplying each of the closing prices by its corresponding index weighting and summing the results produces a total of $1,110.75. Dividing the total units by 6 yields a front-weighted three-day moving average of $3.6938 in comparison to $3.6908 using the simple average method.

A 10-day weighted moving average can be calculated by assigning the most recent day's price a weight of 10 and multiplying the next oldest day's price by 9, the third oldest by 8, and so on to the tenth oldest by 1. The sum of the weighted values is divided by 55, the sum of the weighting units.

Weighting gives greater emphasis to more recent price action, but it is a more complicated calculation. Since calculations are completed each day with new information and can be very time-consuming, programmable calculators and microcomputers greatly simplify the procedure and reduce the time requirement for calculation.

Moving averages are plotted on charts just like bar charts, with price on the vertical axis and time on the horizontal. Figure 5-15 illustrates price activity of corn and sugar markets, along with moving averages of the prices. Several observations can be made in reading the charts. First, the moving average smoothes out the price action. No one day's price can have an extreme effect on the average value, since the last day's price is always averaged in with several others. Using an average eliminates or reduces the distraction (market noise) caused by sudden and extreme daily fluctuations and enables the user to observe a smoother definition of trend rather than market reaction. This feature is the major reason for the widespread use of various types of averages as market analytical tools. Some chartists use 3-9-18-day averages; others use daily-10-40-day ones. The number of days for calculating the average depends upon market performance of which one works best or possibly personal preference. Reasons for using different lengths of time are to improve sensitivity to changes in direction, but at the same time to eliminate false signals. The time period chosen should be sufficiently short to be sensitive to true trend changes, but sufficiently long to filter out market noise.

Farmers who decide to use them should be aware of the features that characterize moving averages:

1. They reflect only what has occurred in the past and do not attempt to project into the future;
2. They change direction after the prices from which they are derived change;
3. They smooth the price waves of market action;
4. The time lag that occurs before moving averages signal trend change depends on the time span covered by the average and on the velocity of price movement after the change occurs;
5. Averages are always less than the prices from which they are derived during advances and greater during declines;
6. The more days of prices included in the moving average, the longer the time required to realize crossovers;
7. During sideways price action, moving averages cannot clearly indicate trend direction, since no trend exists at such times.

Moving averages provide a method for detecting changes in price trend. When plotted as lines on charts, the crossing above and below one another provides signals of change in price direction. In using moving averages as a market decision tool, the supposition is that once a trend has been established, it will continue for some time.

The pattern in Figure 5-15 uses daily, 10-day, and 40-day moving averages. The crossover of the daily and 10-day average below the 40-day average signals a down trend; likewise, a double crossover above the 40-day average signals an up trend. In the sugar chart, a down trend was in force at the beginning of the chart, and an up trend detected on September 13 continued throughout the time period covered, even though all averages had turned down after February 1. The averages were very effective in detecting major trends. By comparison, the corn market bounced along in sideways action rather than staying on a prolonged trend. The market changed from down to up on September 11 but turned down again

Figure 5-15: Moving Averages for Corn and Sugar, Along with Respective Daily Price Charts.

on October 27. It changed to an up trend November 17, a down trend on December 19, and an up trend on January 11. In a sideways trading pattern, moving averages whipsaw in and out of up trends and down trends.

Incorporating moving averages into a marketing decision framework relies on crossovers to signal changes in trends. A change from up trend to down trend would signal farmer-producers to place hedges on their production and tell input buyers to exit long market positions. A change from down trend to up trend would signal input buyers to place long hedges on items they planned to buy and producers to lift hedges or delay taking market positions.

SUMMARY

Is technical analysis of charts a form of magic or witchery? (See Figure 5-16.) Certainly the charts seem far removed from the fundamentals of production — weather, usage rates, and government policy — that farmers are familiar with as price-making forces.

The fundamental forces of supply and demand come together in the market place to generate prices. But when prices start to move, charts often show the activity before fundamental reasons for it are revealed. Professionals who are fulfilling their needs prefer to establish their positions before the information is public knowledge. Often fundamental information is history by the time it becomes public knowledge, since its effects are bid into the market by insiders as they established positions earlier. Technical trading provides a timing device for making purchases or sales before the reasons for making them are known.

Fundamentals change. As they do, the questions arise: How high will prices rise, or how low will they fall? Producers will sell all that is produced at some price, and price will ration supply to avoid empty pipelines. These two conditions create the boundaries for price level fluctuations. But the middle leaves considerable room for price negotiation. Technical analysis is a means to measure the negotiation process.

There is an old investment proverb which says, "It is not what is true; it is what investors believe to be true that counts." Technical market analysis is a method for giving market participants confidence in reading what other people believe to be true. Pure technicians ignore fundamental factors of supply and demand; they pay no attention to planted acres, harvest progress, weather conditions, number of animals on feed, or any other fundamental facts. They simply read the market, knowing that eventually all fundamentals will be registered there. In the end, fundamentals will determine final market value. A chart is simply a road map of how and when the prices set by fundamentals come to be realized.

There is no economic reason that price charts should establish price direction or price level. But so many technicians are watching the same pattern that their collective analysis causes self-fulfilling prophecies. Charts work because chartists make them work. Charts are important to farmers because they indicate how high or how low prices are, or may go. In an up trend market, traders move the market up to a point where no buyers are left. In a down trend, they move it to where there are no sellers. A producer with a perspective from charts can use his or her knowledge of fundamentals to know when the market has gone beyond logical reasoning. Over-bought and over-sold markets create marketing opportunities. At this time farmers who are familiar with chart analysis can "harvest" the speculators. If technicians have driven prices higher than logical levels determined by fundamentals, farmers and ranchers should take advantage of the situa-

Figure 5-16: Magic or Witchery? Examples of Many Chart Formations in Price Action.

tion and price their production. Likewise, when downward pressure drives prices to low value, or below cost of production, farmers and feeders should lock in low input costs. Speculators provide opportunities for both sellers (wanting high prices) and buyers (wanting low prices) to be satisfied by the market. Technical analysis signals when to cash in on those opportunities.

Questions for Study and Discussion:

1. What problems of fundamental information cause some market participants to turn to technical analysis of the market?
2. What is the logic behind use of technical market analysis?
3. Explain the three factors upon which technicians base their analysis.
4. What is the psychology of a trend in commodity prices?
5. Why are volume and open interest important technical factors?
6. What is the foundation of chart formations?
7. List and explain the chart formations that define price objectives. What are these rules of measurement?
8. List and explain chart signals of market extremes.
9. What are the rules to follow in bar charting for market decision making?
10. What are confirmation and reliability in connection with chart reading? Why are they important?
11. Explain moving averages as a tool of technical analysis.
12. Is technical analysis magic or witchery?

Exercise: Price Charting

Bar charts are an example of the adage "A picture is worth a thousand words." Once a person learns how to read them, charts provide pictures of what is happening in a market.

Commodity charts should be kept on professional graph paper that can be purchased at office supply stores. Chart paper should have five divisions, one for each trading day with a heavy line to indicate start of the week. Preferably it should have room for 52 weeks of time on the horizontal axis. Prices are designated on the vertical axis. The increments selected should relate to price volatility of respective commodities in terms of both daily fluctuations and annual range of fluctuation. Soybeans can be plotted in 5-cent increments over a $3.00 range; corn, 2-cent increments over a $1.50 range; wheat, 2 cents and $1.50; cotton, 50 points and 30 cents per pound; cattle, 50 points and 30 cents; hogs, 50 points and 30 cents. Rules change as volatility and price level change.

Accuracy and scale in plotting charts cannot be over-emphasized. Plotting the high, low, and close of each day's market activity at the exact prices at which they occurred is of key importance. Inaccurate plotting can change the angle of trend lines, show false market signals, and in the final analysis give misleading forecasts of future prices.

Most beginners are well served by charting nearby contract options in commodities of most interest to them. As they gain experience and learn charting rules, they turn in time to subscribing to a charting service. Several firms publish weekly or monthly reports which have space for charting until the next report is issued. They supply charts on all options of the various commodities. They typically also provide long-range weekly and monthly charts for long-term analysis. A request to a couple of the following should secure a sample of what they offer.

Charting Services:

Commodity Cycles, P.O. Box 40070, Tucson, AZ 85717.
Commodity Perspective, Investor Publishing, Inc., 327 S. LaSalle, Chicago, IL 60604.
Commodity Price Charts, 219 Parkade, Cedar Falls, IA 50613.
Commodity Research Bureau, 30 S. Wacker Dr., Suite 1820, Chicago, IL 60606.
Commodity Trend Service, 100 E. Kimberly Rd., Northwest Tower, Suite 303, Davenport, IA 52808.
Comtrend, Inc., 25 Third St., Stamford, CT 06905.
Data Lab, Inc., 200 W. Monroe St., Suite 1106, Chicago, IL 60606.
Dunn & Hargitt, Inc., 22 N. Second St., Lafayette, IN 47902.
IBEX Chart Service, Inc., Box 693, 2420 First Ave., Seattle, WA 98121.
The Professional Chart Service, Hadady Publications, 61 S. Lake Ave., Suite 309R, Pasadena, CA 91101.
Security Market Research, Inc., Dept. CM, P.O. Box 14096, Denver, CO 80214.
Spread Scope, Inc., P.O. Box 41221, Los Angeles, CA 90041.
Top Farmers of America, 225 E. Michigan, Milwaukee, WI 53202.
Traders Press, P.O. Box 10344, Greenville, SC 29603.

Footnotes for Chapter 5:

(1) Material for this section comes from the excellent work of one of the masters: William L. Jiler, *How Charts Are Used in Commodity Price Forecasting* (New York: Commodity Research Publications, 1973). Used by permission.

(2) Merrill J. Oster, *Commodity Futures for Profit* (Cedar Falls, Iowa: Investor Publications, 1979) 15/1.

(3) Richard A. Brock and John F. Martin, *Charting Farm Markets* (Milwaukee: Top Farmer Publications, Jan., 1981) 14.

(4) Jiler, *How Charts Are Used in Commodity Price Forecasting*.

(5) Oster, *Commodity Futures for Profit* 15/11-15/14.

(6) Robert J. Taylor, "Technical Personalities of Major Commodities," *Commodities* (Cedar Falls, Iowa: Oster Communications, Inc., Aug., 1972).

Chapter 6

Pricing Strategy

COMMODITY PRICING PROBLEMS	183
PRICING CHOICES	186
Price Takers	186
Routine Hedgers	186
Selective Pricers/Selective Speculators	187
FARMERS STORE GRAIN TO SPECULATE	188
BUILDING A MARKETING STRATEGY	191
Decision Period	191
Role of Futures Prices	191
Futures Market — Workable, but Not Designed for Farmers	193
Farmer Speculation in the Futures Market	195
Dealing with Brokers and Brokerage Accounts	196
PRICING STRATEGIES	197
Calendar Selling	197
Cost-of-Production Models	197
Livestock Sales	197
Grain Sales	202
Shortcomings of Cost-of-Production Pricing	203
Scale-Up Selling, Scale-Down Buying	204
Seasonal Pricing	205
Trend-Following Models	208
KEEPING SCORE	211
PROFIT RATHER THAN PRICE	212
GOVERNMENT PROGRAMS AFFECT PRICING	214
RESULTS OF PRICING STRATEGIES	215
USING THE OPTIONS MARKET	217
SUMMARY	220

Pricing Strategy

Modern farming is a complex production process and an equally complex business process. It is inherently speculative, since the future is unknown. Farmers thrive on speculation; the nature of their business requires them to be risk takers. In addition to insects, drought, and pestilence that create production risks, price risk has become more important in affecting profitability. In the 1970's, the trend toward freer markets and away from government regulation, production controls, and subsidy payments produced greater price uncertainty than previously. In the 1980's, surplus production, economic deflation, and shrinking exports added fuel to the price uncertainty fire. Volatile prices provided the opportunity for both high profits and disastrous losses.

Financial success depends upon management of risk and uncertainty inherent in farming. Improved marketing is one way to manage risk. But two psychological attitudes prevent many producers from being successful marketers: Farmers are optimistic, and they like to speculate. Because of the nature of their business, farmers are the world's greatest optimists. They live by the rule that things must grow. No one else would borrow $150,000, scatter seed, fertilizer, and pesticides across 1,000 acres, and pray for rain. Farmers have an inbred psychology that a crop must grow higher and higher. That optimism extends from their production practices into their attitudes about prices. Prices are always too low and must go higher to give them a greater return tomorrow than the market offers today. Consequently, they often delay pricing decisions until forced to sell. The reality of the market is that prices go down as often as they go up, and often with much more vengeance. Many farmers can accept only up, because higher prices mean so much to their well-being.

Farmers are price and production speculators, and they are risk takers. Crop farmers have no control over the most important factor determining production — weather. Furthermore, they have no control over the most important factor determining price of their product — the market. The nature of their business requires that farmers live with production and price risks; the challenge is to control them as best they can.

Risk-bearing ability varies greatly among producers. Some people are risk takers, psychologically prepared to accept setbacks as offsets to opportunity for significant gains. Others have a risk aversion, preferring to minimize risk. Financial condition greatly affects attitude. A full owner with money in the bank can withstand more risk than a young farmer with high debt payments; but their reaction to risk may be just the opposite. An operation with high cash flow requirements relative to gross receipts is more vulnerable than high-fixed-cost operations. Lenders may place bounds on what they will finance due to risk. While excessive caution may limit financial growth, excessive risk taking may bankrupt the business.

Not all risk can, nor should, be avoided. The problem is to balance the probability of loss against its effects. Large potential losses and their disastrous effects should be avoided as much as possible, but small loss – large reward situations may generate profitable opportunities.

Farmer-producers may attack the problems of production and price speculation on two fronts. The first is to minimize the effects of unfavorable outcomes, and the second is to improve market pricing strategies. To practice the first, they may control risk with techniques that transfer it to others. These may be formal insurance purchases on crops and livestock production and on stored commodities. They may participate in government price support and disaster relief programs. They may build cash reserves or increase equity in the business, and they may use forward pricing methods to lock in prices ahead of harvest. The second step is development of a marketing strategy. Instead of being price takers, controlled by emotions and financial requirements, farmers should develop a price-maker philosophy regarding the products they sell. It takes more time and work but can be very rewarding. A farmer's marketing strategy consists of a set of objectives, plans, and tactics that guide production and marketing decisions. Since the market place presents the farmer with a wide variety of choices, opportunities, and risks, a marketing plan is the blueprint for making decisions and exploiting the market.

The purpose of this chapter is to set the framework for developing pricing strategies and to review several techniques.

COMMODITY PRICING PROBLEMS

Farmers have two commodity pricing problems. The first concerns the basis market which dictates when to move the physical commodity in a cash sale. The second applies to price level decision on when to commit to the price offered by the market.

The first task is relatively straightforward with storable commodities. Farmers weigh the costs of holding grain in storage against the expected cash prices later in the marketing year. Demand for food and fiber is fairly stable from month to month throughout the year, so inventory must be carried forward after harvest. If the price were constant month to month, any producer would have the option of selling the inventory this month, or holding until next month and receiving the same price, while incurring the cost of carrying the product for a month. Rational producers would sell now rather than incur the carrying charge if next month's price were not sufficiently higher to pay storage charges. The larger supply coming onto the immediate market would depress the current price relative to next month. Thus, cash price tends to trend upward over time. If the up trend in price equaled cost of carrying inventory, the producers would be indifferent about which month they made their sale. If late season prices were expected to exceed cost of carry, producers would hold inventory for late season sale. The month-to-month spread in prices tends to settle at levels sufficiently wide to cover the costs of carrying the inventory from month to month. The gradual rise in price, in effect, is the method of charging customers for the service of holding the inventory. The specific details of basis level decisions in a market economy will be discussed in the next chapter.

The second task in pricing is picking the day to set price level decisions. As illustrated in Figure 6-1, farmer pricing has one objective — "Top the Market." As the market price level fluctuates up and down throughout the year, farm profit rises and falls relative to a fairly constant cost of production. Only at the highest

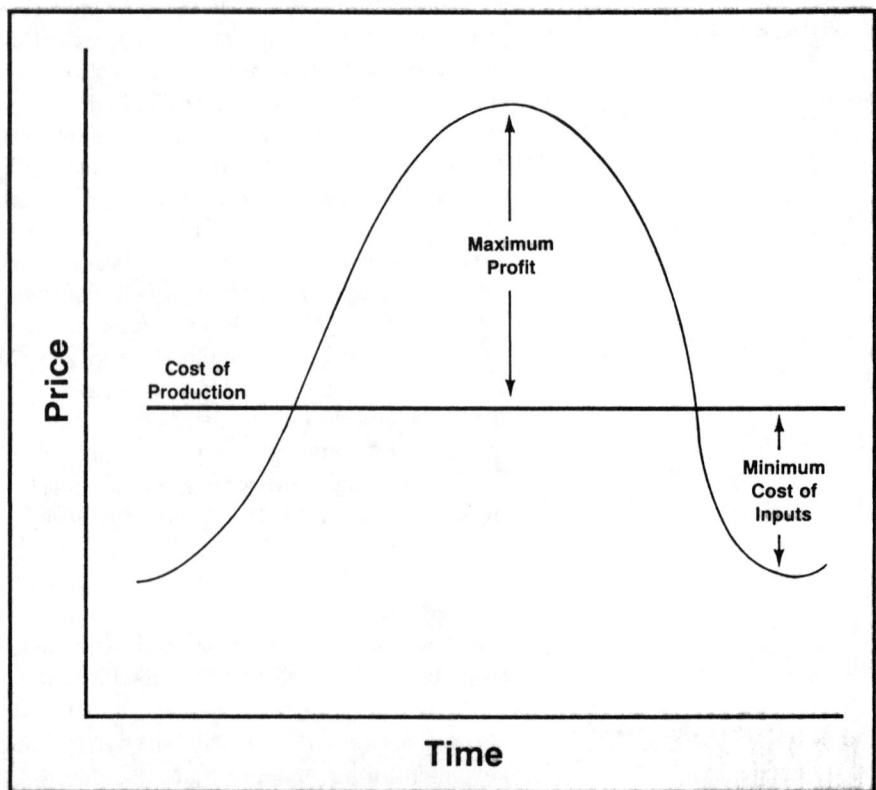

Figure 6-1: Farmer Pricing Challenge — Top the Market.

price level is profit maximized on products being produced. Similarly, a farm operator purchasing feed and feeder animals seeks to buy inputs at the lowest price level of the year. His or her cost of production is minimized with low-cost inputs.

There is a wide array of possible events that influence price on any given date. A wheat farmer being offered a current price of $3.00 per bushel realizes that the following possibilities exist:

1. Crop failure, which could be minimal or extensive, resulting in price of $4.25 or up to $6.00, with smaller probability of extensive failure and extreme prices.
2. A future bumper crop, also of variable size, with corresponding price decline from $3.00 down to $2.00.
3. A business depression, leading to a decline in demand and lower price.
4. A war, leading to reduced output and a higher price.
5. A prospect for either increased or decreased export demand.
6. A shift in consumer taste away from bread toward meat.

Each of these forces will affect the available supply and its rate of use. The role of price is to provide an adequate but not excessive inventory carryover. The expected increase or decrease in carryover sets the tone of price level in the market. As forces from increased demand or reduced supply feed into the market, the price level rises; as decreased demand or increased supply enters, the price level falls. The only thing a holder of wheat can be certain of is that Murphy's Law will prevail and something unusual will happen to change price from its current level.

In making pricing decisions, each market participant, in effect, estimates the probabilities of various conditions of supply and demand at a future date and their effects upon price. These likelihood estimates may be assembled into a frequency distribution of various prices, such as:

Expected Price Possibilities		Probability of Occurrence		Product
$4.00	×	0.05	=	$0.20
$3.50	×	0.10	=	0.35
$3.25	×	0.20	=	0.65
$3.00	×	0.35	=	1.05
$2.80	×	0.20	=	0.56
$2.60	×	0.10	=	0.26
			Expected Price	$3.07

The average expected price is derived by multiplying each price by its respective probability of occurrence and adding the results, giving $3.07 in this case. The producer of wheat would make the decision to continue speculating on price or accept the market offer, depending on the current offer relative to expected average. The producer will sell if market bids are greater than $3.07 or continue speculating if they are less.

The speculative game of pricing based upon fundamentals and price probabilities is to delay making a price commitment when current pricing opportunities are too low. As price rises, the other part of the game is to attempt to identify mistakes bid into the market and to profit from them by making pricing decisions when high price opportunities occur. Some use technical analysis, such as bar charts, in an effort to identify when over-bought and over-sold opportunities exist.

Each seller or buyer has a different set of price expectations and different probabilities attached to them, even though he or she may not go through the rigors of the example in calculating probabilities. In the end, the producer is aware that market price will react to events to prevent accumulation of excessive carryover or, likewise, prevent depletion of stock before year end. Thus, the second pricing decision for a commodity producer depends upon **expectations** of PRICE LEVEL changes. When sales decisions are made depends upon one's assessment of whether events will add strength or weakness to price as determined by fundamental analysis. Whether one's timing turns out to be correct or not depends upon one's skill in accurately assessing future events.

Price projections based on fundamentals and technical analysis might go as follows:

1988 Soybean Forecast

We look for a possible sideways trading pattern to develop over the next three months, with prices trading in the 50-to-60-cent range from current $6.00 level in the first quarter. Based on technical chart patterns, July beans could hit $6.40 in December and, potentially, could reach $7.20 in January, at which point they could encounter overhead resistance. Failure to penetrate that resistance will mean a setback of some magnitude. Furthermore, if no serious problems develop in Brazil, July could easily test the $5.30 level. However, in view of falling interest rates and a weakening dollar, any serious problems in China might test the $9.00 and usher in a push at $12.00 by July. Expectations for a long-term up trend are contingent on too many factors that are still uncertain at this time.

In saying that price level could be $12.00 a bushel or $5.30, the analysis says nothing. It simply shows the frustration of trying to base a pricing plan upon fundamental analysis of uncertain events. Nevertheless, a farmer must make pricing decisions at some time during the production or storage period. The producer did not grow the crop to keep it around forever.

PRICING CHOICES

Farmers are faced with three choices in making pricing decisions:

1. *Price Takers* who deliver their product and ask, "What's the price today?"
2. *Routine Hedgers* who initiate pricing in advance of production without analysis of price. Or they deposit grain in storage, hedge against price change, and hold until time of sale.
3. *Selective Pricers* (or selective speculators) who proceed with production decisions in hopes of making profitable pricing decisions sometime during the year.

Price Takers

Price takers sell when bills must be paid, loans are due, or storage bins are needed for the next crop. They make no pricing decisions but take what is offered when some situation forces them to make a sale.

The other alternatives require a more active role in the timing of pricing decisions. The lure of increased profits encourages many producers to pursue those alternatives.

Routine Hedgers

For routine hedging in advance of production, fixing prices is the final selection criterion of the production plan. Land productivity, equipment, facilities, past production methods, and future production plans limit the alternatives in enterprise selection. Credit availability, cash flow requirements, and risk-bearing ability further limit the number of production alternatives. Out of many combinations of livestock and crop possibilities, a few viable alternatives remain. Farmers can switch planting among a few crops and produce livestock or leave lots empty. To make their production decision, they prepare budgets to determine which production alternative offers the highest net return between variable costs and gross receipts. Routine hedgers rely on price signals during planning periods before production commitments are made and decide what to produce based on relative profits. They fix prices by pre-pricing production. Market price quotes are fulfilling one of their important roles in signaling farmers what should be produced. In effect, currently available price quotes determine the production plans that farmers select.

The Pricing Ahead of Production strategy dictates that a producer either set profitable hedges before initiating production or avoid production. Few producers practice such a strategy, for a variety of reasons. Seldom are producers satisfied with pre-production pricing opportunities; they hope for higher profit opportunities and are willing to speculate on that hope. Many are reluctant to price what they do not have for fear that production problems will occur. Drought or disease could wipe out production committed to forward contracting. Often, a single crop production alternative is not distinctly profitable. Futures prices among several related commodities tend to rise and fall together. High soybean prices encourage high prices in competing crops such as corn in the Midwest and cotton in the South. Seldom do profits of one alternative greatly out-distance others when all limitations are weighed. Finally, agricultural production is characterized by high fixed costs relative to total costs. Production decisions are typically based solely on variable costs and enterprise net margin. What is at risk is their return to labor, management, and capital. Consequently, producers prefer to practice selective pricing, or conversely selective speculation, by going ahead with production hoping that a profitable opportunity will present itself before the cash crop must be sold.

The common mistake of pre-production pricing is to read futures prices, base production plans on them, but avoid making pricing commitments ahead of planting crops or filling feedlots. High futures prices elicit increased production that in turn produces low market prices when production is ready for the cash market. Producers feel betrayed. Their production decision was correct, but they did not complete the strategy by fixing price. They failed to fully appreciate the role of futures prices as presenting a pricing opportunity but chose to view them as a price prediction, which they are not.

Farmers who delay pricing until the crop is in the bin could use routine hedging to fix a future price for their inventory held in storage. However, most farmers do not hedge stored production. They store grain to speculate that price will improve. Sometimes they are correct and profit accordingly; often they are wrong and sell stored grain late in the season for less than what it was worth at harvest, after incurring the costs associated with carrying commodities.

Selective Pricers/Selective Speculators

Producers of grain and livestock who do not pre-price make a pricing decision every day that markets are open for trading. Most days, their decision is to not make a pricing commitment and wait until the next day. Planting seed, filling feedlots, or storing unpriced grain all create speculation on price until some form of pricing commitment is made. With a selective speculation/selective pricing strategy, the decision as to when to commit to pricing commodities for sale is a decision based on controlled speculation. The challenge is to manage price speculation in the way most advantageous to the individual producer.

To initiate a selective pricing strategy, objectives must be established and restrictions placed on the scope of permissible speculation. Every day does not necessarily offer a profitable pricing opportunity, because price often falls below total production cost. Some days offer more profitable prices than others. One producer may set an objective to sell at the high price of the year or buy inputs at the lowest price. If he or she is successful, the strategy will yield maximum income. But such an objective involves high risk because of the difficulty in picking highest or lowest price and unwillingness to commit 100 percent of production. There is only one way to pick the high price of the year. It is called the 365-Day Hindsight Method — look at prices over the past year, pick the high day, and say, "That's when I should have sold." Unfortunately, that system of pricing does not work in the real world, since the opportunity is history.

Price uncertainty affects each producer differently in forming a pricing strategy. Some avoid making pricing forecasts and submit to the Greed-Hope-Fear strategy. GREED keeps them from committing to high profit opportunities when they arise. They refuse to sell at bonanza high price levels because they want only the top price. After a market slump, they refuse to make sales at profitable prices because they HOPE prices will return to their previous levels. Eventually, they do sell at low price because they FEAR it will go even lower. Or they must sell because a loan payment is due.

Another objective might be to get the average price of the year. They sell equal amounts out of storage every Wednesday of the marketing year, or equal amounts on the 15th of every month. This seems rather dull to many operators, but, in actuality, they fail to achieve comparable results.

A different objective might be called income averaging, whereby the objective is to reach the same income level year to year. When price in a marketing year reaches the level of previous years, pricing commitments are made at that time. But there is no guarantee that prices will be the same year to year. Some years prices are high; other years, low.

In order to better understand the process of selective pricing/selective speculation, we need to delve into farmer psychology toward pricing production and into results of various strategies.

FARMERS STORE GRAIN TO SPECULATE

At harvest, farmers store corn, soybeans, and wheat for one reason — they expect price to rise. No one would store grain if he or she expected price to fall lower throughout the storage year. Nor would selling at the same price next July, following eight months of storage, make sense. During that period the grain producer could have had money in the bank earning interest rather than tied up in grain that did not rise in value. Thus, farmers store grain to speculate on a price rise. Common sense tells them that price is too low at harvest and, therefore, it can go only one way — UP!! This is true whether corn price is $2.00 or $3.50 per bushel. Less timid farmers store grain and shoot for the top cash price of the year. Occasionally these storage speculators succeed with part of their production.

Not to be outdone by these coffee-shop top price pickers, we found the Net Chicago High Cash Price — the top cash price of each year adjusted for the cost of storing grain — for 16 years (Table 6-1); 20/20 hindsight is a powerful tool in this search.

We chose the Chicago cash price because other markets tie to it. The one-day high cash price was then adjusted for carrying charges incurred until the high price day arrived. Carrying cost includes interest on the money tied up while grain is in the bin. We assumed a 12 percent per annum interest rate over the entire period. This is low for some years and high for others, but the discrepancies do not materially affect the conclusions. In addition, we calculated the cost of management, storage bin ownership, and risk to equal 2 cents per bushel per month. These interest and carrying costs were deducted from the top cash price to adjust it to harvest-time value, the *NET* high price.

The highest cash price and the net high price do not necessarily occur at the same time. For example, after harvest in 1971 the high cash price of the year, $1.45, occurred the following fall on September 26, 1972. Table 6-1 indicates that the bins should have been emptied on December 15, 1971, six weeks after the November 1 harvest, when the cash price was $1.26. Holding grain until September would have produced a larger check; but with holding cost of 37 cents per bushel, the net realized price would have been reduced to $1.08.

Table 6-1: Results of Storage Speculation, 1971-1986

1	2	3	4	5	6	7	8	9
Crop Year	Harvest Cash Price Nov. 1	High Price of Year	Weeks of Storage Until High	Storage Cost[1]	Net High Price	Change in Price After High	No. of Weeks for Change to Occur	Average Net Price — Sell Every Wednesday
	....(¢/bu.[2])....			........(¢/bu.[2])........				
1971	105	126[3]	6	4.6	121.4	-15	11	105
1972	132	340	41	38.5	301.5	-103	3	179
1973	232	396	41	48.2	347.8	-52	3	277
1974	364	365	0	.0	365.0	-113	11	274
1975	258	314	32	35.9	278.1	-19	3	245
1976	232	263	11	11.8	251.2	-116	13	200
1977	205	270	29	29.1	240.9	-64	10	205
1978	231	310	17	36.4	273.6	-36	6	271
1979	250	354	22	50.2	303.8	-21	4	258
1980	350	369	4	10.4	358.6	-33	10	297
1981	255	265	0	.0	265.0	-70	14	211
1982	210	377	41	47.2	329.8	-52	9	266
1983	338	346	0	.0	346.0	-57	14	289
1984	260	262	1	1.1	260.9	-13	5	228
1985	258	275	12	19.1	255.9	-69	26	225
1986	220	248	21	20.7	227.3	-20	14	163
Avg.	244	304	18	22.0	282.0	-53	10	231

[1]Storage cost = 2 cents per bushel per month plus 12% interest per annum on the cash value of corn.
[2]Cash prices in Chicago 1971-1981, then Central Illinois.
[3]1971 corn was 145 in September, 1972, but it cost 37 cents to hold grain 11 months, so the net price was only 108.

Capturing the top net cash price each year produced average storage profits of 38 cents per bushel per year (average net high price of $2.82 minus average November 1 price of $2.44). Not bad! The obvious conclusion is "store corn at harvest and sell on the HIGH PRICE DAY of the year." The only obstacle is knowing when the top price day occurs. AND THAT IS THE PROBLEM!!

Column 7 of the table shows how closely those who speculate on grain prices flirt with disaster — and, like the moth, are often consumed by the flame. It shows the drop in price AFTER the high day occurred. Column 8 shows how long it took AFTER the highs were in for disaster to strike. To top the market, the 1985 corn crop should have been stored until the end of January, 1986, and sold for $2.75 per bushel (less 19.1 cents storage cost). Those who waited until August to sell watched prices plunge 69 cents per bushel in six months. After the high, sharp price breaks were the rule, not the exception. During this 16-year period, prices dropped an average of 53 cents per bushel within 10 weeks after the highs had been made. In three years, price dropped over $1.00 per bushel on an average of 9 weeks after the high.

Storage speculators hold grain in hopes of capturing 38 cents per bushel. They then watch 53 cents evaporate from the market in 10 weeks. Without 20/20 hindsight, or a reliable crystal ball, their chance of catching the top of the market is about 1 out of 300 (the number of days elevators are open). There are folks who try to fill inside straights at the poker table. Their odds of 1 in 11 are better than the grain speculator's of 1 in 300. As the man said: "Filling inside straights is fun, but don't try to make a living at it." Trying to catch the top of the cash market is fun — but farmers should not try to make a living at it, because the odds are against them.

In our search for something better, Table 6-1 provides some interesting insights. The most profitable holding period is either very short OR fairly long. In 8 years of the 16, the crop should have been sold at harvest or within 6 weeks. In five years the crop should have been stored for at least 29 weeks, or seven months. Even more important, crops that are held for a long time tend to be followed by crops that are held for a short time. This reinforces the adage "A short crop has a long tail." When corn prices were at $3.50 in 1983, many farmers put grain in the bin in hopes of getting $4.00 — many finally settled for $2.90. When their gamble did not pay off, they complained about low prices.

The table also tells us something about "Summer Rallies." Summer price rallies after June 1 (29 weeks after Nov. 1 harvest) that were big enough to pay storage costs occurred in only 5 out of 16 years.

As an alternative, "scheduled selling" suggests dividing the crop into 50 parts and selling one-fiftieth each week. Column 9 shows the average adjusted price from 50 sales each year. Equal amounts were sold every Wednesday and the price adjusted for storage costs incurred on that portion up to the time of sale. The overall average net price of $2.31 was 13 cents per bushel LESS than the average price achieved by *always* selling at harvest.

Farmers store grain to speculate on price. Like the gambler, they occasionally fill an inside straight; but more often than not, they are beaten by the odds. Unless storage speculators are able to come close to the top of the market (and very few are able to do so consistently), the profits from storage speculation are zero to none.

To succeed at the storage game, farmers need to forget about storing to speculate on price level and learn to speculate on the basis (the difference between their local cash price and a selected futures price). Invariably, a strong basis occurs when price level is low. Farmers are then reluctant sellers because they feel downside price risk is minimal. Their reluctance creates the strong basis. When prices are high, buyers find ready farmer-sellers, and they can get all the grain they need to satisfy demand. Therefore, the basis is weak; there is no reason for buyers to bid a premium. Farmers can become storage space merchandisers by (1) storing when the basis is weak, (2) hedging away price level change, and (3) watching the basis narrow to cover the storage cost. Then storage becomes profitable. Its purpose is to capture basis gain.

Farmers should sell cash grain when the basis market says "move it," whether price level is high or low. If they believe that price level is going higher, there is a better way to speculate. Grain need not be held in bins to speculate. Replacing the cash position with a long futures position is the better way to speculate on price level. Futures and cash market prices tend to move up and down together. If cash market speculation is profitable, paper speculation will also be profitable. Furthermore, the cost of paper speculation is only about 2 cents per bushel per year: 1 cent brokerage commission and 1 cent interest on margin money. The total cost of speculating on paper is 2 cents per bushel per year versus the 52 cents per bushel per year holding costs for grain stored in a bin for speculation.

Price may go down and create a loss in the long futures position held for price speculation. But grain stored in a bin loses value just as rapidly as a futures position in a down trending market. Unfortunately, farmers are less aware of inventory value erosion. No neon sign on the bin flashes daily losses in value, but the futures market margin calls ringing every day capture their attention.

Many farmers (and their lenders) object to speculating in the futures market. Yet they borrow thousands of dollars to build storage bins to SPECULATE that the price of corn will go up 38 cents per year — every year. If the books are properly tallied, the cost of making a mistake speculating with the physical grain is far more significant than the SAME mistake made speculating with grain on paper.

BUILDING A MARKETING STRATEGY

In order to improve on cash market speculation, two important concepts provide a framework for building marketing strategy: the decision period for pricing and the role of futures markets.

Decision Period

The decision period for pricing should be extended beyond one-day selling. The marketing tools reviewed in Chapter 2 allow decisions to be effected long before planting or well after harvest. For corn, soybeans, wheat, and cotton, the marketing period extends from the time of production planning until the end of the storage season. This is generally a period of about 20 months for a single crop. Because the storage year overlaps the next year's growing season, a crop producer is concerned with marketing two crops most of the time, one in storage and one growing in the field.

Fortunately, the futures markets trade contracts for pricing 12 to 18 months into the future. They permit forward pricing of planned or growing production, or sales of stored commidities. As one contract month approaches maturity, a new contract for the same month in the next production year is opened for trading. Crop producers are not limited to harvest-time prices to make pricing decisions.

For livestock, the marketing year is not as distinctively defined, since cattle and hogs often involve a more continuous production process. Key production decisions on hog farms relate to increasing or decreasing the number of sows kept for farrowing. A market-ready gilt can be either sold for slaughter on the current market or retained to increase production in about 10 months. Hog futures are traded 12 months ahead of maturity. With both production and pricing decisions extending a year ahead, the marketing year should correspondingly be at least 12 months long. In cattle feeding, the production period extends from the time of purchasing feeders and feed until the finished cattle are sold. Trading in live cattle futures extends about a year; for feeder cattle, futures trading spans about nine months. With the time overlap in contracts, inputs of feeders and feed and production of finished cattle can be contracted about six months before the feeding enterprise is physically started and extended about six months into finishing. Thus, our cattle marketing period is also one year. In this way, futures contracting serves to extend the period of price and production decision making.

Role of Futures Prices

Futures prices are today's price offers for a future business transaction. The price agreed upon today is what buyers and sellers think is the proper future value of commodities based upon currently available fundamental information. But futures prices are poor predictors of cash price when the future becomes the present. Many events can and do occur in the interim to cause great changes in price. Within any one year, futures may fluctuate up and down by several cents or several dollars per unit of production. Futures prices tend to overshoot their targets;

they are too high in anticipation of shortages and too low in anticipation of surpluses.

It is logical to question why they are such poor predictors. Like any prediction of future events, they tend to be self-defeating. High futures prices, before planting or breeding, entice producers to shift acreage to crops with the highest profit alternatives or to breed more females. The resulting increase in supply depresses cash price when the production reaches the market. Likewise, low futures prices discourage production, and prices eventually rise to ration a reduced supply when it reaches the market. In one sense, futures prices are providing a social benefit by fluctuating to extremes. They tend to blunt realized effects by modifying shortfalls in production and discouraging overproduction.

A bigger problem in using futures price as a cash price predictor is the unexpected events affecting price change. A widespread hailstorm, embargo, political turmoil, or export buying often occurs without warning. Uncertainty of events causes constantly changing forecasts, as reflected by constantly changing futures prices. An example is given in Figure 6-2. The chart shows fluctuating soybean futures in response to the newsletter account of changing fundamental factors.

Figure 6-2: Illustration of Changing Fundamental Conditions and Resulting Changes in Futures Price.

Last month's letter on the soybean market described a very bearish supply/demand outlook but called for a short-covering rally in beans, a short-covering rally in the meal versus oil relationship, and widening new crop bean spreads. Those rallies occurred, but the overwhelming bearish fundamental picture was altered in the process. First, the USDA estimated 1983/84 soybean planted acreage at 63.3 million acres which was over 1 million acres below the lowest trade estimate. Then, weather changed. Abundant subsoil moisture and continuous rainfall through the first half of June suddenly changed to above normal temperatures and far below normal rainfall for virtually the entire belt. The trade has tended to focus attention on weather in the western corn belt, but this time around, the danger may lie in the north-central and eastern corn belt and into the mid-southern bean belt. This letter will concentrate on the prospects and effects of a weather market in 1983 — a dangerous proposition in any year.[1]

The role of futures prices is to provide a quoted price for fixing pricing decisions today. Producers and buyers can use futures contracts to hedge away changes from today's price. Because so many things can and do change in the interim, futures prices cannot be relied upon to predict what cash price will be at a later date. The pricing decision is whether to accept pricing offers day by day as time passes.

Futures Market — Workable, but Not Designed for Farmers

The futures market provides a means for eliminating price level fluctuation and its associated cash market speculation. It was designed for grain merchants, not farmers. It allows grain merchants to hedge away the effect of price level change and to speculate on the basis and profit from the cash price–futures price relationship to earn storage profits. Hedging serves them well on both counts.

Grain merchant hedging, with a short futures position equal in size to the merchants' cash position, eliminates the effects of price level change. The parallel fluctuation in value over time of cash grain held in inventory and of futures prices creates offsetting gains and losses. A rise in price level produces a deficit in the futures account but an equal gain in value of the inventoried grain, in a perfect hedge. Conversely, a drop in price reduces grain inventory value, but a gain in the futures position offsets the inventory loss.

Grain trade intermediaries use hedging to avoid the economic effects of inventory value fluctuation. If a Cincinnati grain merchant can buy corn 10 cents under a futures price and sell it in New Orleans for 5 cents over the futures, the merchant has a 15-cent margin. With a cost of operation of 10 cents per bushel, the grain merchant has a 5-cent profit margin. The possibility of the grain dropping 30 cents in value during the four days of the barge trip is too great a risk, given the small operating margin. The grain company hedges away the effects of cash price change by taking an offsetting futures position. Should cash grain price fall 30 cents in New Orleans before the barge arrives, the futures account will accumulate 30 cents of price protection to offset the loss.

As a grain handler, the grain company is not concerned whether corn price level is $2.00 or $4.00 per bushel. It is concerned only with earning 5 cents for each bushel it floats down the river. The futures market serves the grain company well. It provides a vehicle to eliminate the effects of inventory fluctuation, while at the same time permitting it to collect an operating margin for handling grain.

Similarly, grain storage operators, buying corn, wheat, or soybeans from farmers at harvest, focus their attention on earning storage profits. They need only calculate the price difference between the cash cost of the grain and some distant futures option in order to estimate their storage return. They know that the opportunity to deliver cash grain to satisfy the futures contract in the month of maturity will force the two prices to be approximately equal at delivery points. No matter how wide the basis at time of cash purchase, there is reason for basis to converge toward zero over time. Thus, hedging allows grain merchants to speculate on basis while eliminating price level speculation.

Armed with the knowledge that basis differences converge toward zero in delivery months at delivery points, grain merchants have a handle on what to expect from their basis speculation. They are left with their negotiating skills to buy grain at maximum discounts from the futures price (weak basis) and sell it at a maximum premium (strong basis). Their objective is to earn a storage return to compensate them for holding grain from harvest until later use. As long as carrying costs are less than the amount of basis convergence, grain merchants profit from storing grain.

In both situations, the grain merchants' role is to trade the basis. Whether the futures or cash price level is high or low makes no difference to the merchants. Hedging allows them to eliminate the effects of price level change and allows them to concentrate on basis trading.

In direct contrast, the farmer is vitally concerned with price level. To a grain producer, $4.00 corn returns significant profits and $2.00 corn generates significant losses, if cost of production is $3.00 per bushel. The maximum difference between selling price and cost of production maximizes the crop farmer's profit. On the other hand, the livestock farmer buying feed seeks to capture low-priced corn and protein supplement. His or her production costs are reduced by using low-cost inputs.

The farmer who hedges eliminates price level change, just like the grain merchant. A farmer-hedger who establishes a fixed hedge position eliminates both lower prices AND higher prices. But since the farmer is most satisfied when capturing the highest price of the year, the producer is content only with executing a hedge on the one high price day of the year. With growing crop costs fixed for the year, that price extreme maximizes profit.

The farmer who forward prices production or inputs becomes, by necessity, a speculative hedger in comparison to the grain merchant who is a routine hedger. The elevator manager hedges to eliminate price change, but the crop farmer hedges in hope of capturing the ONE high price of the year on products to be sold. In similar fashion, a livestock farmer buying corn, feeder animals, and soybean meal attempts to hedge at the lowest price level of the year to capture low-cost inputs. As a buyer of commodities, the farmer achieves maximum profit when he or she can lock in the lowest input costs. As a speculative hedger, the farmer proceeds with unpriced production until he or she thinks the price extreme day occurs, then sets hedges.

There is always the possibility that unpredictable events (weather, political decisions, unanticipated demand) might move prices higher or lower than what they were on the day that one makes a pricing commitment. Consequently, the farmer-hedger stands beside the market day after day gazing into various crystal balls and praying for divine guidance in selecting THE day to hedge. With analysis and dumb luck, he or she might sell on the one high day or buy inputs on the one low day — once in a lifetime.

Furthermore, the market does not rise day after day to a peak then decline day by day to a bottom. The market moves in fits and starts, up and down, as fundamental and technical forces pull and tug at it. Clouding the fundamental forces driving the market in one direction or another is a lot of noise incorporated in daily and weekly price fluctuations.

In the meantime, the economic rewards from trying to capture high selling prices or low buying prices attract farmers to the futures market. But since the futures market was not designed for farmer pricing, achieving their objectives is not easy. Knowing when to be in a market position and when to be out of it is the challenge for farmers using the futures market. Very few have used the futures market successfully.

Farmer marketing plans and pricing strategies must be based on the fact that futures markets were NOT designed for farmers. But the baling wire/fodder twine/cutting torch inventiveness of farmers can change non-workable designs to make them functional. By accepting the limitation of futures markets, farmers can develop selective hedging/selective speculation programs that offer hope for controlling cash market speculation. A properly developed marketing plan, executed with strict discipline, has the potential for serving farmers better than the usual practice of cash market speculation.

Farmer Speculation in the Futures Market

Use of the futures markets is touted as a price risk management tool. Many farmers believe what they hear and venture into futures trading without understanding the difference between hedging and speculating. The previously cited Commodity Futures Trading Commission study showed that 22 percent of futures market participants were farmers pursuing price speculation.[2] They neither owned nor intended to own the commodities they were trading. Or they placed hedges and lifted them; they traded in and out of the market on whim and emotion, without plan or discipline, and became grist for the commodity mill. Goaded by brokers who collect a commission on every trade, farmers who could be hedgers drift into small trader speculation. All too often they get in and out of the positions on rumors, emotions, and market excitement, and they join the ranks of losers.

On top of their production risk from drought, locusts, and pestilence, some farmers use the futures markets to increase speculation through trading the markets. Their speculative nature takes over to tempt them into doubling up on price risk rather than minimizing it. Whenever farmers buy somebody else's price risk, they are adding to risk already inherent in their business. Most farmers insist they are hedging. That is what they tell their spouses; that is what they tell their bankers. But deep down in their hearts they know that by trading coffee, gold, and pork bellies they are really speculators and not hedgers.

There is a clear distinction between hedging and speculation, as outlined in Figure 6-3. Hedgers own or intend to own the product; speculators do not.

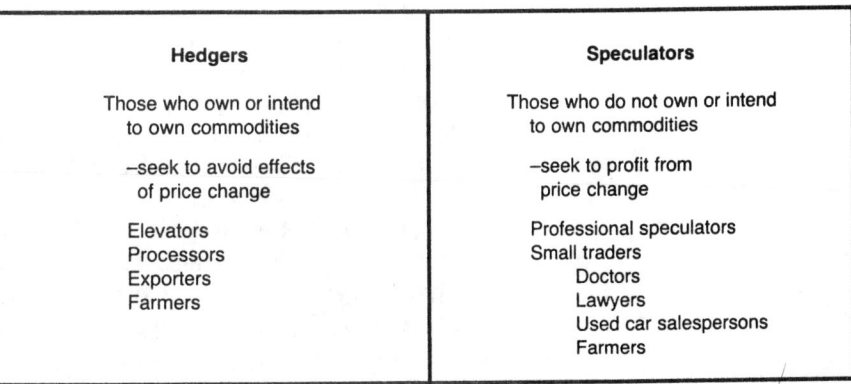

Hedgers	Speculators
Those who own or intend to own commodities	Those who do not own or intend to own commodities
–seek to avoid effects of price change	–seek to profit from price change
Elevators Processors Exporters Farmers	Professional speculators Small traders Doctors Lawyers Used car salespersons Farmers

Figure 6-3: Characteristics of Hedgers and Speculators.

Hedgers seek to shift the price risk associated with owning a commodity; speculators purchase risk. Speculators do not own commodities and assume risk only when they take a futures position. They participate in the futures market and hope to profit by accepting the price risk that hedgers are trying to avoid.

But farmers have no place to hide. Price risk is with them as long as they farm — they always have a cash position, be it a growing crop or grain in the bin. The only way for them to avoid price risk is to sell the farm. Of necessity, farmers must be selective hedgers/selective speculators on the value of their production; fixed hedging in its strict definition does not work for them as it does for the grain merchant. Therefore, their only viable alternative is to learn to manage that ever-present price risk with a marketing plan and pricing strategy.

Dealing with Brokers and Brokerage Accounts

The role of the account executive, or broker, is to fill orders and maintain records of the account. Brokers have contacts on the floors of the various exchanges who can execute orders quickly and accurately. The brokerage office procedures should provide timely and accurate statements. There should be little difference between various firms, since they all must conform to exchange and government regulations as to their practices; however, as in all business, some no doubt execute their responsibilities better than others.

Brokers often provide market information and have trading plans for clients who request them. But their objectives might differ from those of a hedger. Many brokers are biased toward market trading for speculative clients. A hedger with a marketing plan and pricing strategy does not need marketing advice from the broker. He or she is usually better off following the plan rather than mixing several sources of information. Discount firms that do not have large overhead in market research and information services typically charge lower brokerage fees than big firms. Their service is adequate for the producer who has a well defined marketing plan.

A farmer should understand that basic conflicts in objectives may exist between a hedger and a broker. The farmer prefers as few trades as possible to achieve marketing objectives. The broker may prefer as many trades as possible to generate commissions. The broker collects the commission fee when the position is closed out, rather than when it is initiated. Brokers are therefore anxious to have positions close out as soon as possible. The farmer-hedger should be concerned with long-term, major market trends; the broker watches minute-by-minute price change and wants customers to trade short-term action and reaction. Finally, the farmer should understand that a broker has both buyers and sellers as clients and must have a reason to reinforce each in his or her decision as first one then the other calls to place orders.

In daily chats with a broker, farmers may get involved in dynamics of market action and stray from their marketing plans into "playing the market." Few have the time nor discipline to be successful traders, as confirmed by studies of small traders. They must keep in mind that the broker does not know what direction the market is going to take. If brokers did know, they could make larger profits trading for themselves than taking market orders from clients. Farmers should develop their own trading plan and follow it. The broker should be an order taker, not an advisor. The conflicts in objectives between the two are so great that few brokers can be successful trading advisors for farmers.

A farmer-producer should deposit his or her initial brokerage margins as U.S. Treasury Bills, rather than a cash balance with the broker. The interest income on the bills will be credited to the client's account and create an income. Otherwise, the brokerage firm could invest the idle balance and gain the return.

Account surpluses produced by favorable futures positions should be transferred to interest-bearing accounts either with the brokerage firm or with the producer's lender. Properly structured savings accounts with a bank offer additional security by Federal Deposit Insurance that the broker cannot provide.

PRICING STRATEGIES

Many strategies are used by farmers to make pricing decisions. Some involve analysis; some do not. Fundamental analysis and technical strategies were discussed in previous chapters. Other strategies include:

1. Calendar Selling,
2. Cost-of-Production Models,
3. Scale-Up Selling,
4. Scale-Down Buying,
5. Seasonal Pricing, and
6. Trend-Following Models.

Calendar Selling

One method, with several variations, is scheduled selling. A portion of production is priced according to the calendar. Producers may consistently sell at harvest and avoid storage costs, or they may pick July as their sales month, hoping to earn storage profits. Others sell a portion of production each week to achieve an average price. They may vary the schedule by selling larger amounts during the December-January and June-July periods to avoid harvest periods in both the Northern and Southern hemispheres. Those who are comfortable with forward selling may price one-fourth of their anticipated production before planting, another fourth in late summer, another fourth about January 1, and the final fourth in the spring. Livestock producers often close their eyes to fluctuating prices by selling at regular intervals (monthly, weekly). They produce an equal number of hogs or cattle for sale each period. Scheduled selling strategies eliminate the need for making forecasts, while yielding close to an average price for the year, less holding costs.

Cost-of-Production Models

Livestock Sales. A more active program bases pricing decisions on cost of production. Outlined below is a hog production cost analysis and pricing decision procedure. The procedure emphasizes the various "cost of production" figures that can be calculated. When asked, "What is your cost of production?" a farmer should answer, "It depends." Cost depends upon how the calculation is made and what is assumed in estimating fixed costs. This example analysis is based on both price of the product and the major feed cost item, corn. It essentially uses the hog-corn price ratio as a decision criterion to initiate hedges on a larger and larger portion of anticipated production. It includes locking in feed cost as well as product price. Editorial comments in using the procedure are included.

Hog Pricing Decision Analysis

Today, before breeding gilts, go through these steps:

1. Call elevator to get corn price contracted over next nine months.

    $_____/bu.   Contract price on purchased corn for next nine months, OR

    $_____/bu.   Elevator bids if corn produced on farm will be used for feed and producer is willing to wait nine months to receive those prices.

2. $_____/cwt. Buyers offer to buy market hogs in nine months.

3. Calculate your total cost of production coefficient (CP5) in Table 6-2.

Table 6-2: Budget Calculations to Determine Decision Costs of Production (CP's)

	Example Farm[1]	Producer Farm[2]
	(per cwt. hogs sold)	
Purchased Inputs		
Supplies/Med.	$1.29	_____
Utilities	.66	_____
Boar Replacement	.43	_____
Misc.	.46	_____
CP1	$2.84	$_____
Supplements: premix	13.45	_____
meal	7.00	_____
CP2	$23.29	$_____
Corn		
5.43 bu./cwt. sold @ $3.22	17.48	_____
CP3	$40.77	$_____
Labor ($20,788 total)	4.26	_____
CP4	$45.03	$_____
Interest		
Breeding Herd		
$50,000 investment		
x 12% = $6,000 total	1.37	_____
Buildings & Equip.		
$165,000 investment		
x 20% = $33,000 total	6.75	_____
CP5	$53.15	$_____

[1]One year of cost data from Wilmington College Commercial Swine herd, 120 sows, farrowing house, nursery, partially slotted finishing floor. 4,884.00 cwt. produced. 18.95 pigs/sow/yr.

[2]Fill in blanks on your farm from records or 1040F tax forms.

4. Decide.

5. Act.

6. Forget about prices or costs: the profit locked in by the technique will be achieved.

The decision when the buyer's offer is less than CP5 (i.e., price does not cover total costs): Individual producers have the freedom and the responsibility to decide when they will produce and when they will not produce. An economic principle says that farmers will continue to produce as long as the price received covers average variable cost (AVC). From the BUDGET CALCULATIONS each producer can determine what his or her individual AVC is. Depending upon individual circumstances, it may be CP1, CP2, CP3, CP4, or CP5.

If price falls below CP2, all producers will quit production. At this price level, they would be using the buildings free, working free, and giving their corn away, but they would refuse to incur cash production costs and give them away.

Other operators will cease production when price falls to CP3. They are willing to use the buildings for free and work for free, but they refuse to sell their corn through hogs for less than the cash corn market will pay.

Some operators depend upon hired labor. Their quitting price is CP4, since they must pay production cost, plus feed cost, plus labor cost. They will use the buildings without a return because there is no other use for that fixed resource. When the buildings must be replaced, these operators will be forced to quit, rather than rebuild.

Finally, an investor who is considering setting up a pork-producing enterprise does so depending upon whether there is the expectation that price will exceed CP5 (total costs) for the lifetime of the investment. The investor should leave money in the bank to earn interest rather than put it into hog buildings and earn zero return on it.

Farmers who do not develop and use a marketing system to guarantee a satisfactory profit level continue to produce during non-profit periods, waiting for their neighbors to quit. Eventually, a sheriff's auction eliminates the least efficient producers. But there is a better way:

LEARN what to do.

Overcome FEAR of making a nine-month commitment.

And remember, GREED tempts many producers into losses.

Calculation of Satisfactory Sales Prices

CP1 $_____

CP2 $_____

CP3 $_____

CP4 $_____

CP5 $_____

Buyer's offer $_____/cwt. for delivery in nine months.

Buyer's offer of $_____ equals _____ .
(Which CP?)

Are you willing to accept the returns being offered?

YES. Then call your hog buyer and contract to sell _____ (number of) market hogs, nine months from today. Next, call your supplement supplier and contract a nine-month supply of supplements. Finally, call your grain supplier and contract purchase of your corn needs for the next nine months, or agree that the price of your corn as entered in the budget is what you are willing to take for it and that you are willing to wait nine months for the income.

NO. Select Alternative A or B.

A. I expect higher prices for hogs or lower prices for feed. I will take the risk of being naked in the market in hopes of a higher net return.

What do you expect hog price to be? _____

What do you expect corn price to be? _____

What do you expect supplement price to be? _____

Why?

However, based upon my risk-bearing ability, I want to guarantee that _____ percent of my production cost of $_____/cwt. is covered. Make the above calls to contract hog sales and feed costs representing that percentage.

B. I will not operate if I cannot be assured that I will cover cost of production: _____
(Which CP?)

DO NOT BREED SOWS

1. If price prospect is looking up, or costs down, delay breeding until next cycle and recalculate CP's.
2. If price is looking weak, or costs up, take sows to market.

Total Cost + 10 percent pure profit.

Calculation: CP5 × 1.10 = $_____/cwt.

Buyer's offer for delivery in nine months $_____/cwt.

Are you satisfied with covering total costs plus a 10 percent profit?

YES. Then call your hog buyer and feed supplier, and contract sales and feed inputs.

NO. Select Alternative A or B.

A. I expect higher prices and will take the risk of being naked in the market in hopes of higher price.

However, based upon my risk-bearing ability, I want to guarantee that _____ percent of my production covers total cost plus 10 percent pure profit. Call and make contracts to cover that percentage of your output and feed costs.

B. I will not operate if I cannot cover total cost plus earn a 10 percent pure profit.

DO NOT BREED SOWS

Take care, however, that the green-eyed monster, GREED, does not tempt you into being forced to sell at a loss nine months down the road, when your market hogs weigh 220 lbs.

Total Cost + 30 percent pure profit.

    Calculation: CP5 × 1.30 = $_____/cwt.

    Buyer's offer for delivery in nine months $_____/cwt.

      Are you satisfied with covering total costs plus 30 percent pure profit?

        YES. Then call your hog buyer and feed supplier, and contract sales and feed inputs.

        NO. Select Alternative A or B.

          A. I expect higher prices and will take the risk of being naked in the market in hopes of higher price.

          However, based upon my risk-bearing ability, I want to guarantee that _____ percent of my production covers total cost plus 30 percent pure profit. Call and make contracts to cover that percentage of your output and feed costs.

          B. Hog production is profitable at this price level; I plan to expand production _____ percent. I feel I must guarantee that _____ percent of this expanded production earns this profit level, however.

          Remember that at times in the past a 5 percent expansion in pork production has led to a 21 percent decline in hog prices.

Total Cost + 50 percent pure profit.

    Calculation: CP5 × 1.50 = $_____/cwt.

    Buyer's offer for delivery in nine months $_____/cwt.

      Are you satisfied with covering total costs plus 50 percent pure profit?

        YES. Then call your hog buyer and feed supplier, and contract sales and feed inputs.

        NO. Select Alternative A or B.

          A. I expect higher prices and will take the risk of being naked in the market in hopes of higher price.

          However, based upon my risk-bearing ability, I want to guarantee that _____ percent of my production covers total cost plus 50 percent pure profit. Call and make contracts to cover that percentage of your output and feed costs.

          B. Hog production is profitable at this level. I plan to expand production _____ percent. I feel I must guarantee that _____ percent of this expanded production earns this profit level, however.

          Remember that in past hog cycles a 10 percent expansion in pork production has sometimes led to 42 percent decline in hog prices. If GREED convinces you not to contract, what do you expect hog prices to be in nine months? _____

Refinements to the technique.

1. Decide how often to make calculations: annually, monthly, daily, before each breeding, before each feed purchase. Individual interest in marketing and desire to refine expected profits will dictate how often to make calculations.
2. Calculate grinding and storage costs of corn. If you use a fall harvest price on home-grown corn, add storage and interest costs to grinding costs to determine as-fed value of corn when it is converted to cash as hog sales.
3. Hedge cost of feed supplements and live hogs as separate activity in futures market rather than cash contracting.

Grain Sales. A similar method of cost analysis for grain production can be developed from university budget estimates.[3] Separate costs into appropriate variable and fixed cost categories, as illustrated in the hog model, to calculate the various CP's (costs of production) in your operation. Determine which costs must be covered or how much profit is required before hedges are placed.

Another method of cost of production is to calculate cash-flow break-even.[4] A positive cash flow requires that sufficient cash must flow into the business from the sale of grain to meet cash expenditures. The grain price needed to cover expenditures is computed by dividing cash expenditures by the amount of grain produced. The procedure is outlined in Table 6-3. The first expenditure item is cash production expense for seed, pesticides, fertilizer, insurance, fuel, etc. Production expense for 200 acres of corn and 200 acres of soybeans is given as approximately $26,000 and $18,000, respectively. With yields of 125 bushels per acre for corn and 38 bushels per acre for soybeans, variable production costs are $1.05 for corn and $2.42 for soybeans. As long as price level exceeds these costs, production will be initiated; larger losses would result by leaving resources idle.

Prices must be higher to meet the other cash flow requirements. One other cash expenditure is the annual payment on a machinery loan. The payment of $10,800 includes interest and a portion of the unpaid principal. It amounts to 22 cents per bushel of corn and 71 cents per bushel of soybeans. The third cash expenditure item includes all the costs of controlling land, such as cash rent payments, real-estate taxes, and payments on land debt. In the example farm, it was

Table 6-3: Example of Cash-Flow Break-Even Method of Cost Calculation on a Grain Farm

Item	Total	Corn (200 Acres)		Soybeans (200 Acres)	
		(per acre)	(per bu.)	(per acre)	(per bu.)
Direct Production Costs					
Corn	$ 26,154	$130.77	$1.05		
Soybeans	18,364			$ 91.82	$2.42
Machinery Payment(s)	10,800	27.00	.22	27.00	.71
Land Payment(s)	50,100	125.25	1.00	125.25	3.30
Family Living & Taxes	17,150	42.88	.34	42.88	1.13
CASH FLOW REQUIREMENTS	$122,918	$325.90	$2.61	$286.95	$7.56

Source: Futrell 27.

assumed that 200 acres were owned free and clear, 100 acres were cash rented, and 100 acres were recently purchased with 100 percent financing. The land expenditure includes cash rent on 100 acres, principal, and interest payment on 100 acres. The total land payments exceed $50,000. Spread evenly over all 400 acres, the expenditure is $125.25 per acre, or $1.00 per bushel of corn and $3.30 for soybeans.

The last cash flow expenditure incurred covers family living needs. The figure of $17,150 also includes income and self-employment taxes. This labor income adds 34 cents and $1.13 to the respective per-bushel production costs. Total cash expenditures are $325 per acre of corn and $287 per acre of soybeans. The figures translate into cash flow expenditures per bushel of $2.61 for corn and $7.56 for soybeans.

If producers can price their grain at these cost levels, cash inflow will exactly equal cash outflow. Higher prices would produce a surplus to increase cash reserve, retire debt, expand the operation, or increase living expenditures. If the grain were sold at prices below these expenditure levels, a cash deficit would occur. The deficit would have to be covered by drawing upon previously accumulated cash reserves, increasing indebtedness, decreasing family living expenses, or selling assets.

Highly leveraged farmers may well use this type of cash flow analysis to calculate break-even price objectives. It gives them a handle on price level that would keep them out of financial difficulty. Note that depreciation charges are not accounted for, and principal payments are included. Thus, the method does not reflect total cost of production; it only accounts for cash flow requirements.

These examples provide two methods out of many to calculate cost-of-production price decision criteria. Other methods may serve individual producers in their particular situations. However, all cost-of-production models have serious logic problems.

Shortcomings of Cost-of-Production Pricing. The cliché "Know Your Costs of Production" has been quoted so often to farmers that some believe cost of production can be a legitimate marketing strategy. It cannot! And those who have practiced it know that it can lead to bankruptcy instead of guaranteed profit.

First, the market does not care what it costs a farmer to produce corn, hogs, or anything else! If the market were sensitive to costs, why would it have offered farmers $1.50 for corn (when costs were $2.75) or $60.00 for hogs (with $37.00 per hundredweight costs) during 1987? The market is oblivious to farmers' production costs; it has more important roles. In one situation, the market's role is to get prices high enough to ration a short supply among highest-return users to avoid a diminished supply from being prematurely exhausted. Conversely, its role is to drop prices low enough in the face of over-supply to encourage increased usage and discourage further production. Those are legitimate roles of the market; it does not fret about farmers' cost of production.

Secondly, basing a marketing plan on cost of production will win practitioners a quick trip to the poorhouse because of the way it works. Assume that total production costs for corn are $2.75 per bushel (excluding land charge).* What happens if the market rises to $2.70 and then falls to $1.50? The cost of production strategy never signaled a sale. The producer slides along with the market,

---

*The calculation is based on variable production costs of $150.00; labor, $50.00; machinery, $100.00; and overhead, $30.00 per acre; total, $330.00 with a 120-bushel yield.

holding unpriced grain, into financial difficulty. On the other hand, suppose the market rallies to $2.80. "Time to price," rings the strategy. "You can't go broke taking a profit," even if the profit is only a nickel per bushel. Then the market rallies to $3.80 (as it did in 1983 and 1988). The price move creates lots of margin calls. And the salve of "selling at a profit" wears thin when hedgers calculate what might have been. Their problems are often compounded further in high-priced markets by pricing too early. Widespread summer drought characteristically creates bonanza high harvest-time prices; droughts also reduce yield. Few situations are worse than having 125-bushel yields hedged at $2.80, but harvest produces only 70 bushels per acre and a price level of $3.80. The textbook hedge definitions of offsetting gains and losses do not match.

Over time, a cost-of-production marketing strategy is guaranteed to create financial problems. The mechanics of the strategy causes farmers to sell too late (or never) in down trending markets or too early in up trending markets. Cost-of-production hedgers miss out on bonanza high prices by selling too early and they go unpriced in low, declining markets. They suffer from $1.50 corn prices and eliminate $3.50 prices. Their long-term average price will typically be less than cash market speculation.

Scale-Up Selling, Scale-Down Buying

Producers can use scale-up selling in conjunction with other analysis to provide decision points for committing to pricing opportunities. The objective is to capture a higher than average price. Scale-up selling involves setting a price objective at which the first sale will be made and then committing larger and larger amounts at selected price increments. Scale-down buying involves making larger and larger purchase commitments at successively lower prices. Both price increment and percentage of production to hedge at each increment must be selected along with the initial entry price.

Individuals may use cost-of-production estimates, chart objectives, or historical prices to determine their first sale price. Pricing increments should be based on the price volatility of each commodity. They may be 20 cents in corn and wheat, 40 cents in soybeans, and 2 cents in cotton. The percent of production at each sale point dictates how many sales decisions will be made. One producer choosing large price increments might price larger percentages of production each time. Smaller price increments would fit with smaller percentages each time. The procedure is outlined in Table 6-4 with two combinations of increments and production percentages.

Two problems plague scale-up selling (or scale-down buying). First, market price may never reach the first sale point, and if it does, it may not follow through to signal sales of the remainder of the production. The method guarantees the first sale is wrong if the price does happen to rise further. There will have to be liquidity to carry hedges to the top of the market and sufficient fortitude to suffer the agony of having sold a portion too early. Second, the original selling point may be set too low so that all sales will have been made long before a strong price move finally exhausts itself. Should bonanza high prices occur, the producer will have no remaining production to commit at the very high levels. An average price of $3.40 looks good unless the market reaches $4.00. At that point regret sets in.

A variation to scale-up selling is to use a scale-down selling procedure. The first sale is made after a market has peaked and has made a predetermined decline. Larger commitments are made on subsequent declines. Should the market rally before passing through all the decision points, the same procedure is used

Table 6-4: Illustration of Scale-Up Hedging at Two Percentage Rates and Price Increments

Percent of Production		Pricing Points		Results
		($/bu.)		
Option I				
5	initial x	3.00	=	.15
10	x	3.15	=	.32
20	x	3.30	=	.66
30	x	3.45	=	1.03
35	last x	3.60	=	1.26
100			Average Price	$3.42
Option II				
20	initial x	3.00	=	.60
30	x	3.30	=	.99
50	last x	3.60	=	1.80
100			Average Price	$3.39

on the next market decline after a second peak. This method has the advantage of causing the seller to wait until a major market thrust has run its course and reduces the chance of selling too soon. It suffers from the dual problems of being able to detect market tops and farmer psychology against selling in a down market. The agony of knowing they missed the top makes it extremely difficult for many producers to make pricing decisions at lower levels. Hope of a rebound and capturing market tops springs eternal in many producers.

Seasonal Pricing

Seasonal pricing has formed the foundation of many market pricing plans. An analysis of cash price indicates that there are times of the year when higher and lower prices can be expected. Logic would suggest that end months of crop years should have higher prices than harvest time due to the reduced availability of supply. Furthermore, on a particular day grain futures prices are typically higher in distant contracts than in nearby months.

Seasonal price variations within a crop or marketing year are due to seasonality of production, demand factors, or marketing patterns. Fall holiday demand for turkeys is a classic example of demand-induced seasonal price effects. Reduced sow farrowings during winter months lead to reduced supplies in late summer and produce supply-induced seasonal price effects in hog prices. The seasonal change in livestock prices is due largely to traditional production practices and seasonal eating habits. Seasonality of crop prices originates from the dual effects of short harvest periods and the costs of storage over longer consumption periods. For most crops, prices are lowest at harvest, when the supply is most abundant. Charts of seasonal tendencies of cash prices are presented in Figure 6-4.

Producers who use a seasonal pricing strategy find that it often does not work. They store grain at harvest, waiting for a summer rally, only to sell grain at a lower price in July than existed at harvest, while incurring storage costs for the period. Livestock producers who put animals on a feeding floor on a schedule to fin-

Figure 6-4: Cash Price Seasonal Tendencies.

ish them at historically high seasonal periods find that the market forgot to read the history book when the anticipated seasonal price rise fails to materialize.

An analysis of historical cash grain prices leads one to conclude that there was a consistent improvement in late-season price compared to harvest price. This relationship has been dubbed "the summer rally" in grains. During the time period of the 1960's and early 1970's, the futures price level was fairly constant due to government price support and storage programs. The historical rally in price was basis level gain of cash price relative to futures price level. To emphasize the point, the 13 years of recent futures prices depicted in Figure 6-5 discount the

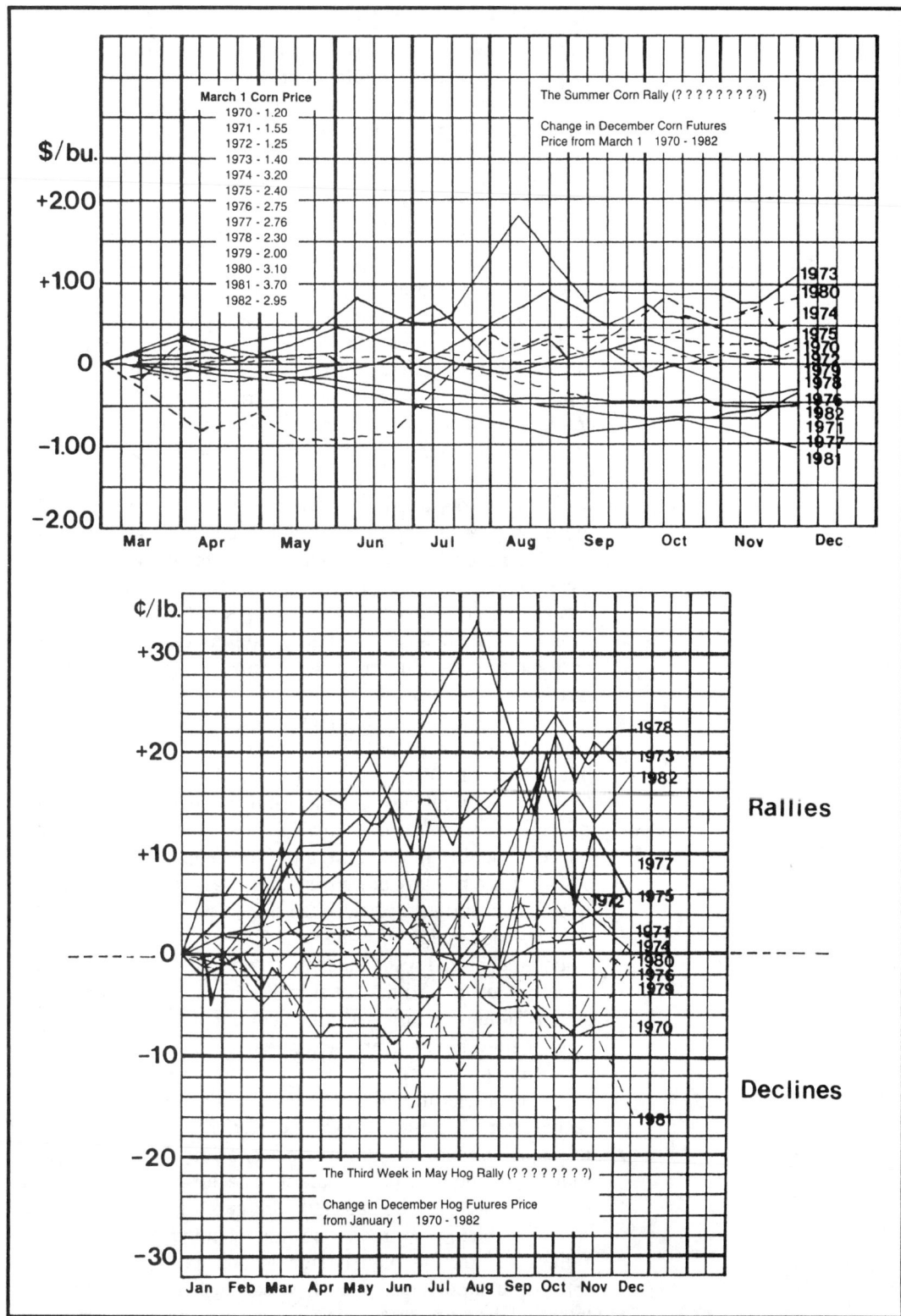

Figure 6-5: Change in Corn and Hog Futures Prices Over Time, 1970-1982.

existence of either a summer corn rally or spring hog price rally. While some years did show such rallies, about an equal number had price declines. The conclusion from analysis of futures prices is that there are no seasonality patterns in level of futures price upon which to base a pricing plan. The effect all too often identified as seasonality is actually basis level gain. Seasonality pricing is typically an unworkable marketing plan because it mixes the two pricing effects, price level and basis level.

Trend-Following Models

Picking market tops and bottoms can be an exercise in futility. Preconceived ideas of what is a high price or a low price can be harmful to a successful marketing plan. As most farmers have experienced, prices can, and often do, move in both directions a lot further than anticipated. Cheap prices have tendencies to become cheaper, just the same as expensive ones become more expensive. To incorporate a pricing strategy more sophisticated than calendar scheduling and not as frustrating as picking extremes, some producers adopt a trend-following technique.

Trend following does not attempt to predict price action; it merely interprets what occurred in trading pits and contains rules for reacting to signals generated. As markets move up, down, and sideways, the purpose of trend following is to analyze the current major direction. Is it turning down from up, or simply settling back for a continued upswing later? To decide the question, various methods of trend analysis are used and rules applied for a clear-cut decision about when changes in direction occur. Popular techniques are moving averages, price channels, linear regression, oscillators, and directional movement. Each one has its criterion to determine the extent of price change required to signal a change in direction.

The change in trend dictates market action. A change from up trend to down trend signals farmer-producers to place hedges on their production and input buyers to avoid or remove market positions. A change from down trend to up trend signals input buyers to place long hedges on items they plan to buy and producers to lift market positions or delay placing them. As a pricing strategy, a farmer-producer should have all of the crop or livestock production hedged or sold during a primary down trend, and he or she should own or be long on all the production during a primary up trend.

The advantage of trend following as a pricing decision technique is that trades are entered systematically. Trend signals provide an automatic method for dictating reaction to market signals and eliminate human judgment and indecision. All that is required to implement the plan is to execute the appropriate position whenever signals are generated.

An important characteristic of trend following is that it will not signal initiation of positions at market tops nor market bottoms. By definition, a trend is detected after the extremes have occurred; the change in direction is confirmed by the signals generated. With a trend-following strategy, producers will never price their production at the top market price nor buy their inputs at the bottom of the market. They must be satisfied with capturing a part of major moves.

A problem with trend-following techniques is that they often produce false signals. An up trend might be signaled, but within a short time a down trend signal follows. An input buyer would have set buy hedges only to be signaled a short time later that the market was going lower. The same situation arises as falsely signaled down trends fail to follow through. These false signals are called whipsaw. They arise because the criterion for detection of trend change is set too close to market price action and corrections within a trend touch the signal criterion, but then the true trend again exerts itself. The action dictated by a whipsaw is to lift the original hedge position and replace it when another change in trend is detected. The canceled position may, however, have generated a loss in the futures transaction, depending upon how soon the switch occurred and extent of price movement in the interim. If the whipsaw occurs within a short period without much price change, the loss can equal the width of the criterion boundary.

No trend-following system can avoid false signals. The success of these systems must be measured by relative frequency of correct calls and the profits they produce as compared to number of incorrect calls and amount of losses. An analogy can be drawn to flipping a coin which has a 50/50 probability ratio of producing heads or tails. If heads yields a 20-cent gain and tails a 10-cent loss, a farmer would spend all day tossing the coin because the overall result would be profitable. With a 2-to-1 reward-to-risk ratio, a participant would need be correct only one-third of the time to break even. Thus, for an evaluation of trend following as a market decision technique, the bottom line is the number of correct calls times their profitability compared to the incorrect calls times their losses.

We analyzed three types of moving averages to illustrate the procedure and to search for a workable one for farmers to use in executing a marketing strategy. Results of two of the analyses are attached at the end of this chapter, and Table 6-5 reports the results of a popular three-way (4-9-18-day) moving average.

A simple, one-variable moving average consists of M days of prices being included in the average, where M can be any number of days. For example, to calculate a 10-day moving average, we summed the closing prices of the previous 10 days and divided by 10. On subsequent days the oldest closing price was subtracted and the newest closing price added. The criterion for analysis of the trend was the comparison of today's price relative to the average. As long as the daily closing price was greater than the 10-day moving average, an up trend was in effect. When the daily price was less than the 10-day average, the market was in a down trend.

To execute a marketing plan with a moving average pricing strategy, long positions were initiated in up trends, and short positions in down trends. One position was liquidated and the other initiated upon signal of trend change. A long or short position was held in the nearby contract at all times.

We analyzed farm commodity prices over two periods from 1970 through 1987. The trading results reported in tables at the end of the chapter include all simple moving averages from 2 days to 28 days in span. The top diagonal numbers of each table report the result of each respective moving average analzyed.

Few positive results occurred, and many large losses did occur. In soybeans, for example, the best simple average for the 1970-1981 period was a 22-day average. It yielded a $65,145 annual profit. But in the 1982-1987 period, the 22-day average yielded a $15,160 loss. Only the 20-day average yielded positive results for the second period.

We concluded from these results that a daily price versus a simple moving average was unsuitable for detecting market trends in executing a marketing strategy.

The second analysis consisted of the crossover of daily price and two moving averages, defined as follows:

1. The average of M days' closes and N days' closes were calculated, where the number of days in the M average (the faster-moving average) was less

than the number of days in N (the slower-moving average). For example, if M = 10 and N = 20, we analyzed a 10-20-day combination.
2. If the short-term (M-day) average was greater than the long-term (N-day) average, and the daily market closed above the M-day average, a long position was established on the opening the next day.
3. If the short-term (M-day) average was below the long-term (N-day) average, and the daily market closed below the M-day average, then a short position was established on the open the next day.
4. If the high and low of the next day were equal, it was assumed the market was locked the limit and no trade was made. The trade was made the first day the market was not locked the limit.
5. The model was always in the market in selected nearby contracts with either a long or short position.
6. The model was tested on November or December, March, and July grains and cotton contracts; February, July, and December hog contracts; and February, June, October, and December cattle contracts.

The tables at the end of the chapter report the results of trading one contract in farm commodities over two periods, 1970-1981 and 1982-1987. They show results of all combinations of the faster-moving M-day average (from 2 to 28 days) against the slower-moving N-day average (from 2 to 60 days). The intersecting cell of the respective row and column reports the trading results of each combination. At the bottom of each table, the return from the most profitable combination is reported. For example, trading corn in the 1970-1981 period, a 12-60-day combination yielded the highest return. It averaged $24,065 per contract ($4.81 per bushel) over the 11-year period (almost 44 cents per bushel per year). Over the 1982-1987 period, however, the 24-30-day combination was best for corn, with $8,240 of returns. In the second period the 12-60-day combination yielded only $970.

We concluded that trading profits from a two-crossover moving average model were very elusive. Furthermore, the best combination in one period did not guarantee that it would be the best over the next period. Some other combination of moving averages worked better in every second period tested.

Optimization of a pricing model in one test period does not guarantee that it will perform satisfactorily next year, or over the next several years. The difference in results between periods in our test should alert farmers who are offered marketing schemes to ask how consistently the scheme performed over two or three different marketing periods. A marketing strategy is of no value if it works only one year in five. And fantastic average results from a 10-year test period may turn to disastrous losses in the eleventh year of actual market trading. The farmer who expects to follow a marketing plan should look at both the average results of the test period AND the extremes in profits and losses for the period.

The results of the third moving average model are presented in Table 6-5. This one uses 4-9-18-day averages. When the 4- and 9-day averages exceed the 18-day average, an up trend is in effect; when the 4 and 9 averages are less than the 18, a down trend is in effect. The crossover is signaled when both the 4 and 9 averages are either greater or less than the 18. In every commodity and in both periods tested, returns from the 4-9-18 scheme were less than the best of the two-day moving averages. A simple 4-9-18-day average thus proved to be unsatisfactory also. We leave the problem to researchers to find a more market-sensitive method for calculating moving averages for farmers to use in a pricing strategy.

Trend following is a systematic, technical marketing plan that rules out human judgment in making decisions to let profits run while cutting losses short. As long as the number of profitable positions and their amount of profit exceeds the number of loss positions and their amount of loss, the overall result will be profitable. The problem is to find a consistently profitable moving average combination.

Table 6-5: Trading Results of a Three-Way (4-9-18-Day)
Moving Average, Major Farm Commodities,
1970-1981 and 1982-1987

Commodity	Time Periods	
	1970-1981	1982-1987
	(total dollar returns per contract over each period)	
Corn	$ 6,180	$ -770
Soybeans	68,785	14,420
Wheat	25,085	-4,580
Cotton	39,240	12,875
Soymeal	28,580	-1,840
Live Cattle	-6,912	-8,776
Hogs	2,237	15,156

KEEPING SCORE

Every farmer likes to brag about making a $12.00 soybean sale, a $60.00 hog sale, or a 95-cent cotton sale. Each year every coffee shop has its high-selling hero. But because prices realized at different times of the year do not have the same value, there must be some procedure to calculate price equivalency irrespective of when sales are made during the year. A procedure for keeping score is suggested as follows. Use late harvest-time price as the base price. Since wheat harvest in Oklahoma, South Dakota, and Montana is completed on different dates, the base month will be different depending on location. But that makes no difference in calculating the results for a particular farm. The objective is to have a specific initial pricing date year to year. The second step is to add profits from forward sale commitments to the harvest price. If a contract price for fall corn delivery was set at $3.50 early in the season and harvest price was $2.85, the forward pricing profit was 65 cents per bushel on that portion of production so committed, say one-third. Next, add storage returns to the harvest base price as sales are made during the year, but deduct costs of storage up to time of sale. Assume another third was sold in January at $3.00, and the final third was sold in June at $3.40. They represent additions of 15 cents and 55 cents to the base harvest price. However, storage costs must be deducted from the storage sales. Using the carrying cost of 7 cents per bushel per month, the January gain must be reduced by 14 cents and the June sale by 56 cents on the portion of sales made on each date. The summary calculation follows:

Price per Bushel

	$2.85	Base Nov. 1 Cash Price
+	.22	Pre-harvest Hedge (65 cents × 33% of production)
+	.05	January gain over Nov. 1 base (15 cents × 33%)
+	.18	June gain over Nov. 1 base (55 cents × 33%)
−	.05	Storage Cost to January (14 cents × 33%)
−	.18	Storage Cost to June (56 cents × 33%)
	$3.07	Effective Price

This method produces an effective net price of $3.07 per bushel. This compares to an average of $3.63 from the three sales when carrying charges are not accounted for.

This method can be used to evaluate the bottom line of various marketing strategies suggested in this discussion or offered by various advisors. It was used in the analysis of marketing strategies cited at the end of this chapter and in the analysis of 730-Day Marketing explained in Chapter 10.

PROFIT RATHER THAN PRICE

Specialized production in farming means that many producers buy commodities as production inputs and sell finished products. Poultry, hog, and cattle producers buy corn and soybean meal as feed ingredients. Feedlot operators also buy feeder livestock. To the extent production inputs are traded on commodity exchanges, these producers can hedge their input purchases as well as sell contracts to cover sales of the product.

A viable strategy is to hedge in major production cost items while simultaneously selling against production. This technique is effectively a futures spread that focuses on relative prices rather than price level. Product prices do not have to be high to lock in profits as long as input prices are low. The objective is to lock in a profit spread rather than being concerned with price level per se. Since feed costs are 60 to 70 percent of costs of hog production, the pork producer should focus on the hog-corn price ratio as the decision criterion. In cattle feeding, it is called the feeder spread.

To produce finished cattle, two major inputs are required, feeder animals and feed, primarily corn. Other cost items are supplements, interest, veterinary expenses, death loss, buildings and equipment, labor, and a profit margin. Most costs are stable or fixed; feeders and feed are the most variable. The objective is to maximize the difference between the two major variable costs and price of finished cattle.

Balancing the number of contracts among the three items is essential for an accurate spread. Feeder cattle contracts, traded on the Chicago Mercantile Exchange, call for 42,000 pounds of feeders per contract. Using an average weight of 700 pounds, each contract would represent about 60 head. Live cattle, also traded on the CME, represent 40,000 pounds, or about 36 head of 1,100-pound cattle. Corn is traded on the Chicago Board of Trade in 5,000-bushel contracts. Producing 400 pounds of gain with a conversion ratio of 7.5 pounds of corn per pound of beef would require 53.6 bushels of corn per animal. Putting these figures together in the approximately correct proportions produces a trading unit of three feeder cattle and two corn contracts on the input side and five live cattle contracts on the output side. The unit would represent 180 head. Smaller-size contracts on the MidAmerica Exchange allow other combinations. To calculate the proportions for any size volume of operation, first determine the number of head of finished cattle to be sold and their weight (180 head at 1,100 lbs. = 198,000). Divide by the size of contract to determine the number of live cattle contracts to sell (198,000 divided by 40,000 = 5 contracts). Next decide weight of feeders to be bought and the number of contracts to cover them (180 head at 700 lbs. = 126,000 divided by 44,000 pounds per contract = 3). To calculate how much corn is required for the gain on the animals, subtract purchase weight from selling weight and multiply by the number of pounds of corn per pound of gain and number of animals (1,100 lbs. − 700 lbs. = 400 lbs. gain, times 7.5 pounds of corn per pound of gain = 3,000 lbs. of corn times 180 head, or 540,000 pounds). Finally, divide the corn weight by 56 pounds per bushel to calculate the number of bushels and number of corn contracts required (54,000 lbs. divided by 56 = 9,643 bushels, or about 2 contracts).

Using closing prices for the respective contracts permits calculation of the profitability of the spread. Assuming December corn futures are $2.70, November feeders are $65.00 per cwt., and April live cattle (five months later) are $59.00 per cwt., the dollar differential is:

Fats	59 cents × 40,000 × 5 =	$118,000
Less Feeders	65 cents × 42,000 × 3 =	−81,900
Less Corn	$2.70/bu. × 5,000 × 2 =	−27,000
	Spread Margin	$ 9,100
	Margin per head (180 hd.)	$50.55

If this margin is sufficient to cover other production costs, the feeder can set the hedge to lock in this spread. If it is not satisfactory, he or she can delay setting the hedges and/or delay initiating the enterprise until profitable opportunities exist.

When market conditions do not allow profitable feeder spreads to be established, placing opposite futures positions in each respective contract permits a feeder to speculate on the possibility of improved price relationships without placing cattle in the lot. It is called a reverse feeder spread. The opposite action is to sell feeder and corn contracts and buy live cattle contracts. If a feeder fills his or her lots when the spread profit is abnormally low, that farmer-speculator is betting that the spread will widen as the cattle mature. It may not do so, but setting a reverse feeder spread in futures is no more speculative than placing cattle in the lot and hoping relative prices will improve.

Setting feeder spreads when it is profitable to do so eliminates the effects of price level change and reduces feeder and corn buying and fat cattle selling to basis considerations. The feeder who can buy corn and feeders on basis below the futures contract prices and sell fats at a premium to futures will improve on the results of the spread. But cash buying and selling at poor basis can eliminate profits hedged into the spread. In effect, futures contracting eliminates price level speculation while allowing basis level speculation.

GOVERNMENT PROGRAMS AFFECT PRICING

Government programs exert an influence on market pricing. The U.S. Food and Agricultural Act was instituted in 1977 and was revised in 1981 to extend through the 1985 crop year. In December of 1985, the 1985 Food Security Act was signed into law. The major justification of the program was to protect farmers from severe price declines and to assure consumers of adequate food supplies without sharp increases in food costs.

The grain reserve portion of the act affects wheat and feed grains through three components: (a) the farmer-owned reserve, (b) rules for managing government-owned stocks, and (c) a program of low-interest government loans to farmers to encourage construction of grain storage facilities. The program encourages farmers to put grain in the farmer-owned reserve when stocks are abundant and market prices are low. When the market price becomes relatively high, disincentives are applied to encourage farmers to remove the grain from storage and place it on the market to keep price from going too high. If total stocks of grain become too large, production is reduced through cropland set-aside and diversion programs. If grain prices are low, producers receive deficiency payments to maintain income levels.

The farmer-owned reserve applies to wheat, rice, corn, sorghum, barley, and oats. Farmers who participate in the program have to comply with restrictions on planted acres. They are then eligible for non-recourse loans and storage payments. With the grain as the only security against the loan, the farmer can forfeit it rather than repay the loan, even if the value of the grain is less than the loan balance plus interest. The government's Commodity Credit Corporation (CCC) has no recourse to claim additional payments if grain value is less than the loan (hence, a non-recourse loan). The loan rate in effect places a floor under grain prices. Whenever the loan rate exceeds market price, the producer can place grain in reserve and take a loan rather than sell it on the open market.

Once farmers put grain into the reserve, it cannot be removed without penalty unless the market price exceeds a release price. The supply of grain in the reserve is thus withheld from the market to remove its effect on market price. Price support is achieved. When the market price is above the release price, farmers can repay the loan, remove the grain from the reserve, and sell it at the higher price, or they can leave it in the reserve. If they leave it in reserve, they no longer earn storage payments, but they can retain the low-interest loan on the grain. If market price increases to a higher level, the government can call the loan, requiring the farmer to repay the loan within a specified time. At still higher price, government stocks owned by CCC can be released from storage into the market. The reserve stocks thus place a ceiling on how high prices rise as larger and larger supplies are released into the market. Figure 6-6 illustrates the bounds placed on price fluctuation by provisions of the program. Between loan and call levels, price fluctuates in response to supply and demand forces. In effect, the farm program exerts control on price levels through economic incentives and disincentives.

Farmers choose to participate in the program by complying with its provisions in order to become eligible for its price-supporting benefits. They may choose not to participate and produce for an open market price. Thus, a farmer's market pricing strategy is consequently affected by provisions of the farm program. If the farmer expects low prices, he or she can use the program to hedge against them. If high prices are expected, the farmer could stay outside the program, but he or she has to keep in mind the upper bounds placed on prices by it and have a marketing plan to guarantee those higher prices.

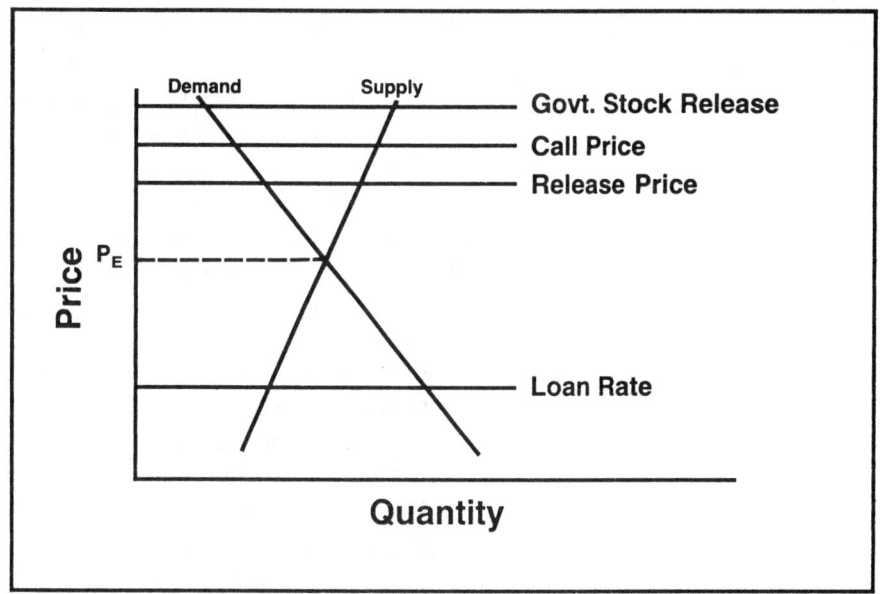

Figure 6-6: Price Bounds Established by Government Grain Reserve Program.

The various loan and release prices, interest rates, storage payments, and land use diversion payments can be changed by the secretary of agriculture to raise or lower the free market price band and control the amount of grain in the reserve. Starting in 1983, Payment-In-Kind Certificates (PIK Certs) added another variable into price effects of governmental action. In 1986, early redemption of Certs and County Posted Prices created a new currency for farmers and added to their market decision options. Because of changing government programs and changes in implementation, it is necessary for farmers to keep up to date on announced programs and fit their provisions into marketing plans.

Crop producers, livestock feeders, and others have to consider not only the usual market uncertainties as they make their plans, but also how agricultural programs might be changed. Government policy has added another consideration to be weighed as part of a marketing strategy.

RESULTS OF PRICING STRATEGIES

An analysis of grain marketing/pricing strategies by Kentucky researchers provides some factual results of various marketing strategies for corn and soybeans.[5] Nine grain marketing strategies were defined with respect to pricing mechanisms that triggered sales, timing of sales, quantity sold, and the delivery period. The strategies were separated into two categories: (1) five which required no market information, only timing; and (2) four which required information. The producer who required no information delivered and cash priced production in a specific and predetermined harvest or post-harvest month; he or she made no attempt to forward price. Strategies requiring market information instituted sales based upon price action in the futures markets. In each instance where applicable, the variable costs of marketing and storage were deducted from market price to determine net price in terms of a harvest season equivalent. Definitions of the various strategy alternatives:

Those Requiring No Market Information:

1. Entire crop priced and delivered at harvest. No carrying or storage costs were incurred with this strategy.
2. At harvest, grain was stored and equal amounts ($1/12$) of production were priced and delivered for cash each month. With this strategy a producer received the season's average price, net of carrying charges.
3. Grain stored at harvest with half of total production delivered and cash priced in historically high price months, August for corn and June for soybeans. The remaining half of production was sold equally in each of the remaining 11 months beginning at harvest.
4. Grain stored at harvest and one-third of total production priced and delivered for cash in June, July, and August. With this strategy the producer attempted to capture the season's high gross price by making sales in three historically high cash price months.
5. Grain stored at harvest and delivered for cash in December after the harvest glut was safe in storage and cash basis had improved.

Those Requiring Market Information:

6. Scale-up selling. Grain was stored at harvest, and 25 percent of the production was sold each time the average monthly price rose a selected percentage above the projected season's average price. Corn sales were made at 5, 10, 15, and 20 percent above the projected average; soybean sales at 10, 20, 30, and 40 percent levels. If sales had not been triggered by the targeted prices by September following harvest, the remaining inventory was sold at September's average price.
7. Trend analysis. The producer was assumed to sell the commodity and/or hedge when the closing price for two consecutive days fell below an up trend line as defined by chart analysis. Grain was stored at harvest, and one-half of the remaining inventory was sold for cash each time a sell signal (as defined above) occurred in the futures market.
8. Pre-harvest hedging of fall sales. The producer hedged one-third of expected production in March, July, and September for harvest delivery by using December corn and November soybean futures contracts. The producer bought back the contracts in October and delivered the total production for cash at harvest. Timing of the sales decisions during the selected three months depended upon the trend analysis and use of sales signals as defined in option 7. If the producer did not price during any one specified month due to a lack of sell signals, the cumulative quantity was priced in the next scheduled pricing month that contained selling signals. Any portion not hedged by September was sold for cash at harvest.
9. Pre-harvest hedging for sales from storage. The producer hedged one-third of expected production in April, July, and September according to the same signals used in option 7 for delivery in February. The March contracts were used to establish the hedges.

Average results over the 10-year period 1972 to 1982 are given in Table 6-6. Few strategies produced better results than selling at harvest. Alternatives 5, 7,

Table 6-6: Average Net Prices Received from Nine Corn and Soybean Marketing/Pricing Strategies, Kentucky, 1972-1982

Marketing/Pricing Strategy	Average Net Price[1]	
	Corn	Soybeans
	($/bu.)	($/bu.)
Calendar sales:		
1. Delivering and pricing the crop at harvest.	2.37	5.80
2. Selling stored grain in equal amounts each month.	2.36	5.80
3. Half of crop sold in historically high-priced months (Aug. corn, July soybeans); other half sold 1/11 per month.	2.37	5.88
4. Thirds of stored crop sold June, July, August.	2.36	5.92
5. Stored at harvest and sold for cash in December.	2.45	5.87
Scale-up selling:		
6. Sales when monthly price exceeded projected season's average price by defined percentages.	2.14	5.58
Trend analysis:		
7. Grain stored at harvest and 1/2 of remaining volume sold on trend violation signals.	2.47	5.95
8. Hedging thirds of production in March, July, September on trend violation signals for harvest delivery.	2.38	5.86
9. Hedging thirds of production in April, July, September on trend violation signals for February delivery.	2.57	6.18

Source: Riggins, Skees, and Reed 61.
[1] Net of carrying costs.

and 9 produced better results in corn. In soybeans, strategies 4, 7, and 9 were better than harvest sales.

The analysis emphasizes the importance of storage and carrying costs in affecting net return. Selling one-twelfth each month, hitting both high- and low-priced months, produced no better results than selling at harvest after deducting carry charges on stored grain. Nor did selling in historically high-priced months improve on harvest sales. It appears that early season basis gain may be more important in improving cash sale results than trying to pick high prices.

USING THE OPTIONS MARKET

Using the options market for price protection appears to offer an attractive alternative to use of the futures market. Farmers dread the possibility of margin calls, and they are reluctant to lock in a price when prices could go higher. With options, a minimum price level can be established for a premium payment, without eliminating the potential of higher prices. And, options supposedly eliminate the threat of margin calls.

The hog market provides an example of use of the options market. In 1986 hog prices rose to approach the record levels set in 1982. Hog producers seeking a way to capture those high price levels and lock in excellent profits could have turned to the options market.

Various prices bid into the markets on July 21, 1986, are given in Table 6-7. A $54.00 October Put gives the hog producer the right, but not the obligation, to market his or her hogs at $54.00 (less the basis and option premium) at any time prior to the option expiration date in mid-September. The option "insurance" premium for the right to sell $54.00 hogs was $3.00 per cwt., and the $56.00 option was $4.72, while the $44.00 option was "only" $0.20. Respective net prices

Table 6-7: Futures and Options Prices, July 21, 1986

October Futures	Options			
	Strike Price	Terminology	Premium	Price Level Insured by Option[1]
	($/cwt.)			
$54.00	56	in the money	$4.72	$51.28
	54	at the money	3.00	51.00
	44	out of the money	0.20	43.80

[1] Less local basis on day of delivery. The producer faces the same basis risk in using the futures market.

"insured" by the options were $51.28, $51.00, and $43.80, less local basis on the day of delivery.

A farmer who purchased a $54.00 October Put paid $3.00 per cwt. for the right to sell hogs in a $54.00 market at any time prior to expiration of the option in mid-September. If hog prices had continued to rally, say to $60.00, the right to sell hogs for $54.00 was worthless. The farmer's net price was $55.00 ($60.00 cash less $3.00 option premium less $2.00 assumed basis). On the other hand, if prices fell to $44.00 by September, the right to sell $54.00 hogs made the option worth approximately $10.00 per cwt.* The farmer's net price would be about $49.00 ($44.00 cash market, plus $10.00 option value, less $3.00 premium, less $2.00 basis).

The $54.00 October Put was "at the money" since the October futures price also traded at that same price on July 21. The entire $3.00 premium was thus attributed to the "time value" inherent in the remaining eight weeks of life left before the option expired. The $56.00 October Put was "in the money," in that the price level being locked in exceeded the futures price of $54.00. The right to sell $56.00 hogs in a $54.00 market was worth $4.72; $2.00 of the premium constituted "intrinsic value" (the difference between $56.00 and $54.00); the remaining $2.72 ($4.72 total value minus $2.00 intrinsic value) was attributed to "time value" of the option's remaining eight weeks of life. The $44.00 October Put was "out of the money" since the option strike price was below the futures close. Its premium of only $0.20 was all "time value."

Were the $4.72, $3.00, and $0.20 respective premiums high or low? The answer depends upon how much value the producer places on the $51.28, $51.00, and $43.80 price levels insured by the respective options. The producer eliminates the possibility of lower prices but does not eliminate the opportunity of higher prices by purchasing options. Cost and value are relative.

Some hog producers might have concluded that the $3.00 per cwt. premium on $54.00 options was an exorbitant price to pay for eight weeks of time. Others might have felt that the 20 cent premium on $44.00 options was a better buy because it was cheap. It was cheap, but it was not inexpensive. In addition to the 20 cent premium, the $44.00 option cost $10.00 per cwt. in value given up before the option had value. The purchaser was guaranteed a price level of only $43.80 per cwt. ($44.00 strike price less 20-cent premium). In order to buy low-cost disaster insurance, the purchaser gave up $10.00 per cwt. from the begin-

*For the option to be worth $10.00 in a market price move of $10.00, the delta must be 1.0, that is, a 1:1 change in option value versus change in market price. A constant 1.0 delta is not a valid assumption.

ning. Out-of-the-money options are like insurance with a deductible clause. The $44.00 October Put had a $3,000.00 "deductible clause" (300 cwt. times $10.00 per cwt.) and therefore had a low-cost premium of only $60.00 (300 cwt. times 20 cents per cwt.). The $54.00 at-the-money October Put covered all the downside risk (no deductible clause) at a cost of $900.00. If the $44.00 option were bought and price fell to $44.00, the hog price being protected was below $43.80. The option buyer regretted not having purchased the $54.00 option that netted $51.00 in a $44.00 hog market. Hindsight is always 20/20.

Options are often touted as having a distinct advantage over the futures market because there are no margin calls. That claim is largely erroneous when options must be carried over time in order to guarantee long-term price level. A comparison of futures and options supports the point. For example, if an October hog futures contract is sold at $54.00, the 30,000-lb. hog contract requires an initial margin deposit of $500.00, but it is eventually returned to the hedger. If the price level goes up, each $1.00 per cwt. increase requires an additional $300.00 to maintain the hedge position. By comparison, an option with a $3.00 per cwt. premium costs $900.00 initially. Futures prices require a $3.00 move against the position before margin calls exceed the initial purchase cost of the comparable option. Furthermore, if hog prices are constant or increase by the mid-September option expiration date in the example cited, the $54.00 option expires worthless. The $900.00 "insurance" premium would not be collected upon and would become an out-of-pocket expense. Thus, in September, the producer is faced with another decision. In order to again lock in an attractive price and guard against lower prices, an at-the-money $54.00 December Put option might be purchased for, say, $3.00 per cwt. Should prices remain constant, or higher, until mid-November, the December option will also expire worthless. In mid-November the decision of whether or not to buy additional price protection must be made once again. A $54.00 February at-the-money Put might cost $3.00. In mid-January the decision to continue extending price protection might be made once again, if the anticipated price break has not yet occurred. Use of a July $50.00 at-the-money option might cost $3.00.

If by mid-June hog price falls to $40.00, the producer will have achieved the long sought-after objective — price protection — but at what cost? The four options for almost year-long price protection on one month's worth of production cost $12.00 per cwt. ($3.00 October + $3.00 December + $3.00 February + $3.00 July). Hogs marketed in June realize a price level of $38.00 ($50.00 July option price less $12.00 cost of options). That price is $2.00 per cwt. less than the cash price. During the time period, the options buyer enjoyed an average cash price of $52.92 for all hogs marketed during the year. In addition, the options hedger enjoyed peace of mind knowing that he or she had locked in minimum prices in case of price disaster. Neighbors who were cash market speculators enjoyed the same average prices on cash hog sales plus the $12.00 less cost for hogs marketed in June. They did not have the peace of mind associated with knowing what minimum price to expect when the price drop materialized. The annual insurance premium for peace of mind cost $12.00 per cwt. in the example given. Is it worth it? Ask the options buyer.

The point is that the assertion that there are no margin calls with Put options is not necessarily true. If the market does not "cooperate" by falling significantly lower shortly after the first option is purchased, additional options (equivalent to additional "margins") must be purchased to maintain the price protection.

Buying options for price protection is like buying fire or life insurance. The best time to buy is the day before it is needed. The worst case is to cancel the insurance the day before it is needed. Not buying the July Put in the above example was the worst possible case. It is not until then that the price protection was needed.

The alternative form of price protection is to sell a hog futures contract. If the market follows the same pattern noted above, the hedger enjoys the same cash market prices and the same price protection at a cost of $0.40 per cwt. ($60.00 commission* for a 30,000-lb. futures contract plus $60.00 interest for the $500.00 margin money at 12 percent). Additional margin is required if the futures market moves higher than the $54.00 hedged price. However, cash hog prices will usually go higher as well to offset the futures loss. Higher prices often generate consternation when margin calls have to be met. What has really happened is that the hedger has developed regret. What started as a $54.00 hog market can become a $64.00 market. A $54.00 hedged price level loses its attraction in a $64.00 market — regret sets in. Then the hedger laments: "If I had just waited." The problem with waiting is that $64.00 possibilities can quickly turn into $44.00 realities.

Options appear to solve the problems caused by futures. But Put options are not cheap, and they are no panacea. Unless the buyer is purchasing Put options that are well in the money and therefore have a lot of intrinsic value (e.g., $64.00 hog puts in a $54.00 market), he or she is paying a premium price for time. But time is being given away for nothing in the futures market. Therefore, if one can successfully trade the futures market, one does not need the options market. If one cannot trade the futures market, the options market may not be all that it has been publicized to be.

SUMMARY

Pricing decisions have become more important than production decisions in their effects upon farming profitability. Economic effects of price variation have become relatively more severe than production variation in items within control of the farmer to decide. Unstable prices force operators to develop risk management and pricing strategies. Some strategies are active; others are passive. Some producers intend to capture market extremes; others are content with averages.

The purpose of this discussion was to emphasize the time period available for price decision making, the role of futures prices in pricing and production decisions, the choices available for mixing speculation with hedging, and techniques used to make pricing decisions. We suggested a method for keeping score of the results of a pricing plan and suggested it could be used to evaluate proposed strategies. Finally, we introduced the concept of locking in profits irrespective of price level whenever differences between prices of input and output items presented profitable opportunities.

We reviewed the features of current agricultural price legislation. Rather than removing uncertainty, current agricultural policy has added another element of uncertainty because of its impact on total supply, available supply, and potential changes in the program.

These ideas were intended to form a foundation of thinking to help an individual define a marketing plan suitable to his or her situation. A Kentucky study provided results of average prices for comparing various strategies for making price level decisions. And an effort was made to focus agricultural Put options into perspective.

---

*Commissions total $300.00 if the hedge stays in the near month and rolls from October to December to February to April to June.

Corn 1970-1981
Total Futures Profits from Trading Various Moving Average Combinations
Long and Short

No. of days	2	4	6	8	10	Faster Moving Average 12	14	16	18	20	22	24	26	28
2	-100835													
4	-78045	-55225												
6	-47475	-46705	-44515											
8	-39315	-27695	-24645	-26705										
10	-29875	-15555	-7905	-20095	-16675									
S 12	-19545	-7315	565	-11125	-13735	-11985								
l 14	-6145	-7465	2095	2885	885	-11155	-13085							
o 16	-6005	135	5575	9415	-5335	-9285	-13495	-13795						
w 18	-3495	1545	7275	11795	4725	1215	-5405	-11335	-1195					
e 20	95	3525	5325	9985	5535	2305	-7555	-4535	-13865	-435				
r 22	-1065	5095	8285	8435	7745	7255	1315	6425	745	-4765	-7175			
24	-7175	4635	5065	13945	13145	15465	9025	9285	4105	-75	-9805	-4905		
M 26	735	6995	11205	15575	11585	8015	11875	9775	4585	915	2955	-4655	2725	
o 28	3755	6925	8185	12145	12795	9855	5325	7125	1295	6215	3445	-1605	-11305	1555
v 30	9395	9355	8585	9385	14495	9455	8355	6285	1515	1185	4465	-3935	-5295	-5295
i 32	5975	14495	11485	14105	12825	7745	10715	9205	3335	7395	2055	-4025	-1675	-125
n 34	9055	9915	11715	12975	11625	12875	8125	6855	11775	6505	3675	8575	-105	
g 36	8825	6015	16285	15375	14095	15885	13325	12635	12025	8425	4785	4475	8255	2635
38	9815	8605	20325	15875	14495	17805	16615	13015	15185	11715	10305	10905	6105	6445
A 40	11705	14675	17485	16345	15825	18185	19215	20575	15675	14335	8815	10155	3395	3475
v 42	11955	15585	17625	12865	12655	17885	17185	50645	15015	12625	8705	7775	7075	8475
e 44	11875	16755	11835	9845	10195	18865	13285	16845	9345	11785	11685	9305	11345	6835
46	15595	14245	14305	9275	5955	14645	12915	11245	12365	15785	11365	9655	10295	9625
48	14955	16035	12255	10555	13235	16685	16645	15725	13875	15435	9655	10545	9715	8715
50	14085	13125	11105	12305	16485	17765	14245	11265	14425	12505	12325	10575	13735	11375
52	15085	15085	12915	15915	20895	16595	18725	18045	16895	16235	11885	13335	11965	15365
54	11265	14655	15755	18995	22065	20065	18995	17315	16185	19725	12835	11635	13765	15355
56	12245	17275	16505	22655	22915	22685	20665	20295	17845	16985	15975	11835	12885	15685
58	11705	16765	19035	23875	22005	22035	23215	20255	19225	21145	18905	18925	13325	14535
60	12255	18795	18245	23395	22375	24065	22195	21325	19655	20195	19205	21105	16855	17115

Max. profit = $24065 @ 12-60 days

Corn 1982-1987
Total Futures Profits from Trading Various Moving Average Combinations
Long and Short

No. of days	2	4	6	8	10	Faster Moving Average 12	14	16	18	20	22	24	26	28
2	-51550													
4	-39570	-28430												
6	-32107	-32350	-31250											
8	-27650	-25810	-25370	-16040										
10	-20970	-17050	-12580	-10410	-7510									
S 12	-17510	-7210	-5670	-2910	-10130	-7790								
l 14	-13550	-8830	-470	-1910	-2630	-12150								
o 16	-10010	-10910	-5190	-4850	-5410	-11150	-13050	-8890						
w 18	-7270	-9570	-5210	-2650	-4470	-4390	-6870	-10590	-10170					
e 20	-5390	-5310	-6550	-2130	-2010	-2980	-3610	-2870	-3750	-4010				
r 22	-4890	-7130	-4910	-1750	1610	-710	-2750	-1690	210	-2270	-5310			
24	-4830	-5210	-3550	350	-170	-90	350	310	-950	-1610	-9470	-4870		
M 26	-2550	-4970	-1370	910	30	390	-890	670	-2550	1010	110	-6730	-150	
o 28	-4130	-4410	1430	870	4310	2330	-710	1810	-1330	2150	3990	3250	2090	2670
v 30	-2210	-3090	3590	4330	2170	4290	1290	2410	-1570	1150	4530	8230	6090	1350
i 32	-2550	-1550	2490	2590	530	2790	-330	670	-490	4410	7230	6910	5290	610
n 34	-2290	-190	2810	3130	1730	50	1750	1770	770	5070	7150	6450	4450	1650
g 36	1810	1890	6430	6730	3130	1710	1810	3090	3250	3750	6770	6270	2190	530
38	-1450	5090	7270	6410	4190	2610	2050	2450	2390	2990	5110	2290	310	2350
A 40	-1590	6310	5370	4990	2710	2870	1230	2510	1530	1530	2730	1910	1070	2250
v 42	-1830	3790	6330	6290	2150	990	830	2610	3390	870	890	-210	-770	1570
e 44	-770	4650	4170	5370	2970	2090	1210	2530	930	1630	2190	-590	190	430
46	-2070	3290	5690	4970	2010	690	-390	610	1250	2490	1550	1350	1490	870
48	-890	2930	4830	3270	490	-990	-1050	570	2450	3190	1710	2130	930	2150
50	-1430	1690	4430	2010	-290	-570	-230	1070	2170	2550	3330	2450	2550	1630
52	-2570	2290	3390	210	1010	-1370	-1830	-830	2170	3410	2890	3070	2250	2370
54	-430	2520	3250	1250	910	-1230	-530	670	4610	3410	3930	4110	4090	1570
56	-1950	3490	4050	2630	-70	-2390	-410	2510	2650	3550	4470	3850	2470	1090
58	-430	2150	4110	1650	1370	-1910	570	3750	3050	3730	4370	3710	1390	2610
60	1470	2350	3930	2930	1710	970	1490	4630	4450	4370	3810	2070	1170	910

Max. profit = $8230 @ 24-30 days

Soybeans   1970-1981
Total Futures Profits from Trading Various Moving Average Combinations
Long and Short

```
=========================================================================================================================
                                          Faster Moving Average
No. of    2      4      6      8     10     12     14     16     18     20     22     24     26     28
days
-------------------------------------------------------------------------------------------------------------------------
     2 -74490
     4 -61115  22040
     6   -865  14805  26300
     8   7645  44465   1355  22305
    10   -575  60075  12915  26165  23845
S   12  17365  45045  14045  18585  12425  42630
l   14  28365  57685  18575  20696  19575  49445  49370
o   16  61195  50345  64105  80845  33865  76245  75655  59005
w   18  53255  80335  55745  71745  65435 106265  85795  34785  57925
e   20  66165  82685  74635  77425  84285  72575  61385  34795  28265  54240
r   22  55355  89895  77755  73535  80005  77105  63785  53245  46055  52185  65145
    24  67395  85585  94795  91465  71905  70125  55245  53495  45185  39085  29475  64935
M   26  80955  91525  81325  86945  59025  61555  52595  42925  42235  48725  36555  55165  51765
o   28  82665  84545  78655  76415  52985  47485  45585  46365  47655  36215  44625  51125  29665  53900
v   30  85765  76905  81275  64745  56835  55215  53155  43095  23505  37725  47055  67315  58795  41495
i   32  77295  67895  70575  75375  64245  62635  62395  46735  57695  67295  53995  52005  43135  43295
n   34  83505  71125  64215  72575  62025  68355  62405  56615  69045  70515  75065  62355  56395  53235
g   36  68645  73405  62855  70345  67755  69135  65215  69705  77075  82475  77585  74695  64895  55385
    38  66325  77595  74635  81595  75345  77515  77355  71555  80545  86975  90915  85965  59985  47745
A   40  63395  75625  74265  77295  79615  75985  71205  73815  81755  82165  89385  66855  48235  50975
v   42  68965  75055  66785  86075  84465  76355  67905  86195  78465 100055  78685  69125  50565  55025
e   44  58105  76325  73895  85925  82145  79425  79045  80375  92745  81635  73195  61075  58345  50765
    46  50535  75705  70855  76125  82525  83215  81045  89155  88875  74995  62025  63365  53685  57075
    48  62015  67665  70235  71765  85435  78875  80215  87095  77605  76505  61775  58695  60255  59005
    50  75065  68395  73775  74675  76115  79455  83135  83165  67235  78875  62475  59955  65605  57315
    52  77065  81675  79515  88305  85995  72915  77885  83465  80275  66995  73765  61325  70465  56835
    54  76295  84855  92565 100365  85785  86005  87565  69355  75735  60255  47965  64795  61835  66425
    56  91945  93325  85135  96105  91435  83405  85415  67505  65535  61215  57885  72165  73305  67205
    58 100495  86775  89455  86635  77405  83435  81785  71195  58595  59605  66015  76345  66535  68715
    60  92275  86005  86625  82855  81235  84065  75045  71345  70285  68615  73435  83195  81065  72285
=========================================================================================================================
```

Max. profit = $106265 @ 12-18 days

Soybeans   1982-1987
Total Futures Profits from Trading Various Moving Average Combinations
Long and Short

```
=========================================================================================================================
                                          Faster Moving Average
No. of    2      4      6      8     10     12     14     16     18     20     22     24     26     28
days
-------------------------------------------------------------------------------------------------------------------------
     2 -72720
     4 -51700 -18840
     6 -24380 -26400 -46970
     8 -25180 -40740 -34520 -28240
    10 -17620 -25980 -19260  -3760  -6900
S   12 -13460 -23340 -11100   9560   7160  -2120
l   14 -14520 -12160  -2080  19180  11040   1680  -8600
o   16  -6780  -9960  -4560   2520   7400 -11640 -15020  -8120
w   18  -2080  -3240    120  10360  -9900 -20980 -11360 -24720  -3900
e   20   3900  -1500   3280   3240   -880 -14760  -8360  -6840  -8120     10
r   22  -2460  -3320   9920   4390   1360  -5320  -5280   4180  -5180 -15160
    24  -2060  -7080   8800   4480   6060  -3930 -10920   6440  13520   -940  -8480  -4540
M   26  -2780  -5320   1820   6800   1740  -5960 -10620  -1460  11780   -200  -1420 -14120 -13680
o   28  -2020  -3980   7360  10440  -3480 -15780  -5600  -2560  10780  -1520  -8840 -15300 -26260  -3240
v   30    160   -840  10220   5120  -8360 -15080    460  -6660     60  -7340 -10620 -21280 -19020 -26540
i   32  -7760  -7640  11660   1680 -12440  -6320  -4860   1700  -5720 -17740 -11440 -18100 -21860 -21960
n   34  -5000  -3960   8800  -1740 -10000  -6720    900  -7840 -12860 -11160 -11360 -11540 -10040 -10440
g   36  -6860  -3980   -160  -5640  -4180  -5100  -4780  -9460 -12040 -16740 -12080  -3560  -3680 -12620
    38  -3100  -4880  -1820  -8100  -4400  -7540    180 -16660 -22160 -19480  -6080    880  -8820  -7860
A   40   -200  -7000   1160  -5900  -8960 -12080  -6300 -15040 -22460  -8940   -960   2220  -8160  -8780
v   42  -1260  -3580  -3440  -6680  -7280  -7440 -11960 -16500 -10380  -8260  -3440  -3920  -6120  -5460
e   44   5960   4140  -6880  -9120  -5760 -13020 -21900 -14040  -8080  -1620   5060  -3260   -940  -9960
    46   3080   1000  -1120  -6760  -4880 -10740 -16340 -12620  -2260   -200   3860   1280  -6840  -8880
    48   4740   4100  10280  -1980  -5660  -8720 -11760 -10980  -1440   2400   4820    120 -10120 -14500
    50  17140   9160  13280   4320  -2200  -8940 -13360  -7820   5060   5300   1020  -5000  -7500 -10580
    52  11300   7900  11580   3300   1540  -7260  -5240  -1360   7080   9000   3420    940  -2580  -5280
    54  11400  13980  14720   7640    580  -4100 -12200  -3120  10140   8140   2520   1340  -3380  -5420
    56  13120  14120  13140   9640   2160  -3320 -13420    680  10660  10900   4700      0  -7180  -6080
    58   4520  15980   9780  10740   3300  -3840 -10460   2080   7960   6740    760  -2760  -8780  -8100
    60   4560  10100  11480   9460   6620  -4900  -5120   3020   8180   6020   -180  -4540  -9820  -6940
=========================================================================================================================
```

Max. profit = $19180 @ 8-14 days

Wheat   1970-1981
Total Futures Profits from Trading Various Moving Average Combinations
Long and Short

```
                                    Faster Moving Average
No. of    2      4      6      8     10     12     14     16     18     20     22     24     26     28
days
------------------------------------------------------------------------------------------------------
     2  -113070
     4   -96940 -53780
     6   -51880 -43560 -54410
     8   -41720 -21230 -32200 -46550
    10   -37355 -27810 -23160 -11800 -17210
S   12   -27740 -10200 -11800    110   1890  -8230
l   14   -32800    340   5210  -1190   1540  -8490 -10570
o   16   -25050   1820  11330  13670  23890  12330   5260  -2730
w   18   -12390   3050  18140  21390  15280  17720  11080  -6030  -2760
e   20    -6060  13800  19660  33120  18730  21190  16040   9400   3020  -3223
r   22     7900  21340  26850  29230  18580  22340  18690  13830   8530  -3280  13030
    24     8470  21610  32230  22120  25890  25780  26700  19820   7300   4890   1250  13860
M   26    11110  17840  29510  27050  27590  28500  24650  19340  11130  16030  12140  16590  13080
o   28     9310  24430  36450  32560  29810  37340  29760  18830  19260   9640  16800  19330   8330  11240
v   30    14490  19720  34670  33900  39060  40030  33690  26250  23470  11070  18270  23200  24530   6350
i   32    18430  27120  40150  33500  39970  38530  38150  34530  25530  15860  19510  24850  18730  14860
n   34    17050  34680  32250  30190  35230  39630  41470  36130  22470  17700  19500  27960  25550  20980
g   36    16960  33060  30230  34910  37900  43500  45470  28400  20860  17800  19070  23600  21270  19830
    38    15680  28960  34230  38640  41850  44540  36150  26230  15950  16650  21430  23020   8500  10730
A   40    14480  27020  26010  33660  35070  38430  28790  20050  18360  18940  14360   5490  -1690
v   42    16300  22760  28440  28580  31560  38340  21110  21020  19280  12490  13430   4800   -980  -5210
e   44    12110  27720  32790  29160  40300  32230  28190  22040  12460  11990   9850   3370   -650  -5610
    46    20200  26450  32100  33650  36620  38400  30500  18110  14640   7780   6090   1950 -10270  -9110
    48    25740  27830  36370  34120  37260  36440  26540  25610  13660   8090   4810   -400  -7700  -1020
    50    26540  30970  36230  33590  36490  28650  27100  21280  15170  10390  10780  -3020  -8210   1570
    52    28450  33810  27740  32760  34390  31870  20260  20420  19530  10960   1440   2200  -3450
    54    31830  34020  30040  31970  26940  23530  19660  16880  22160  15480   5890    560   6440   6210
    56    35050  34310  30440  30080  28390  20510  20640  23550  17450  12990   7630   4920  11930   7390
    58    32790  35610  33730  28130  27350  20580  16440  16540  16520  13360   7280   6760   5400   2310
    60    25260  35860  33030  27420  28360  18050  12330  15140  12280   9960   9780   6980   4010    -30
```

Max. profit = $45470 @ 14-36 days

Wheat   1982-1987
Total Futures Profits from Trading Various Moving Average Combinations
Long and Short

```
                                    Faster Moving Average
No. of    2      4      6      8     10     12     14     16     18     20     22     24     26     28
days
------------------------------------------------------------------------------------------------------
     2   -52770
     4   -44420 -32500
     6   -32360 -30580 -25560
     8   -28800 -26560 -24740 -14750
    10   -19080 -19420 -19860 -20000  -8800
S   12   -13440  -9160  -9880 -12720 -16120  -5400
l   14    -5560  -6420  -1200  -5780  -6560  -2600  -5950
o   16    -9840  -2600  -3200  -5040  -1760  -1780  -9260  -2140
w   18   -12200  -1820  -4400  -6780  -3680    100  -7840 -17080  -3760
e   20    -6480  -3400  -5060  -6360  -1060  -3300  -7800 -12960 -11340  -8220
r   22    -5520  -1280  -5800  -6000  -4020  -4240 -10820 -13860 -16920 -18500 -11840
    24    -3980  -3520  -3840  -6100  -4400  -9160 -10320 -17360 -14560 -11480 -16280 -11220
M   26    -1240  -1200  -5620  -5280  -5840  -6100  -8820 -10200 -12800  -9420  -5660  -9480 -11580
o   28    -1680  -2520  -4600  -5120  -4580  -7200 -10020 -11920 -12220  -9720 -12200 -14000 -15340  -6580
v   30    -3900    140  -4900  -7700  -4480  -7700  -6640 -10260  -8040  -7040 -13640 -13660  -9060  -9660
i   32    -4080     20  -5500 -10360  -6700  -9040  -8060  -9880 -10780  -8500 -11400 -10820  -7620 -12080
n   34    -1660    -80  -6860  -8520  -6900  -7560  -6120  -8740  -9500  -9780 -10000  -6560  -5780  -9720
g   36     -320  -3680  -8540  -7460  -5660 -10240  -8280  -8240  -6580  -7160  -9580  -5040 -10360 -13940
    38    -1480  -4440  -6820  -7360  -7900  -8560  -7960  -6700  -7080  -8340 -12320  -8120 -10660 -11880
A   40    -3220  -7420  -8340  -5900  -9660  -9940  -7320  -7340  -5500  -9760  -8820  -9280 -12240 -13000
v   42    -7860  -9200  -7040  -8640 -11000 -13180  -9500  -8620  -6860 -10620  -7800  -9880 -11500 -11960
e   44    -9120  -8260  -9460 -14340 -12380 -10580  -9540  -7820 -10620 -10620 -10220 -10420 -12540 -15140
    46    -9920 -10100  -7960  -9200 -14720 -14460  -9800  -7980 -10160 -10340 -10320 -12880 -14420 -14780
    48   -10260 -12740  -7780  -9620 -14640 -11300 -11300 -10600 -10500 -11620 -11800 -12160 -14700 -14280
    50   -11520 -13080  -8920  -9080 -15080 -11980 -10620 -12420 -11680 -11320 -11880 -11840 -13240 -10280
    52   -10720 -12180  -8700  -8380 -13660 -10580 -10580 -13880 -12520  -8360 -11140  -9960 -11940 -11180
    54   -10640 -11520  -7720  -8860 -13280 -14060 -13480  -9720  -9760  -8300 -10020  -7860 -10677  -8600
    56   -11040  -9240  -6020 -10720 -10920 -13000 -12820  -8720 -10720  -6900  -6240  -7600 -10940  -7880
    58    -9040  -9640 -56801 -11240  -9780 -11360  -9620  -6120  -7480  -3860  -7120  -8300  -8680  -8120
    60    -9040  -8480  -2440 -10000 -12940 -11020  -7900  -7637  -6420  -5780  -5560  -7140  -7960  -7040
```

Max. profit = $140 @ 4-30 days

Hogs 1970-1981
Total Futures Profits from Trading Various Moving Average Combinations
Long and Short

No. of days	2	4	6	8	10	Faster Moving Average 12	14	16	18	20	22	24	26	28
2	-119519													
4	-96934	-42204												
6	-50248	-31014	-37334											
8	-21830	-16822	-30228	4656										
10	-4858	2086	9242	1816	10902									
12	11298	20860	21952	14074	1182	-4508								
14	12500	22780	18886	8856	188	-21972	-11494							
16	16736	17118	11754	6374	-1808	-13358	-3394							
18	16534	17394	6502	4606	-1838	-3712	-15484	-25386	9320					
20	14042	15714	8322	2806	6642	3322	-2086	-15618	-24278	5358				
22	12968	13584	10764	6686	9698	4532	-2514	-9532	-17668	-14806	-10758			
24	11548	17610	12764	10530	9380	-2228	1310	-4508	-5142	-15806	-16242	136		
26	9840	17732	11670	12600	11942	-3217	-4090	-4780	-8694	-9052	-5938	-11990	-6835	
28	11356	12756	8776	8534	6410	-2860	-296	-6430	-7316	-2604	-2040	-2172	-14494	774
30	12780	12048	11910	5448	868	-3604	-2882	-7338	-4508	1146	-1774	-3196	-13602	-21214
32	11990	15438	12206	6856	2238	-3006	-5766	-8382	-1374	1450	-1202	-5800	-11664	-11428
34	13600	19196	12216	5824	4822	-1182	-3420	-2460	-2734	1034	-4936	-5356	-5854	-6352
36	16202	14878	8072	9106	6546	1182	-2330	-750	-622	-5378	-2412	-6666	-4780	
38	13112	13000	10260	9882	9954	5812	-1032	1058	408	2020	-2712	-5674	-5244	-6708
40	15004	17664	12372	13948	13144	4402	-1908	-530	-2188	6034	-440	4022	-1950	-1630
42	16084	19466	16254	14374	8694	4076	-186	708	2814	9260	4064	1656	22	1198
44	14054	17832	15230	15566	7792	4258	3016	8950	8008	11464	6894	1554	2858	8178
46	14074	21538	18204	15002	11532	8462	7318	10088	8804	15536	15838	7358	11114	13922
48	12434	22822	19718	14778	9168	4282	6670	10580	13744	16502	14486	12778	16608	12926
50	12150	21010	18460	15706	4492	7382	7854	13942	18136	17900	17642	14362	14842	13630
52	13172	20692	13600	13196	9068	8126	10554	12158	20252	19372	18836	12546	14132	14740
54	14722	21402	16682	10546	9078	8284	11132	14940	20264	20934	17028	10916	14032	13218
56	17305	17493	15319	10627	10855	14171	13685	14555	19975	15283	16463	14185	17207	12477
58	18400	15610	21502	13704	11266	13254	16080	12158	20146	13860	12660	16582	18662	14020
60	19805	16249	16407	15213	14427	15353	12737	12657	19969	13997	12285	15437	16209	11889

Max. profit = $22822 @ 4-48 days

Hogs 1982-1987
Total Futures Profits from Trading Various Moving Average Combinations
Long and Short

No. of days	2	4	6	8	10	Faster Moving Average 12	14	16	18	20	22	24	26	28
2	-67412													
4	-51022	-27416												
6	-33548	-22192	-25516											
8	-22039	-13048	-4810	-13162										
10	-13362	-6810	1636	-10118	-12538									
12	-8466	-6914	-4708	-8506	-8664	-7752								
14	-6664	-7284	-5166	2176	6858	1024	-934							
16	-3528	230	1600	5568	3374	2322	-1772	-118						
18	-452	3894	2918	7156	3024	3778	3140	-6306	1868					
20	570	3760	7822	11040	8008	7356	6642	566	-2426	788				
22	-1396	2122	7662	9874	7992	8600	79786	3880	2632	510	9268			
24	-4216	3202	8616	7204	8216	7202	10736	3776	3652	5116	-1092	9948		
26	-2614	8638	8968	6908	7964	7340	10966	9750	4638	6012	1778	2012	230	
28	-1172	8234	11956	10512	9256	11758	10954	8930	8316	8086	6734	3668	-934	4752
30	7188	8586	12200	12366	10164	13474	12766	11084	11020	39718	7384	5472	2290	422
32	2864	5482	11622	10100	10242	11694	12784	13160	11098	8390	6164	3994	4784	3264
34	4432	6388	10114	10286	9076	14592	11158	11942	11702	7488	4846	5910	4784	3264
36	3726	8704	8406	10682	12308	13598	11866	12219	8312	4162	1210	4118	3562	262
38	5128	8560	8158	5534	9428	13146	13708	11326	5180	3904	4564	3412	3154	260
40	5576	9528	10846	6618	7342	10458	13094	7416	4908	2518	2416	3246	2908	2810
42	5844	12570	9890	7212	10630	10020	10632	5454	3276	40	1870	2190	3112	3608
44	6996	12858	8022	5630	8370	9178	7590	2884	1944	-456	1240	4366	5982	5870
46	6198	9604	7410	6082	4680	7558	4302	2192	1392	-54	2946	2458	2186	2494
48	6340	8702	6750	4916	5402	7990	4130	356	1360	-172	2186	972	-1650	-1600
50	3656	7560	5748	5140	6306	5868	2858	-808	758	-1094	498	1086	-284	-854
52	48	6566	4114	3744	6116	5328	64	-3164	-658	-890	-566	664	730	42
54	844	4874	3240	3818	6152	2226	-1796	-4970	-760	-1646	310	1750	610	-1484
56	-2470	1748	2742	3636	3094	1460	-1790	-3762	-2100	-1292	-404	2452	154	-1916
58	-3668	-44	752	2604	90	2418	-4316	-3300	-2412	-548	586	2038	-1942	-2698
60	-3252	-1240	350	2430	-2264	-182	-3232	-3354	-2346	-1568	700	2938	-1756	-510

Max. profit = $79786 @ 14-32 days

Live Cattle 1970-1981
Total Futures Profits from Trading Various Moving Average Combinations
Long and Short

	Faster Moving Average													
No. of days	2	4	6	8	10	12	14	16	18	20	22	24	26	28
2	-66140													
4	-41304	16996												
6	6660	15224	3752											
8	4230	24948	-11892	16564										
10	13180	17348	2192	18112	17712									
S 12	19960	10032	15820	12644	6600	26844								
l 14	22480	8500	11312	19516	19304	-1276	15020							
o 16	22564	12796	17444	17340	8472	-1036	-8264	9112						
w 18	28448	20640	20308	22140	12860	7548	-1140	-15172	8268					
e 20	23784	15060	14136	16351	3288	5348	-312	-644	-6096	18436				
r 22	26480	22940	18832	12676	6300	8952	1412	4004	3920	-2168	14512			
24	25940	25272	14394	7624	1976	2764	-296	2488	844	5772	-2388	14376		
M 26	25044	26984	14268	19040	14304	948	-4224	264	1688	-604	-5032	-18048	432	
o 28	29484	23492	13144	15616	6912	-1476	-1816	-1384	5124	1260	-4844	-17952	-24440	-12136
v 30	29312	20672	14668	18000	8003	2768	5348	-144	-3976	-4248	-12196	-20908	-24248	-34184
i 32	25792	21008	11172	18792	7328	528	-444	-436	-1196	-10396	-12284	-18212	-24588	-24848
n 34	23068	20676	10268	9876	6232	-968	-6580	-7452	-8132	-8440	-14988	-22036	-31796	-21292
g 36	27780	22204	15424	10696	1032	-4372	-6412	-7156	-9816	-9480	-14576	-16528	-16584	-17044
38	30708	21492	18384	9228	-80	-3488	-6504	-9536	-6684	-10484	-15336	-18444	-14224	-9448
A 40	28536	22956	16412	9936	8840	-5820	-8284	-6804	-5556	-5904	-18996	-9112	-8964	-11600
v 42	26584	23120	10424	10096	9984	-4924	-2216	-3984	-8012	-11164	-5656	-8708	-9476	-9864
e 44	24788	18880	12416	9848	7344	332	4208	616	-4468	-10012	-3980	-9072	-9640	-12560
46	21796	18376	11092	10212	3636	2512	4068	-3024	-520	-8548	-5876	-8108	-10564	-19644
48	19568	14392	10208	11356	2652	2888	3404	-1020	-3192	-3352	-1636	-9276	-11500	-11180
50	15536	10972	9076	12336	8760	6592	1980	2512	-6824	-1892	-5360	-10612	-12568	-11960
52	14692	9680	9700	9192	9412	11844	4508	-612	-556	-6264	-6576	-8896	-9700	-14992
54	10116	9420	8056	9368	8960	5528	812	-484	644	-4236	-8568	-8904	-13360	-24779
56	9656	7236	2312	6088	6084	-460	-4968	2064	-1708	-3108	-10112	-11104	-19816	-22484
58	11696	2224	551	4464	-812	-4276	-6548	-1832	-3024	-2088	-8376	-16352	-20720	-20572
60	11348	6788	-4924	-924	-3984	-6052	-60	-2288	-5092	-5412	-5612	-19204	-22132	-24024

Max. profit = $30708 @ 2-38 days

Live Cattle 1982-1987
Total Futures Profits from Trading Various Moving Average Combinations
Long and Short

	Faster Moving Average													
No. of days	2	4	6	8	10	12	14	16	18	20	22	24	26	28
2	-51936													
4	-30344	-28660												
6	-19604	-22204	-20156											
8	-9308	-15776	-25600	-12424										
10	-17868	-16924	-18732	-22040	-17584									
S 12	-15032	-16524	-18260	-25896	-31140	-11180								
l 14	-8028	-10260	-20224	-18440	-20700	-24648	-18680							
o 16	-8736	-12184	-19532	-15664	-19240	-9116	-8768	-12892						
w 18	-4456	-8568	-8660	-11268	-10336	-8272	-6492	-1408	-13160					
e 20	-4544	-10536	-7976	-7888	-6424	-9176	-9156	-160	-136	-3408				
r 22	-1000	-10736	-7296	-5876	-6628	-1848	-3328	-1028	7316	6524	-1184			
24	-1872	-13040	-10604	-5600	-992	-32	416	3136	3308	5836	-1480	-13444		
M 26	-5872	-12364	-15340	-6844	-596	1256	3872	2844	3368	-1264	-5312	-13612	-12004	
o 28	-5984	-9184	-12544	-5020	-2336	-2440	2552	-1220	-3436	-7436	-8032	-10748	-14832	-16920
v 30	-3128	-11416	-12488	-7864	-4396	108	-1000	-6504	-976	-6540	-6782	-11612	-16680	-14388
i 32	-4864	-8144	-8236	-9300	-3192	-4176	-8004	-3820	-6988	-6788	-4896	-8524	-9408	-8544
n 34	-5840	-3024	-4324	-6444	-6092	-9388	-8212	-2996	-8276	-2944	-9456	-13084	-9256	-6868
g 36	-2928	-2616	-452	-6276	-7176	-9172	-8204	-7948	-5300	-7880	-11200	-8540	-8160	-4488
38	-3456	-264	584	-5508	-8296	-8876	-6740	-8264	-6476	-8596	-9848	-9916	-7472	-2808
A 40	-104	384	2560	-5616	-4320	-7104	-4004	-5776	-6288	-6448	-6192	-6004	-8128	-7132
v 42	3828	1280	8	-3096	-4152	-5248	-6168	-3984	-6112	-3404	-6556	-10540	-8860	-6808
e 44	416	816	456	-1620	-4856	-3940	-6416	-5136	-5856	-1164	-4636	-7156	-9328	-8452
46	-464	3008	-424	-1204	-4432	-4596	-6456	-3680	-3808	-260	-3912	-6112	-9804	-10480
48	-5392	2336	-1036	-1092	-4012	-2820	-6380	-5336	-48	-860	-2660	-4904	-5600	-10604
50	-6688	1248	-632	420	-4240	-2124	-7004	-3440	104	-168	-4524	-8156	-6628	-6248
52	-9352	368	180	716	-2160	-2900	-5916	-4696	-2044	640	-8112	-5324	-9460	-11424
54	-10296	-2084	-1696	2848	144	-3300	-6764	-5892	-3408	-2352	-7440	-6384	-8700	-6264
56	-8432	-164	-2704	2792	1880	-1908	-2884	-4456	-892	-6376	-5728	-5332	-7208	-7576
58	-8384	-1644	-5060	-348	180	-4080	-6040	-316	-1860	-5096	-6680	-7920	-7056	-6660
60	-8408	-3960	-5868	-1928	2636	1036	-1904	-844	-3164	-2088	-3968	-8800	-8136	-6456

Max. profit = $7316 @ 18-22 days

```
                                        Cotton     1970-1981
                  Total Futures Profits from Trading Various Moving Average Combinations
                                         Long and Short
=================================================================================================
                                       Faster Moving Average
     No. of    2       4       6       8      10      12      14      16      18      20      22      24      26      28
     days
     -------------------------------------------------------------------------------------------
        2   -86240
        4  -105225  -34855
        6   -45225  -33015  -16335
        8   -29535  -14235  -59105  -70645
       10   -32715  -46855  -57065  -69455  -48010
   S   12   -19315  -45805  -57045  -48175  -15835   -2125
   l   14   -11635  -28355  -35275  -18245   -5675   -1615   12375
   o   16    -6925  -13465  -15835    7175   32585   25055   27135   32545
   w   18     4715   24475   12785   25645   33465   29335   49915   32805   30135
   e   20    20735   33175   25925   34835   31165   38765   62675   59665   57955   30415
   r   22    31975   26835   37665   38385   38205   44185   64735   63715   58715   27565   22345
       24    34665   36815   38125   39675   37335   57125   58025   55965   44565   14485   12915    3035
   M   26    32825   38155   38385   34685   43315   53365   48275   61205   31405   22465   31915    6025   17935
   o   28    22335   31945   19585   30605   37565   45805   47675   45705   34855   31605   35695   20165    4245    4815
   v   30    22445   27055   19905   40615   42755   33875   53595   53975   44535   22205   27605   24865   17965   24115
   i   32    18235   19995   14585   41955   42045   46255   46285   54635   42295   24125   28705   30035   22845   21405
   n   34    18105   16965   26455   35775   41895   45805   55145   65445   36275   29445   36375   35165   23855   20165
   g   36    13995   26585   37145   39925   49395   51555   55705   54185   37575   27925   28375   40615   18515   24815
       38    27605   23345   36075   41165   45435   57795   61055   55595   39055   28395   29665   39605   20845   33015
   A   40    19365   25165   27145   37525   37965   60655   55905   54285   29995   34805   32845   17395   27005
   v   42    34145   28985   26565   37855   37595   51455   52795   44675   34305   35305   35355   34155   28915   38385
   e   44    26255   36425   33226   32775   26805   45195   45235   38525   36375   40155   44035   38385   31905   43885
       46    25095   32445   39365   28355   26325   36085   42365   37485   39015   41435   43685   42645   32835   48165
       48    28455   23825   36735   24555   26765   40585   38965   42905   38665   45355   44075   45115   42865   48645
       50    25705   22125   29095   22185   32785   37005   37905   42755   50865   54425   51145   57875   57045   59495
       52    17535   29965   28935   39305   34145   35925   39895   43275   49915   55695   57155   54235   55885   61815
       54    20545   37815   32785   41415   35335   42145   43855   47745   51225   48035   52736   51385   56235   55555
       56    19395   40015   37065   43045   40475   44335   39985   45635   49275   51775   52845   46385   58935   42615
       58    16635   34175   42265   39735   43765   44285   43595   39665   45655   50575   46355   49685   44325   40805
       60    15725   37065   42015   37785   48335   47365   47705   42125   42205   49025   37925   44685   46325   44615
=================================================================================================
                                Max. profit = $65445 @ 16-34 days


                                        Cotton     1982-1987
                  Total Futures Profits from Trading Various Moving Average Combinations
                                         Long and Short
=================================================================================================
                                       Faster Moving Average
     No. of    2       4       6       8      10      12      14      16      18      20      22      24      26      28
     days
     -------------------------------------------------------------------------------------------
        2   -50430
        4   -43830  -47840
        6   -39950  -45060  -34700
        8   -25380  -21790  -31990  -21490
       10   -12320  -17770  -28100  -17770   -7570
   S   12    -4410  -11320  -25080  -19260  -27110   -2780
   l   14    -2740  -13720  -13800   -6450  -24960  -18060   -2420
   o   16    -3600   -4400    -440    5580    1160   -2170    3820   21260
   w   18     5000    5130    8330   14970   14870    9600   18100   12460   10220
   e   20    13440    9640   11590   13910   17620   11640   12690    9610    9740    1250
   r   22    12780   13380    8900   18050   12170    7660   14350    5070    -840   -7110    -940
       24    13440   11580   12720   19600   17020   11060   10150    4290     320    2510   -3360     730
   M   26     8450   10340    6600   14200    9760    7340   10580   -1690    1730   -4240     340    2450     -90
   o   28     7370    9380    7150    7640    6180    9490    9440    2380     300   -3990    2220     910   -5460   -3240
   v   30     6380    7960    9520    6330    3030    9730    9140   -2330    3430   -5120    3410    4110   -5920   -5860
   i   32     4800    5870   10780    4890    4140    6150    1040    1770    3850    1370   -1190     320  -11190   -3710
   n   34     1260    9520   15150    2250    4260    5270    -130    3620    3550    -870    -290   -1740   -7500   -3090
   g   36    -1210    4490    7920    5720    2750    2340    2320    -600    2530    -760    3350   -2430   -1730    1470
       38     2510    9850    6760    6640    6070    1100     620   -1600    1730   -1330    -140    1440      90    1690
   A   40     4970    5260    6440    5400    3950    1540   -2250   -4870   -1760   -1000    2870    -340    -970    -140
   v   42     3520    6730    6740    4150    6430     320   -1620   -5610    -950   -1580   -2140   -1870   -3070    3500
   e   44     3350    7030    5640    5550    2250    -250   -3060   -4960   -3780   -3010    -530   -2260   -3860    2240
       46     4160    4170    4650    5980    2030    2480   -3220   -3290   -2830   -2080    1330   -1960   -2720    8390
       48     5980    3630    3870    6240    4410    4270   -2840    -590    -510   -2020     590   -7330    3180    9740
       50     3770     210    3100    6500    7270    3620    -990    3650    -670   -3250     240   -5280    4750    5000
       52     1000    1590    4550    4600    8540    6430    -980    3550     690   -5170   -1410     330    5330    -210
       54     1300   -2620    5450    8720    9420    8130   -1720    3360     100   -4160   -1050    2230     810    1890
       56     5800   -3050   -1450    8760    7710   10230    1950    2580   -1080   -1010    1000     820     680    4830
       58     8680   -1560   -2890    8930    3750    7710     260     170   -1880    2750    3740   -1370    2560    6210
       60     8250    -520   -2690    4820    3960    2610    2020    2240    2440    6260    3300     470    8000   11510
=================================================================================================
                        Max. profit = $21260 @ 16 day single move. ave.
```

Soybean Meal 1970-1981
Total Futures Profits from Trading Various Moving Average Combinations
Long Only

No. of days	\	Faster Moving Average												
		2	4	6	8	10	12	14	16	18	20	22	24	26	28
	2	-49190													
	4	-35620	300												
	6	-11280	1510	-2070											
	8	-9230	14560	16050	8360										
	10	-7390	17360	14510	13470	3170									
S	12	-800	32010	18160	9950	8190	-4830								
l	14	790	27260	21180	11340	15700	14700	16020							
o	16	2410	31680	24230	22610	21730	24170	28430	17520						
w	18	6270	26280	21550	32240	28630	27210	28400	11830	18230					
e	20	18080	33050	28630	34010	24630	29630	34280	21020	23970	19910				
r	22	24500	36110	30010	29530	25590	32180	30900	18540	22040	22610	28780			
	24	24240	35330	37170	33750	28150	30200	26280	18160	30500	25990	18600	24690		
M	26	31120	39730	33490	27990	28770	28330	21570	22880	28860	23530	26760	15620	19490	
o	28	35570	38490	32610	26970	24440	27440	24960	22650	21960	22740	21140	15000	1640	20580
v	30	32520	37760	31920	31530	25800	23740	28650	21910	22510	15840	16190	15600	5830	10310
i	32	32020	36000	29940	26520	23970	21540	23590	25100	20670	19710	21120	17660	14210	11310
n	34	25970	32560	28380	24280	24040	25230	25530	26080	20490	22920	20050	21390	16520	19320
g	36	19220	29530	27990	25140	25270	26220	27320	23970	24000	24210	25160	18590	22810	21080
	38	14810	26850	24160	23210	26610	30030	29910	23170	27400	25530	22430	25810	28600	23990
A	40	14990	26700	25610	25050	26150	30200	26810	24280	26140	23380	23380	26450	25120	25540
v	42	18380	22530	27890	25910	27100	30250	27040	23780	18050	20310	21350	23440	26810	24380
e	44	15210	24190	24950	26660	29880	29380	21850	19390	21730	21890	25790	21830	20680	24600
	46	19480	24830	24630	25710	26980	24660	27550	21750	19020	22570	22760	21200	20530	23260
	48	18520	23290	23480	23700	24800	25450	24860	19590	21930	24030	24060	22060	22530	23270
	50	19760	21050	24960	24140	26980	24690	25150	20350	24120	27150	21770	21350	22650	24940
	52	19860	22800	22830	24350	26580	26510	26630	21810	22550	24990	21820	22220	26470	23690
	54	20350	25670	22670	29000	28630	27840	24830	22850	24480	23130	24260	24820	23200	22540
	56	24670	26760	22440	28070	28700	26780	22800	22550	23510	24610	25510	23740	23040	20920
	58	26310	28740	21020	23200	26830	26760	24240	21020	22890	21740	21250	24070	24210	20920
	60	30990	26730	21750	24880	26230	21550	24370	25200	21410	24220	21210	21550	22720	22710

Max. profit = $39730 @ 4-26 days

Soybean Meal 1982-1987
Total Futures Profits from Trading Various Moving Average Combinations
Long Only

No. of days	\	Faster Moving Average												
		2	4	6	8	10	12	14	16	18	20	22	24	26	28
	2	-28880													
	4	-27880	-20370												
	6	-20530	-23340	-20560											
	8	-18350	-18270	-20450	-14460										
	10	-15550	-17890	-14440	-10130	-7090									
S	12	-14930	-10970	-6400	-500	-5210	-11580								
l	14	-15940	-10360	-7980	-2350	-3600	-8630	-8740							
o	16	-14870	-9000	-5600	-4780	-4080	-8680	-11150	-11070						
w	18	-9320	-5330	-4500	-3280	-3910	-4060	-6490	-7670	-5810					
e	20	-7870	-4570	-4420	-3710	-5310	-4680	-5230	-5800	-9290	-1890				
r	22	-8630	-5420	-4030	-2460	-3240	-5450	-5720	-5510	-7300	-6100	-1640			
	24	-5850	-5020	-2880	-2500	-3190	-3340	-5250	-4330	-4370	-7320	-6720	-4300		
M	26	-4650	-4250	-2030	-3780	-4960	-4650	-5370	-5560	-3470	-4850	-4200	-3630	-3710	
o	28	-4560	-3730	-3200	-4210	-5430	-4030	-5810	-4880	-3200	-2580	-3140	-1810	-4060	-3920
v	30	-7720	-3330	-3520	-3900	-5430	-6810	-5910	-3380	-2260	-1750	-2560	-1240	-4310	-5940
i	32	-6730	-3180	-2490	-2140	-5830	-7210	-5530	-3490	-3460	-2430	-3690	-4660	-4800	-7700
n	34	-7940	-3410	-4570	-3540	-4070	-4820	-5140	-3120	-4660	-3670	-4650	-5590	-4660	-5280
g	36	-5150	-4620	-1380	-3900	-4560	-4780	-4170	-3500	-3840	-5120	-4490	-5700	-5750	-6010
	38	-4580	-5210	420	-3260	-5540	-5880	-5780	-3420	-2980	-4770	-4480	-7530	-7630	-6780
A	40	-3640	-4500	-420	-3130	-6530	-6510	-5820	-3500	-4270	-3680	-5360	-7310	-7570	-7650
v	42	-5500	-3400	-110	-5320	-5960	-6300	-5930	-4190	-5110	-5470	-5910	-6810	-7230	-6750
e	44	-5280	-3420	-1850	-5450	-6320	-6090	-4090	-4630	-4650	-5850	-4840	-4860	-5680	-4030
	46	-4530	-3460	-690	-4630	-5500	-5290	-2900	-2590	-3590	-6130	-5460	-5350	-5790	-4010
	48	-2460	-990	-940	-3440	-5660	-3200	-1310	-1560	-2850	-3490	-1050	-3400	-3320	-2380
	50	-4860	-370	-50	-4300	-5670	-3530	-1810	-1300	-2370	-1920	-390	-1500	-470	-2060
	52	-980	-960	-820	-3900	-4840	-1590	-1490	-2590	-3100	-700	-720	-1380	-1380	-1250
	54	-510	370	-1280	-3290	-3940	-1430	-540	-2560	-560	150	-460	-1180	-1190	-240
	56	560	590	-430	-2060	-170	-1280	260	-1310	-950	-80	-380	-1130	-130	-340
	58	-1550	1160	-40	-1830	-1300	-610	-1600	-910	350	270	-430	20	180	420
	60	-820	340	80	-1420	-1460	-1180	-1530	800	740	860	800	540	590	1480

Max. profit = $1480 @ 28-60 days

All Ag. Commodities 1970-1981
Total Futures Profits from Trading Various Moving Average Combinations
Long and Short

```
                                    Faster Moving Average
No. of   2      4      6      8     10     12     14     16     18     20     22     24     26     28
days
     2 -609484
     4 -515183 -146728
     6 -200313 -122755 -124612
     8 -129755    3991 -140665  -92015
    10  -99588    6649  -49271  -41787  -26266
S   12  -18777   44627    1697   -5697    2477   37796
l   14   13555   80745   41983   43858   51517   19637   57636
o   16   64925  100429  118603  157429  113399  120745  101363   98263
w   18   93337  173719  142305  189561  158557  185581  153161   21497  119923
e   20  136841  197009  176633  208532  174275  173135  164427  104083   68971  124701
r   22  158113  215799  210161  198477  186123  196549  178323  150227  122337   77341  125879
    24  165083  226857  234543  219109  187781  199231  176289  154705  127357   74341   33805  116127
M   26  191629  238961  219853  223885  196531  177496  150651  151609  111209  102009   99355   58707   98592
o   28  194475  222583  197405  202845  170917  163589  151193  132861  122833  105071  114821   83891   -6359   80728
v   30  206707  203515  202933  203623  187816  161479  179911  144033  107051   84923   99615  102941   63975   21577
i   32  189737  201951  190113  217103  192621  174227  174925  161387  146955  125439  111899   96513   60993   54469
n   34  190353  205117  185499  191495  185869  184805  187055  182483  144269  144949  137571  123153   93245   85951
g   36  171627  205677  198001  205497  201993  203105  198293  180989  161097  145253  135021  143035  112495  101921
    38  178055  199847  218069  219595  213609  230009  213549  181087  171859  160801  156697  161187  104567  105769
A   40  167475  209805  199299  213759  216800  212367  196483  187291  168471  155933  155933  145575   88721   92075
v   42  192413  207501  193983  215755  212053  213437  183633  223039  159917  178881  155933  132243  102931  112389
e   44  162397  218127  204342  209779  204461  209685  194829  186741  176195  168907  167469  126447  114843  116093
    46  166775  213589  210551  198329  193573  207979  205761  184809  182199  169553  155887  138065  107625  123293
    48  181687  195859  209001  190829  199315  205205  197299  200485  176287  182565  157225  139517  132773  140361
    50  188841  187647  202701  194937  202107  201539  197369  195269  183127  199353  170777  150485  153099  156365
    52  185859  213707  194659  223023  220485  203295  198457  198561  214921  196553  187845  156205  171417  154003
    54  185123  227837  218553  241659  216793  213397  206849  188601  210693  183323  152146  155207  162147  154524
    56  210266  236414  209216  236670  228854  211426  198222  196154  191882  179750  166196  162126  177486  143808
    58  218031  219899  227558  219743  207809  206073  198807  179001  180007  178197  164089  176015  151737  140733
    60  207658  227492  213148  210624  216978  204396  194322  185504  180712  180600  168228  173748  165052  144560
```

Max. profit = $241659 @ 8-54 days

All Ag. Commodities 1982-1987
Total Futures Profits from Trading Various Moving Average Combinations
Long and Short

```
                                    Faster Moving Average
No. of   2      4      6      8     10     12     14     16     18     20     22     24     26     28
days
     2 -375698
     4 -288766 -204056
     6 -202479 -202126 -204712
     8 -156707 -161994 -167480 -120566
    10 -116770 -121844 -111336  -94228  -67992
S   12  -87248  -85438  -81098  -60232  -91214  -48602
l   14  -67002  -69034  -50920  -13574  -40552  -58384  -57474
o   16  -57364  -48824  -36922  -16666  -18556  -42214  -55200  -21970
w   18  -30778  -19504  -11402    8508  -14402  -24224  -17812  -55314  -24712
e   20   -6374  -11916   -1314    8102    9944  -15900  -14824  -18454  -25322  -15480
r   22  -11116  -12384    4446   16228    9244   -1308   66238   -8958  -13062  -32126  -26806
    24   -9368  -19088    9262   17434   22544    1710   -4838   -3738     920   -7888  -46882  -27696
M   26  -11256   -9126   -6972   12914    8098    -384    -282   -5646    2696  -12952  -14364  -43110  -40984
o   28  -12176   -6210    7552   15112    3920   -5872     806   -7460    -790  -15010  -19268  -34030  -64796  -26478
v   30   -3230   -1990   14622    8682   -7302   -2188   10106  -15640    1664   13078  -18278  -29980  -46610  -60616
i   32  -18320   -9142   20326   -2540  -13250   -6112  -12960     110  -12490  -21288  -19222  -30880  -44804  -50120
n   34  -17038    5244   21120   -4578  -11996   -8576   -5794   -5364  -19274  -15826  -23760  -26154  -27830  -31798
g   36  -10932     188   12224    -144   -3388  -11644   -9438  -14439  -13668  -29748  -26020  -14882  -23928  -34796
    38   -6428    8706   14552   -5644   -6448  -14000   -3922  -22868  -29396  -35622  -23194  -17544  -31028  -25028
A   40    1792    2562   17616   -3538  -15468  -20766  -11370  -26600  -33840  -25780  -13316  -15558  -33090  -31642
v   42   -3258    8190   12378   -6084   -9182  -20226  -24328  -30840  -22746  -28424  -23086  -31040  -34438  -22300
e   44    1552   17814      98  -13980  -15726  -22612  -36206  -31172  -30112  -21090  -11736  -24180  -26176  -29042
    46   -3546    7512    7556   -4762  -20812  -24358  -34804  -27358  -20006  -16574  -10006  -21214  -35898  -26396
    48   -1942    7968   15974   -1706  -19670  -16610  -30510  -28140  -11538  -12572   -6204  -24572  -31280  -31474
    50      68    6418   16956    5010  -13904  -17656  -31156  -21068   -6628   -9902  -11706  -28240  -20822  -23392
    52  -11274    5574   14294     290   -3454  -11942  -25972  -22970   -8382   -2070  -15638  -10770  -17050  -26932
    54   -8332    5520   15964   12126     -14  -13764  -37030  -22232     362   -4758  -12210   -5994  -18437  -18548
    56   -4412    7494    9328   14678    3684  -10208  -29114  -12478   -2432   -1208   -2582   -6940  -22154  -17872
    58   -9872    6402  -50149   10506   -2390  -11672  -31206   -4646   -2272    3986   -4774  -14582  -22328  -16338
    60   -7240   -1410    4842    6292   -1738  -12666  -16176   -1145    3880    8074   -1098  -14462  -17912   -7046
```

Max. profit = $66238 @ 14-22 days

Questions for Study and Discussion:

1. How may a farmer attack the problems of price risk?
2. Discuss the factors which should be considered in framing a market strategy.
3. Are futures prices accurate predictions of cash price?
4. What are the two problems of commodity pricing?
5. What choices do agricultural producers have in making pricing decisions?
6. Explain selective hedging/selective speculation.
7. List and explain several pricing strategies.
8. Why is cost-of-production strategy inadequate as a pricing strategy?
9. Give examples of scale-up selling and scale-down buying.
10. Why did seasonal selling appear to be a viable strategy prior to 1972, and why has it proven unsatisfactory since 1972?
11. What is the reasoning behind a trend-following market strategy?
12. Explain a method for "keeping score" so that various pricing strategies may be evaluated on a common basis.
13. Explain the concept of "profit rather than price."
14. Do government programs increase or reduce market uncertainty?
15. Explain non-recourse loans and their effect on price level.

Exercise: Making Pricing Decisions

A group exercise in committing to pricing decisions can be conducted by using playing cards to generate changing prices. Students can be put into a position of having to make pricing decisions by assuming that they have 100,000 bushels of commodity in storage and are planning production of an additional 100,000 bushels in the new crop year. They can pick any day to price all or a portion of each crop. The first day's price should be the current local price.

Face cards are removed from two decks of playing cards. The cards are shuffled, and each draw represents weekly change in price. A red card represents decline in price equal to the number on the card; a black card is a price increase, again equal to the number on it. Jokers represent limit day price moves in the same direction as the previous color drawn. (The instructor can bias the price trend which will evolve by removing either some black or red cards. This will also serve to confound "card counters.")

The exercise begins when the instructor asks participants whether they want to price any of their inventory or planned production at the current market price. Those who do write down the amount and the price. A card is drawn, and price rises or falls according to the card's color and number. Students are again offered the opportunity to price their production by duly noting quantity committed and the new price.

The exercise continues with successive card draws and time lags for students to record their decisions as they wish to make them. After 52 draws, one year of time for making pricing commitments will have lapsed. The students should average together their individual chosen prices and respective quantity commitments. The high- and low-price sellers might be recognized.

The exercise is intended to emphasize the difficulty in committing to price offers as prices randomly change. It emphasizes the necessity of developing a pricing strategy based upon some foundation rather than having to react to changing market price.

Footnotes for Chapter 6:

(1) Quoted from "Monthly Grain Report" (Chicago: Stotler & Co., July, 1983).

(2) Blair Stewart, *An Analysis of Speculative Trading in Grain Futures*, Tech. Bulletin No. 1001 (Washington, D.C.: Commodity Futures Trading Commission, Oct., 1949).

(3) Allen Lines, et al., *Ohio Crop Enterprise Budgets* (Columbus: OSU Cooperative Extension Service, 1986).

(4) Gene Futrell and Robert N. Wisner, eds., *Marketing for Farmers* (Doane Information Services, 11701 Borman Drive, St. Louis, Mo. 63146), 1982, 27.

(5) Steven K. Riggins, Jerry R. Skees, and Michael R. Reed, "Analysis of Grain Marketing/Pricing Strategies," *Journal of the American Society of Farm Managers and Rural Appraisers*, Vol. 50, No. 1 (Denver: ASFM and RA, April, 1986).

Chapter 7

Basis: The Economics of When and Where to Sell

PLACE AND TIME PRICE DIFFERENCES	232
Components of Basis	233
RELATIONSHIP OF CASH AND FUTURES PRICES	234
Parallel Fluctuation of Cash and Futures Prices	235
Convergence of Cash and Futures Prices	236
BASIS AND HEDGING	236
The Pre-harvest Hedge: Protecting Price Level	237
The Storage Hedge: Earning a Storage Profit	238
CASH AND FUTURES PRICES OVER TIME	240
Basis Level Negotiation in a Market Economy	242
Cash Versus Nearby Futures	243
Farmer Psychology and Basis	243
Contract-to-Contract Futures Spreads	243
Local Basis Year to Year	246
BASIS RECORDS AND CHARTS	246
Plotting Basis Charts	247
Historical Charts and What They Show	250
BASIS SIGNALS FOR CASH SALES AND PURCHASES	252
Contract Month to Place Storage or Production Hedge	253
Rolling the Hedge	255
Relationship of Futures Price Between Marketing Years	256
TO OWN OR NOT TO OWN GRAIN STORAGE	258
LOCATION — WHERE TO SELL	262
SUMMARY	263

Basis: The Economics of When and Where to Sell

Each commodity traded on a futures market has a parallel market for trading the physical commodity, or actuals. We refer to the latter as the cash market, or spot market. However, it is not a single place, like the futures market. Instead, cash markets are scattered across the landscape at the nearest elevator, auction barn, or buying station, down the road from every producer.

The two markets have related, but distinctly different, prices arising from their respective price-making forces. These price differences and their relationship to each other constitute the study of BASIS. An understanding of the relationship and factors affecting it is one key to profitable marketing. Basis signals tell a commodity producer **when** and **where** to sell or buy the physical commodities on the cash market.

PLACE AND TIME PRICE DIFFERENCES[1]

Price differences cause commodities to move from areas of surplus production to areas of deficit production, in both location and time dimensions. The Southeastern poultry production region depends upon Midwest grain production as a feed source. Export markets in Toledo, New Orleans, and Portland, Oregon, depend upon interior production areas to supply commodities for export shipment. Price differentials between production and consumption or export locations must be sufficiently wide to pay costs of transportation and handling. Otherwise, there is no incentive to move commodities between locations.

In a time context, concentrated harvest periods create surplus supplies relative to immediate consumption needs. During later periods, consumption exceeds production (there is no production until the next harvest). Price differentials over time encourage storage until the supply is needed for consumption; they ration supply over time.

On a given day, location and time factors cause many prices to exist for a particular commodity. For example, on a day in April, some prices of No. 2 yellow corn were:

By Location $/bu.		By Time $/bu.	
Gulf	$3.625	Spot, Chicago	$3.515
Chicago	3.515	May futures	3.615
Cincinnati	3.345	July futures	3.7075
Decatur	3.415	Sept. futures	3.7625
Des Moines	3.135		
Omaha	3.200		
Iowa farm	2.930		

These individual prices were functionally related to each other and actually constituted the corn price level as it existed that day. As such, they should be viewed as one price. Differences existed with respect to place and time of delivery.

Basis is the name of the spreads between prices. It is defined as "the number of cents per bushel that, on a particular day, a local cash price of a commodity is above or below the price of some futures price or the cash price at another market location."[2] On the particular day cited, corn prices per bushel were: May futures $3.615; Chicago cash (spot) $3.515; farm price in central Iowa $2.93; and Gulf, $3.625. In basis terms, Chicago cash was 10 cents under May, Iowa farm price was 68.5 cents under May, and Gulf price was a half-cent over the nearby May futures. Additional basis definitions are given in Table 7-1.

In Chapter 2 we discussed two elements of marketing: space, reflecting locational differences; and time, reflecting costs involved in holding storable commodities from short periods of harvest into longer periods of use. Both concepts are important in studying basis.*

Table 7-1: Basis Definitions

Basis: The difference between a futures contract price and a local cash price.

```
    Futures Price      $3.615 May Corn CBOT
  - Local Cash Price   $3.135 Des Moines Cash
    (Spot or Flat Price)
         = Basis        .48 Under Nearby
```

Negative Basis: Futures contract price is higher than cash price. Quoted as *under*.

Positive Basis: Futures contract price is lower than cash price. Quoted as *over*.

Strong: Price difference is relatively small (5 cents, 10 cents).

Weak: Price difference is relatively large (70 cents, $1.00).

Option or Option Month: The specified futures contract from which the basis is calculated.

Nearby: The next option month during which delivery on a contract could be made. On Feb. 10, March is nearby corn contract month.

Distant: Some option month later in the year. On Feb. 10, July is a distant corn contract month.

Old Crop: Futures contracts on the crop that has most recently been harvested.

New Crop: First futures trading month of crop not yet harvested: Dec. corn, Nov. SB, Dec. cotton, July wheat.

Components of Basis

An explanation of the difference between a futures contract price and a local cash price of storable commodities requires understanding several different components. The first component is the amount that Chicago cash price is below the nearby futures contract. This difference arises from handling costs in loading and unloading, the handler's profit margin, and the costs of storage until grain can be delivered against a futures contract.

The second component is the amount by which cash price in the country is below the cash price in Chicago because of transportation cost. If transportation were free or if the cost of transportation were not related to distance, then the price of a commodity would be the same in all locations. Transportation is not

---

*Futures markets exist for both storable and non-storable commodities. Significant differences in basis concepts exist for the two. Price and basis relationships in non-storable commodities will be discussed in Chapter 9.

free, and it is related to distance. Thus, commodity prices on a particular day vary considerably from location to location as a result of transportation costs.

The third component of basis arises from the carrying charges incurred in holding commodities from harvest until they move into use at a future time. Storage is a market function that must be performed by someone who is willing to incur the costs of providing that service. The returns to storage are offered in the market by basis spreads between cash price and different future contract options. As time of delivery on a futures contract approaches, reduction in carrying costs causes futures and cash price to converge. The time relationship of prices is an important part of basis.

The effects of these various components are summarized in Figure 7-1 as they contribute to price differences which existed on any particular day. The value of corn was not one price but several, as different components were bid into the market to account for the place and time differences.

Figure 7-1: Components of Basis on a Particular Day in an April 1 Corn Market.

RELATIONSHIP OF CASH AND FUTURES PRICES

Two features characterize the relationship between cash and futures prices over time, as illustrated in Figure 7-2.

1. Movement up and down together, called "parallel fluctuation," and
2. Convergence of cash and future prices.

Each feature has its reason for occurring, and each one is important in understanding hedging as a business tool.

Figure 7-2: Illustration of Fluctuation and Convergence of Futures and Cash Commodity Prices.

Parallel Fluctuation of Cash and Futures Prices

Cash and futures prices tend to move in unison because they are affected by the same fundamental forces of supply and demand. These forces are usually first registered in the futures market. The trading pits offer convenient facilities for trading and are an arena for price discovery. Dealers in the actual commodities, such as local elevator managers, cotton merchants, livestock buyers, or exporters, look to the futures market for guidance with respect to the cash price they bid or ask for a commodity. Stronger futures prices encourage them to raise cash bids; weaker futures often lower cash bids. Even if a local grain elevator manager does not hedge wheat purchases in the futures market, but instead calculates the offering price to farmers on cash price in a terminal market, that local price is still linked to the futures market, since the terminal's bid is based on the futures price. Participants in a trade typically use a futures price as reference in negotiating the cash price. But the ultimate determinant of value is what buyers are willing to pay and sellers are willing to take in a cash transaction in the spot market. Reluctant sellers can withhold supply from the market and force futures prices higher. An example of cash leading futures occurred in the corn market in the spring of 1983. The Payment-In-Kind (PIK) price and production control program provided farmers with an alternative to selling corn at low prices. Their withholding of deliverable supplies from the market strengthened the cash market. In turn, higher cash prices led to a rise in the futures price.

Invariably, the question arises: Which price actually controls which? The answer is neither. The cash price influences the price of futures, and the futures price influences cash. In the PIK example above, cash led the futures, but most often, futures price leads the cash.

The foregoing does not mean that cash and futures price are the same in dollars and cents. It means only that the law of one price operates so that most price differences can be attributed to transportation costs between locations and carrying charges that reflect holding costs over time. If price difference between locations exceeds transportation costs, or bids among different months exceed carrying charges, a speculator looking for a profit will buy cheap in one market and sell dear in another. This arbitrage trading moves cash and futures prices into the "proper" relationship to each other, as determined by fundamental economic factors.

Once relative values are established, the tides of rising or falling price level affect both prices. It is the parallel fluctuation of futures and cash price that makes it possible to hedge for price protection.

Convergence of Cash and Futures Prices

Cash commodities can be delivered to satisfy the obligation inherent in a short futures contract. Consequently, during the period of delivery against a maturing futures contract, the cash price of a commodity and its futures price are forced to be equal, save minor transaction costs and some uncertainty, at the point of delivery. The basis spread between the two prices approaches zero, no matter how wide the difference between the two prices might have been at some other time. Market forces squeeze cash and futures prices together at this time and location.

Consider the situation of a terminal elevator manager in Chicago. The firm has representatives on the floor of the Chicago Board of Trade and a storage facility half filled (Figure 7-3). The elevator can receive grain into storage or make deliveries from it. To maximize profit, the manager wants to buy grain "below the market" or "sell it above the market." If the futures price is higher than cash, the manager will "sell" a contract to establish a delivery obligation at a price higher than the grain is worth in the warehouse. Upon maturity of the futures contract, the manager need only deliver a warehouse receipt for the grain and collect the higher price. On the other hand, if the futures price is less than cash, the manager will "buy" a futures contract and hold it in order to accept grain delivered against the obligation. The manager will have bought grain below the market price. Thus, the terminal operator's practice of arbitrage between the cash and futures market forces the two prices to be essentially equal in months when futures contracts mature. The opportunity to make or take delivery against futures contracts is the key element that forces prices to convergence.

The convergence relationship creates the opportunity to capture a storage profit from basis change, as we shall see shortly.

BASIS AND HEDGING

There are two reasons for hedging. One is for price protection; the other is to capture a storage return. Each use depends respectively on parallel fluctuation or on convergence of cash and futures prices. The first can be illustrated with a preharvest hedge, the second, by the storage hedge.

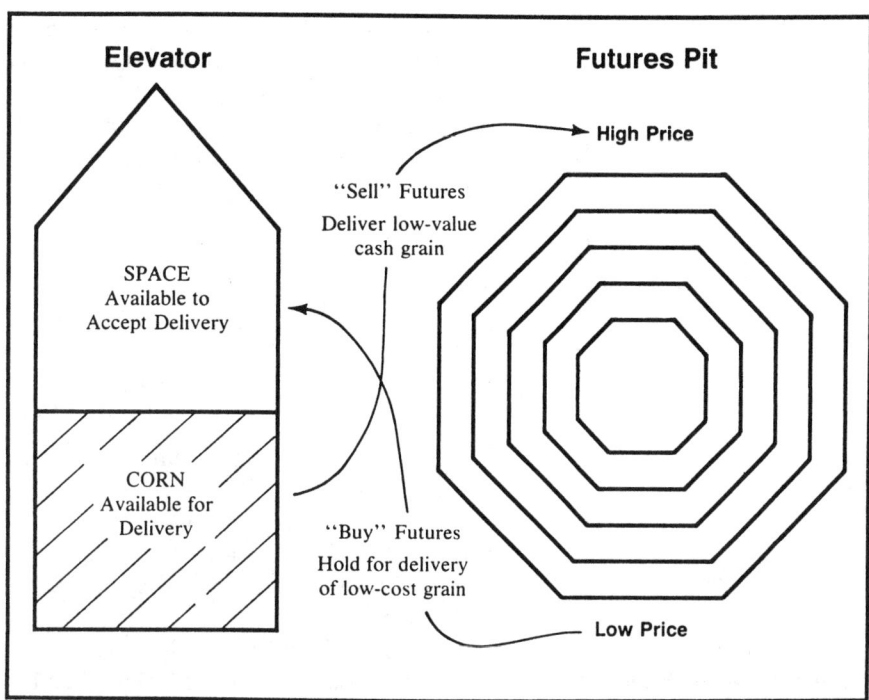

Figure 7-3: Illustration of Arbitrage Between Cash and Futures Markets That Causes Price Convergence.

The Pre-harvest Hedge: Protecting Price Level

A farmer who is concerned that price level might decline during the growing season can protect against that risk by selling a new crop futures contract, that is, by hedging. Suppose on July 1, December corn futures are trading at $3.70 per bushel. If this price level is satisfactory, a producer could hedge away effects of price change. The approximate price of $3.70 is protected by the equal, but opposite, positions in two markets — a long position in the cash, offset by a short position in the futures. The producer will not be in a position to realize the full $3.70 price, since the farm is not located in Chicago or Toledo and he or she would have shipping and handling fees in making delivery to satisfy the futures contract. The hedge will protect a farm gate price equal to $3.70 less delivery costs and handling fees incurred in delivering against the futures contract.

But few producers deliver against futures in Chicago or Toledo. Most sell cash grain on their local market and offset their futures obligation by buying back that contract. They enjoy price protection without the cost and hassle of delivery.

Knowledge of local basis becomes essential in order to translate a given futures price into a probable cash price for local delivery. The producer "localizes" the futures price to estimate the cash price he or she will receive by placing the hedge. The futures price, plus or minus basis in the local market in the month of delivery, is the price that the futures market is offering for grain delivered to the local elevator during a particular future month. A historical basis record might show that cash corn price averaged 10 cents under December futures during mid-December for the producer's location. By subtracting the 10-cent average basis from the futures contract price, the anticipated "localized" price is $3.60.

Actual basis at time of delivery may be greater or less than the average 10 cents and affect the final results of the hedge. If the basis is 20 cents at delivery, the local price will be $3.50. If the basis is zero, the price is $3.70. Thus, a futures hedge removes the effects of price level change but does not remove basis level risk. Since year-to-year basis fluctuation is usually less than price level fluctuation, hedging fixes an approximate price that can be fairly accurately projected.

The Storage Hedge: Earning a Storage Profit

Hedging can be used to earn a storage return. It is done to take advantage of expected convergence of cash and futures prices. In this use, a hedge is a profit-making tool, as contrasted to a risk management tool used to protect a price level.

It is common business practice for a warehouser to use a storage hedge to hold commodities over a period of time when basis is expected to narrow. A storage hedge has two purposes: (1) to protect the value of inventory against adverse price level change, and (2) to capture a storage return during the storage period. The alternative is to store commodities unhedged and speculate on price. Few can afford that luxury. There is no guarantee that price level will rise over the storage period to pay the costs of storage. It may fall dramatically. The storage hedge protects the elevator operator from a price level change. It also serves to lock in a storage return. This tool can be used effectively by farmers who own or rent storage space.

A storage hedge is usually established at harvest. Relatively low-priced cash grain is deposited in storage facilities, and an equal amount of higher-priced futures contracts are sold. The futures position counteracts the effect of price level change while allowing basis change to occur. As long as the difference between cash and futures prices shrinks, or moves to a premium of cash over futures, there is a basis gain. Figure 7-4 shows five different combinations of price changes that occur to produce the expected convergence of cash and futures price. In each case, the narrowing of basis earns a storage return, irrespective of price level change.

For example, at harvest in October the cash soybean price is $6.40, and the March futures is trading at $7.40. From a basis history, a southwest Ohio farmer determines that the Cincinnati basis in January averages 35 cents under the March future. By localizing the March contract, the farmer calculates that the futures market is offering about $7.05 for January delivery. Compared to the harvest bid of $6.40, the farmer expects to capture a 65-cent gross storage return by a storage hedge. The farmer stores and hedges. In January, basis is 35 cents under. The farmer sells cash soybeans and lifts the hedge. The results can be traced as follows:

Date	Cash Market	Futures Market	Basis
Oct. 15	Store SB worth $6.40 for later sale	Sell March SB contracts @ $7.40	$1.00
Jan. 15	Sell Cash SB @ $6.65	Buy March SB contracts @ $7.00	0.35
Gain (loss)	+ 0.25	+ 0.40	+ 0.65

Results: Jan. 15 Cash Price	$6.65
Plus Futures Gain	0.40
Net Price	$7.05
Fall Value of SB	$6.40
Storage Return	$0.65

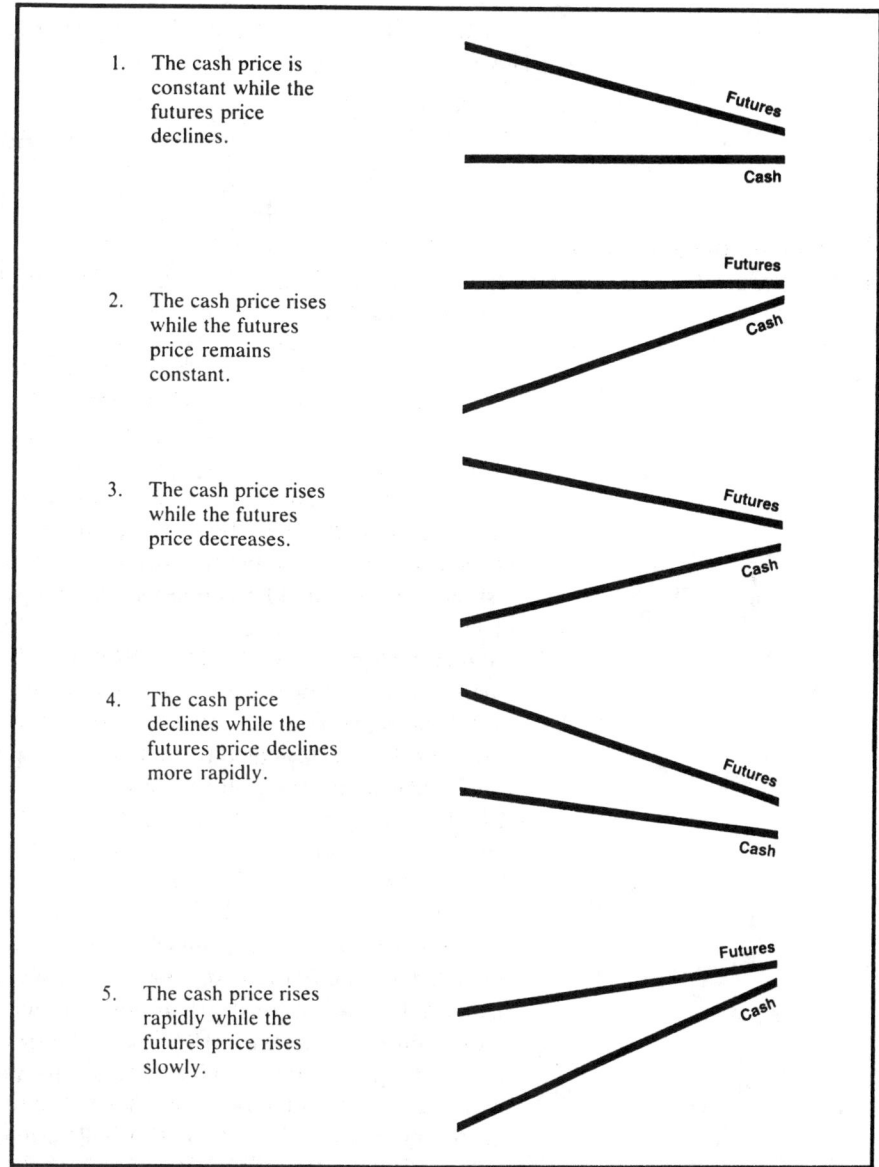

Figure 7-4: **Cash-Futures Relationships Causing Basis to Narrow.**

In the example, futures price level declined and cash price increased. Without the hedge, the farmer would have received $6.65 for the grain, or only a 25-cent storage return. The hedge protected against price decline in the futures market by capturing 40 cents of price protection. These numbers illustrate situation #3 in Figure 7-4. Each of the other four situations will produce the same results if the prices produce a 65-cent convergence in basis.* Situation #5, with a rising fu-

---

*The 65 cents is a storage return — not storage profit. The cost of facilities, risk, management, and money tied up in the grain must be subtracted from the 65 cents to determine net profit from storing.

tures price level, is the only one to have negative results. An example of it is as follows:

Date	Cash Market	Futures Market	Basis
Oct. 1	Store SB worth $6.40 for later sale	Sell March contracts @ $7.40	$1.00
Jan. 1	Sell Cash SB @ $7.20	Buy March SB contracts @ $7.55	0.35
Gain (loss)	+ 0.80	(− 0.15)	+ 0.65

Results: Jan. 15 Cash Price $7.20
Less Futures Loss   − 0.15
Net Price $7.05
Fall Value of Corn $6.40
Storage Return $0.65

The storage return of 65 cents is the same as the other examples because of narrowing basis. However, the cash price is reduced by 15 cents due to the futures loss; the farmer would have netted $7.20 instead of $7.05 had he or she not placed the storage hedge. This example emphasizes the fact that a hedge eliminates price level improvement as well as providing protection against falling price level, while capturing the storage return. If a warehouser were certain that futures price level would rise, he or she should delay setting a storage hedge and speculate on both price level and basis. Only when price level started to fall would the warehouser set the hedge. In actual practice, storage managers cannot afford the risk of not hedging. They cannot predict when price will fall nor how fast. Three limit down moves in price on 1 million bushels of stored grain produce a financial disaster of $900,000.

Profitable storage hedging is based upon the expectation of cash and futures price convergence. A problem of storage hedging is that basis levels cannot be predicted exactly. The farmer or warehouser will know basis level on the day of storage. However, he or she does not know what the basis level will be on a later date. While historical averages provide a guide as to what to expect, basis fluctuation is a totally different market situation from price level fluctuation. A storage hedge instituted in October with the expectation of a "normal" basis of 35 cents under March during January may actually produce a larger or smaller return than anticipated. If basis is weaker than 35 cents when the cash soybeans are sold and the futures contract lifted, the storage return will be less. On the other hand, a stronger basis than 35 cents will yield a larger storage return. The storage hedge removes price level speculation but does not eliminate basis speculation. Variations in basis notwithstanding, basis is normally much more predictable than the general price level because of required convergence. This fact enables hedgers to substitute basis risk for price level risk through use of futures contracts. Skill must be developed in forecasting basis change to be successful with storage hedging, however.

CASH AND FUTURES PRICES OVER TIME

Cash and futures prices are related over time by carrying charges. This relationship depends upon four conditions:

1. Storable commodities are harvested at one time of year and used at fairly

even rates throughout the year. Hence, inventories must be carried forward.
2. Costs are incurred in the process of storing, financing, and maintaining quality of the commodity.
3. There is virtually no cost in holding futures contracts.
4. Cash and futures prices will be essentially equal in the month of maturity at delivery points.

A soybean processor in Decatur, Illinois, has a constant supply requirement each month to keep the plant operating at capacity. If, at harvest, the processor decides to cover the processing needs for next July, he or she must have storage space to hold the inventory. That space has fixed investment, depreciation, and maintenance costs. Additionally, the processor must operate the space and invest capital equal to full value of the grain. On the other hand, July futures contracts provide the right to receive soybeans needed to keep the plant operating in July. The only cost of holding futures contracts is the interest on the margin money deposited and brokerage fees for the transaction. Whether harvest price level is high or low, the cost of holding soybeans in storage is about 2 cents per week, or about 80 cents per bushel for 9 months. If the difference between the October cash and the July futures is more than 80 cents, the processor can hold long futures contracts at lower cost and wait for delivery of soybeans. If the basis spread equaled carrying charge, the soybean processor would be indifferent as to whether the firm owned the physicals or futures. The effective cost of soybeans as a production input for July would be the same.

Figure 7-5 shows the widening price spread, or divergence, between futures and cash as we move back in time from July toward harvest. Theoretically, cash price at harvest differs from the end of a crop year futures price by the exact cost

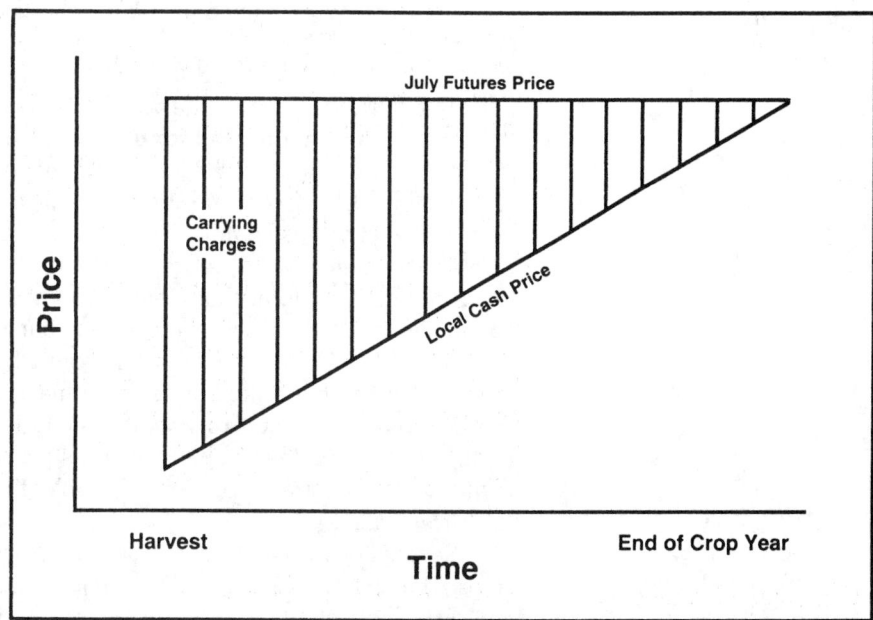

Figure 7-5: Theoretical Convergence of Cash and Futures Prices Over Time from Harvest to End of Crop Year.

of holding the cash commodity over time. Furthermore, prices should diverge an equal amount each day. If actual cash price change followed its theoretical path, carrying charges would exactly equal the amount of convergence.

The theoretical relationship of carrying charge between cash price and futures options forms the benchmark for basis analysis.

Basis Level Negotiation in a Market Economy

In the real world, basis spreads never exactly fit the expected theoretical relationship. The deviations from the expected offer profit opportunities and require analysis of the basis market. Understanding what causes deviations prepares commodity industry participants to take profitable advantage of them.

Rather than conforming to what should exist as carrying charges, actual basis level is a negotiated price. The negotiation process is predicated on the fact that somebody has to pay the costs of holding grain from harvest until it is used. The decision to convert grain into cash (by the producer) or cash into grain (by the user) depends upon the calculation of costs involved and returns expected. If the spread between cash price and a futures contract is greater than holding costs (excess carry), there is an incentive to buy cash grain from a motivated seller (farmer with no on-farm storage), hedge away the effects of price level change, and hold the commodity until it can be sold or is needed for processing. On the other hand, if the spread will not cover holding costs (less than full carry), a commodity processor can protect against the possibility of price level change and at the same time guarantee delivery by buying a futures contract and holding it for delivery. In this case the processor is guaranteed receipt of the grain when needed but does not incur the cost of holding grain.

The relative offerings of cash commodities versus futures contracts determine basis. Heavy cash sales relative to futures widen basis. Light cash sales relative to futures narrow basis.

The carrying charge component of basis becomes the **market price for storage space**. When the available supply of a commodity is large relative to space available, and grain is flowing to the market rapidly, the cash price is weak relative to the futures so that the basis spread is wide. The market is signaling that the commodity is not needed for use at the current time and should be held in storage by the producer until later. The demand for storage is greatest at harvest, and therefore the price for storage is typically the highest, or basis weakest at that time. Once harvest is complete and commodities are in storage, basis narrows quickly. Thereafter, basis level depends upon reluctance or eagerness of sellers to move grain out of storage and into use. When offers to sell are limited and the commodity is flowing slowly to the market, but demand is strong, basis is strong.

The demand for a flow of cash grain into the market, relative to reluctance of farmers and warehouser to sell, sets the tone of basis negotiation. If rivers are frozen, and roads and railways clogged, a New Orleans exporter will offer a premium cash price to avoid demurrage fees on waiting ships. That premium is expressed in the market as a strong basis level. It ripples through the market to entice sellers to deliver grain.

Negotiations to determine basis take place in two different price relationships. The first concerns the relative values of cash grain and the nearby futures contract. The second concerns price spreads between futures contract months. Both are related to what should theoretically exist, but both are modified by current and anticipated conditions.

Cash Versus Nearby Futures

On any given day, the cash price in Chicago is determined by trading between large buyers and sellers of grain. Terminal elevator operators who own grain, and exporters, processors, and merchandisers who need grain, are constantly bidding in cents per bushel under or over the nearby futures prices for grain for immediate delivery. If demand for cash grain is strong and/or the current available supply is small, buyers become eager bidders, and elevator operators holding stocks become reluctant sellers. Without any change in worldwide supply and demand conditions to affect the futures market, local demand for cash trading produces strength in the cash market. The cash market rises relative to futures price and strengthens the basis. Conversely, the demand for cash grain for immediate delivery may weaken because of a decline in processing or export requirements. In this case, cash price in Chicago may fall relative to the futures price, producing a weaker basis.

Farmer Psychology and Basis

An important determinant of basis results from farmer reactions to price level. Farmers are willing sellers at high prices and reluctant sellers at low prices. At what they judge to be low prices, they are willing to defer selling decisions because they feel downside price moves are less likely than improving prices. They are willing to continue to speculate on cash price. At high price levels, they sell more readily because they gauge that downside risk exceeds the probability of higher prices.

Since farmers are typically attuned to local cash prices more than to futures prices, their market actions significantly affect basis levels. When futures prices are high, spot cash prices are likewise strong. Farmers are willing cash sellers; there is no need for buyers to bid a premium to entice farmer selling. On the other hand, when futures prices are low, cash prices are low and farmers are reluctant sellers. Buyers are required to bid a basis premium to entice sellers. Basis is, therefore, wide at high price levels and narrow at low price levels. Farmer reaction to price level and the resulting basis levels are summarized in Figure 7-6.

Elevator operators are well aware that they earn larger profits in high-priced markets than they do in low markets. The opportunity to buy cash grain at wide basis discounts and subsequently lay off hedges to capture potential basis gain is improved when farmers are willing sellers. At low prices, narrow basis does not offer significant storage returns.

Contract-to-Contract Futures Spreads

The second basis negotiation determines the spread between futures prices in different contract months. Prices on various contract options are related to each other by carrying costs over time, theoretically calculated to cover holding costs. In actuality, they are negotiated between buyers and sellers.

Futures contracts trade for delivery in selected months rather than daily. Hence, the convergence is reflected in contract option prices that normally stairstep upward. For example, if carrying costs to hold soybeans were calculated to be 10 cents per bushel per month, various futures option prices should be related to harvest price and to each other, as shown in Figure 7-7. An October 1 harvest-time price of $7.00 per bushel would be exactly equivalent to the following August 1 futures price of $8.00. The $1.00 basis spread between the October cash and August futures would equal "full carry," the $1.00 cost of holding soybeans for nine months. All contract options would be related to the October cash price and to each other by the cost of carrying grain from option month to option month.

In practice, the spreads between futures prices for different delivery months do not always equal full cost of carrying inventory. Price differences between contract months can be more than storage costs, producing a situation called "excess

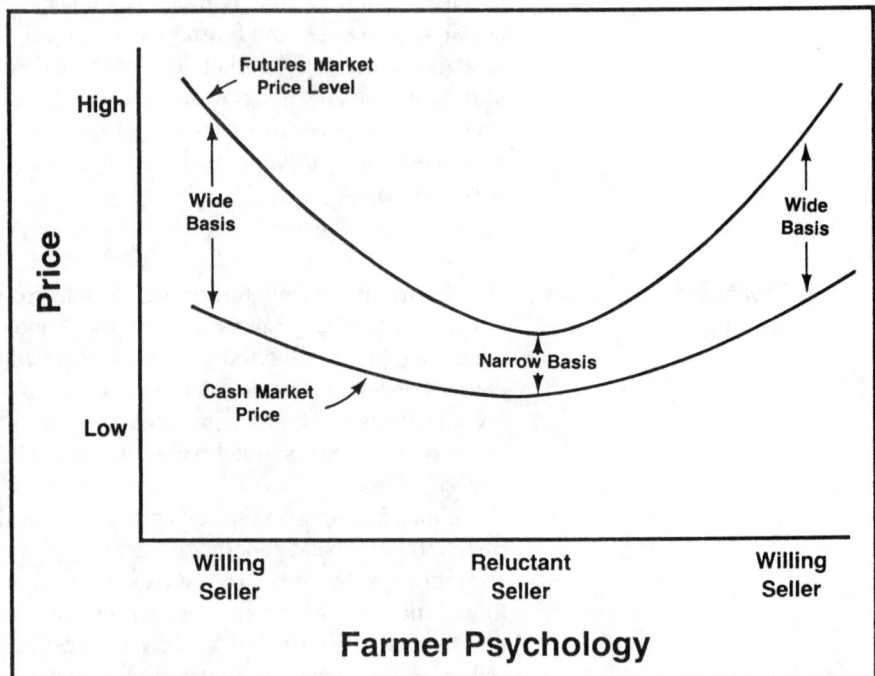

Figure 7-6: Illustration of Farmer-Seller Reaction to Price Level and Resulting Basis Levels.

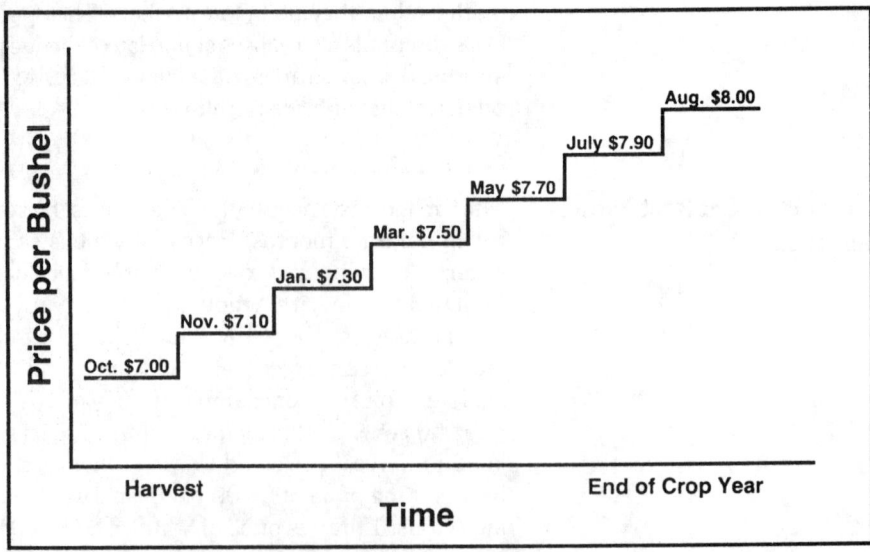

Figure 7-7: Stairsteps in Soybean Futures Contract Prices to Reflect Cost of Full Carry, Contract Month to Contract Month.

carry." The spread in prices is greater than the cost of carrying grain into the future. This situation seldom exists for very long because commodity traders are quick to capitalize on market differences that exceed costs of storage. On the other hand, month-to-month differences between successive futures contracts can be less than full carrying costs. Price differences between contract months are less than the cost of storing grain month to month.

The bidding process that establishes whether futures contract prices stairstep upward to reflect a full carry, less than carry, or excess carry takes place between warehousers (who may be farmer-producers, elevators, or users) and speculators. A warehouser wants to hold the commodity in storage only when he or she can sell a futures contract at a price spread sufficiently wide to generate a storage profit. Otherwise, the warehouser would sell the commodity rather than store it. The opposite side of a hedge is typically taken by speculators who are forced to pay a premium for futures contracts relative to cash price. Without a premium, no hedgers would offer to sell futures; all would be selling cash grain. The shortage of contract offerings causes the price of futures contracts to rise relative to cash price. Thus, the stepwise price rise is established.

In holding actuals and selling futures, the warehouser benefits when the spread between cash and futures in various months is as wide as possible. The warehouser is locking in a differential between what the cash commodity costs and what he or she can guarantee by hedging the price level for future delivery. The warehouser chooses the contract month which offers the widest spread relative to holding costs. Heavy offerings of contracts in one month depress price relative to other months, emphasizing the contract-to-contract steps in price.

Speculators who practice spread trading add emphasis to contract price spreads. On one hand, spread trading narrows contract-to-contract basis as traders bid excess returns out of the market. On the other, spreading widens basis. Speculators can remove the effects of price level change from their trading plan and limit their speculation to basis level change through spread trading. If the price spread between nearby and distant futures is small, a spread trader would establish offsetting futures positions in the market by selling the nearby and buying the distant months. He or she knows that if warehousers are motivated to sell cash commodities early in the season, a shortage of supplies later will lead to higher price. Cash markets will pull the distant futures price upward. Less than full carry between contract months produces a market bias that should produce a profit from the spread position, as calculated in Table 7-2. The trader profits from widening the basis as market forces react to excess supply in the present relative to shortages later. Spread trading adds to the stairstep price effect by increasing selling pressure on nearby contracts and buying pressure on distant ones.

Table 7-2: Illustration of Profitable Spread Trading of Futures Contracts

Set Spread		Lift Spread		Results	
January 20		February 20			
Sell March	376	Buy March	396	Loss	20 cts
Buy May	380	Sell May	416	Profit	36 cts
Basis	-4	Basis	-20	Net	16 cts

There are economic reasons for the nearby futures option to be higher than cash price, and deferred options higher one to the other over time. But each step is a price negotiated in the market place rather than a fixed amount. The cash-to-nearby option and the contract-to-contract spreads in futures develop as a bidding contest between hedgers who wish to widen the spread and speculators who wish to narrow it. The bidding process establishes a "normal" relationship between different futures contract months. Warehousers want to earn returns based on costs calculated from new cost of facilities, full depreciation, and prime interest rates on capital invested in inventory. However, if they have older facilities with lower costs, or if they are willing to accept lower rates of return, they will calculate costs significantly below those of a newly built facility. They then offer contracts at less than full carry. They will not offer zero, as the speculators want, because their alternative is to sell cash commodities and not offer contracts.

In a competitive market environment, "normal" carry is bid into the market place by the offerings of cash or contracts by hedgers and speculators. The market may offer full carry through wide basis, or less than full carry due to narrow basis, depending upon who prevails.

Local Basis Year to Year

Country basis changes do not simply mimic Chicago basis changes. Relatively low country prices, resulting from a widening of basis, may be triggered by higher transportation charges, by a weakening of local demand for grain, by a local supply larger than the storage space available, or by a shortage of railroad cars. Conversely, a short local supply, strong local demand for grain for use or inventory accumulation, and a favorable change in freight rates could result in narrowing basis and raising local prices relative to the futures price.

Just as basis can vary week to week or month to month in any year, it can vary from one year to the next. The reasons may indicate a permanent shift in the relationship or a temporary aberration that will be quickly corrected. A permanent change may result from a change in local demand, as livestock feeding increases or decreases in the area. Likewise, opening or closing of a grain-processing plant in the community, abandoning or restricting use of railroads, and changing freight rates will affect local basis positively or adversely.

Local basis is most affected by size of the crop and available storage space. In years of large crops and storage shortages, basis tends to be weaker, especially during harvest months. The amount of farmer-owned storage relative to size of crop has a significant effect upon harvest-time basis. Long lines of trucks waiting to be dumped at elevators encourage wide basis. Short lines cause more aggressive bidding and strong basis as elevators try to fill their space.

BASIS RECORDS AND CHARTS

Since basis varies from year to year and place to place within a year, commodity producers and users need to know their individual local historical basis patterns. Basis patterns provide the information for timing cash sales or purchases and for deciding whether or not to store a crop, what contract month to use for hedging, and when to turn an unusual basis situation into a profitable opportunity. In addition, knowledge of basis can identify the most profitable option for hedging price level.

Each producer faces a different basis situation depending upon the cash market alternatives available. It is necessary for individuals to maintain their own basis records; a national average is not satisfactory. A producer should at least maintain a weekly record of cash and futures prices. Select a day of the week to record local cash bids and the futures prices. Friday's prices give the producer a longer time span to reflect on the information until the market opens on Monday. But all too often, Friday's trading is distorted by speculators avoiding positions over the weekend or elevators taking protection against possible price changes before trading resumes on Monday. A midweek day (such as Wednesday) is preferable. The extra work of maintaining a daily record during historically strong selling months can better identify the most profitable day to sell.

One handy method is to clip the market report from a newspaper such as *The Wall Street Journal* and paste it in a notebook (Figure 7-8). Record relevant local prices on the same page. This method offers the advantage of recording all contract months. Another way is to create a table with vertical columns to show date, nearby futures, local cash price, and basis (Figure 7-9). A sheet must be maintained for each commodity sold or bought. If the producer has two or three alternative cash markets, it is best to record all cash bids. Also, one may need to record several contract months. The clipping is a handier method than the table for all these details.

Plotting Basis Charts

The daily or weekly prices can be converted into a basis chart useful in visualizing basis patterns. To make a basis chart, set some base price equal to zero and plot basis deviations around it. Several types of basis charts can be drawn. They differ only in which price is chosen as the base. The illustrations in Figure 7-10 show the appearance of each of the methods. One method is to set the local cash price equal to zero and plot futures prices relative to it. This method has the advantage of showing several different futures delivery months on the same chart. It focuses attention on the value of the cash commodity relative to different futures contracts. It is continuous through time and does not come into existence, mature, and expire like a futures contract. Its use might satisfy a hedger who is seeking to determine which futures contract month offers the greatest basis spread relative to cash.

An alternative method of basis charting is to set the last maturing contract of a crop year equal to zero and plot cash prices relative to it. This method shows how cash price converges on futures as expected. It emphasizes deviations from the expected theoretical relationship and flags profit opportunities. The disadvantage of this type of chart is that it focuses on the distant futures contract, while trading is usually conducted on the nearby. Furthermore, it typically spans only 9 or 10 months from harvest to the end of old crop year, not a longer marketing year. This method is the simplest and most straightforward method for farmer-storers.

The third method is to set the nearby futures contract equal to zero and plot daily cash quotes and the other futures options relative to it. This method illustrates the cash gaining on futures, and it focuses on the nearby futures contract commonly used in actual trading. As the first day of the delivery month arrives, the base is switched to the next maturing contract. This method spans an entire crop marketing year and is continuous. The other futures contract months, plotted as deviations above the nearby option, help identify which contract offers the greatest potential basis gain. This last type is most applicable to decision making by commercial warehousers who daily trade basis, since it incorporates the effects of both the first two types.

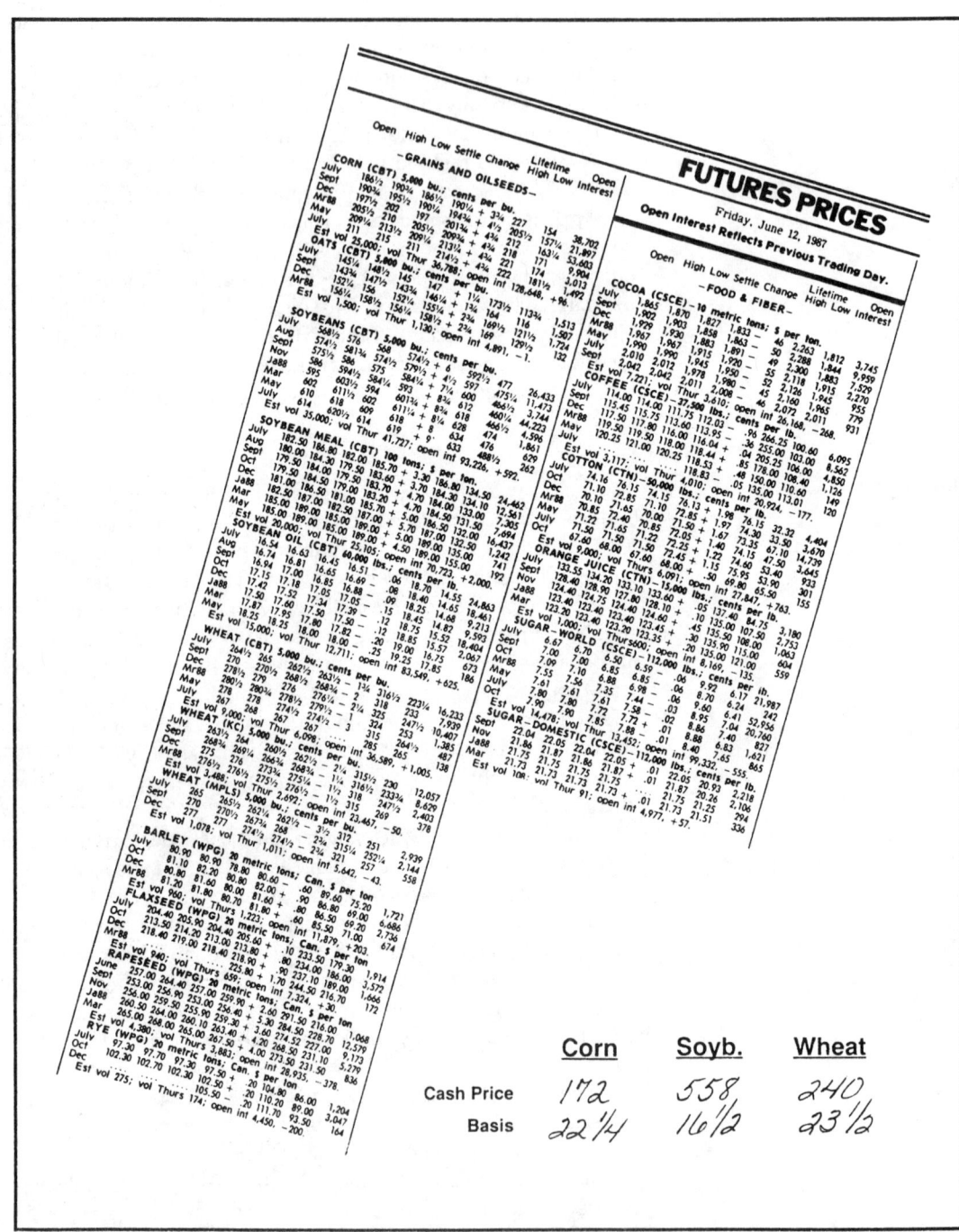

Figure 7-8: Illustration of Method to Record Cash and Futures Prices for Calculation of Basis.

Basis: The Economics of When and Where to Sell

Date 1984	Local Cash Price	July Futures	Basis	Date 1985	Local Cash Price	July Futures	Basis	Date 1986	Local Cash Price	July Futures	Basis
Oct. 1	$2.98	$3.66	-68¢	Oct. 7	$2.48	$3.29	-81¢	Oct. 5	$1.96	$2.53	-57¢
Oct. 8	2.98	3.65	-67	Oct. 14	2.47	3.36	-89	Oct. 13	1.90	2.58	-68
Oct. 15	2.85	3.74	-89	Oct. 21	2.38	3.27	-89	Oct. 20	1.88	2.55	-67
Oct. 22	2.97	3.85	-88	Oct. 28	2.36	3.27	-91	Oct. 27	1.89	2.45	-56
Oct. 29	3.18	3.87	-69	Nov. 4	2.39	3.25	-86	Nov. 4	2.03	2.46	-43
Nov. 5	3.16	3.93	-77	Nov. 11	2.33	3.16	-83	Nov. 10	2.15	2.54	-39
Nov. 12	3.18	3.95	-77	Nov. 18	2.35	3.10	-75	Nov. 17	2.19	2.58	-39

Figure 7-9: Illustration of Method to Maintain a Basis Record.

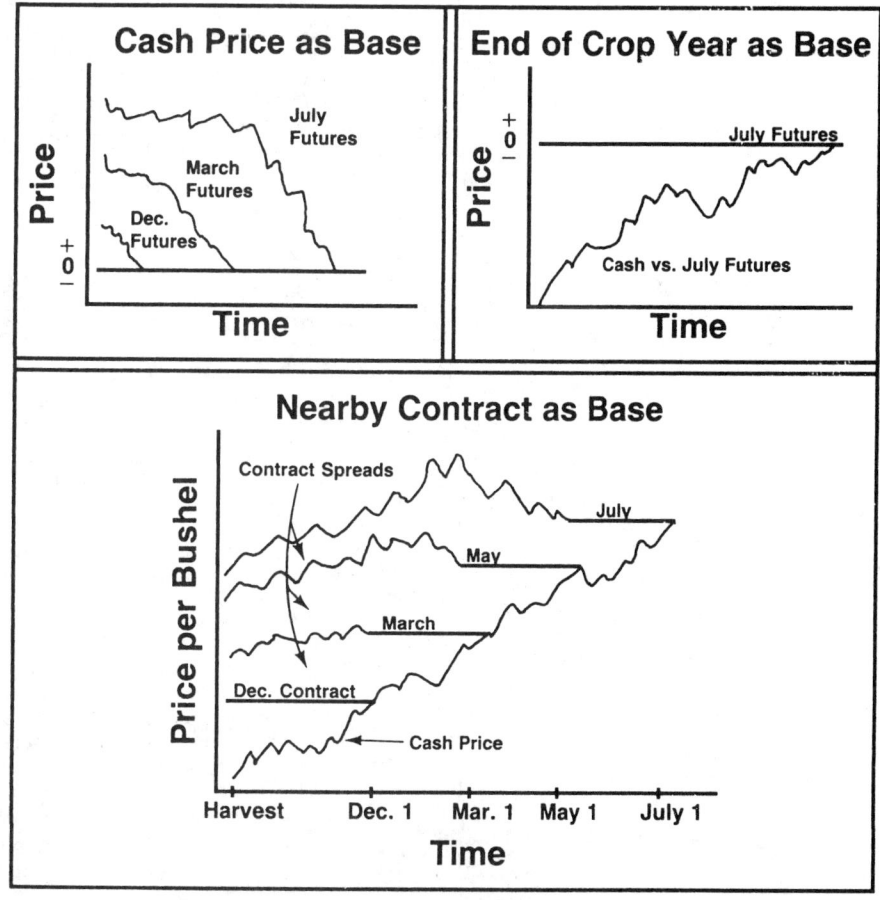

Figure 7-10: Basis Charts Using Different Prices as the Base.

Historical Charts and What They Show

Historical basis charts can be developed from information available from local elevators. Five years of information should provide enough time span to reflect the local pattern. To develop the history, the producer should secure cash and futures prices for the period and calculate the difference between the two prices for one day of each week during each year, as shown in Table 7-3. The next step is to average the weekly basis figures into a long-term average figure. Finally, a plot of the annual and average for the period of years will produce a long-term historical picture, showing when basis has been wide or narrow. The average is of little value in timing cash sales in the current year, because no two years have the same basis

Table 7-3: Historical Corn Basis (Wednesday Prices), Cincinnati Cash Versus July Futures

Week	1983/84	1984/85	1985/86	1986/87	4-Year Average
Oct. 1	.04	-.11	-.21	-.36	-.16
2	.10	-.08	-.25	-.11	-.16
3	.00	-.12	-.30	-.39	-.20
4	.07	-.15	-.25	-.45	-.20
5	.13	-.18	-.19	-.30	-.14
Nov. 1	.05	-.24	-.11	-.27	-.14
2	.08	-.17	-.03	-.18	-.08
3	.02	-.17	.08	-.14	-.05
4	.06	-.21	.14	-.11	-.05
Dec. 1	.11	-.17	.09	-.11	-.02
2	.04	-.15	.05	-.12	-.05
3	.15	-.08	.07	-.15	.00
4	.14	-.03	.05	-.16	.00
Jan. 1	.08	-.01	.05	-.18	-.02
2	.13	-.02	.02	-.14	.00
3	.13	.05	.04	-.09	.03
4	.14	.07	.11	-.05	.07
Feb. 1	.09	.07	.09	-.09	.04
2	.07	.02	.13	-.07	.04
3	.03	.04	.11	-.09	.02
4	.08	.03	.10	-.06	.04
Mar. 1	.12	.11	.08	-.02	.07
2	.13	.13	.08	-.05	.07
3	.14	.08	.12	-.03	.08
4	.14	.09	.15	-.04	.09
Apr. 1	.16	.05	.17	-.05	.08
2	.11	.05	.23	-.01	.10
3	.13	.08	.23	-.01	.11
4	.12	.09	.24	-.01	.11
May 1	.11	.12	.19	.02	.11
2	.14	.09	.19	.07	.12
3	.17	.07	.20	.00	.11
4	.15	.06	.20	.02	.11
June 1	.16	.05	.20	.00	.10
2	.20	.04	.32	.01	.14
3	.22	.02	.18		.14
4	.20	-.01	.34		.17
Jul. 1	.13	.00	.04		.06
2	.07	.01	.05		.04
3	.00	.00	.00		.00

making factors. But it presents a generalized pattern of usually profitable selling months.

The average basis for three years of data for soybeans in the Cincinnati and Wilmington, Ohio, markets is presented in Figure 7-11. Several interesting insights can be gleaned from the picture. The average illustrates the narrowing that is expected to occur, but it is not the straight line that theory suggests. In interpreting the picture with likely events, late December showed a narrow basis because farmer-sellers may have been in Florida or Acapulco, or they were delaying income into the next tax year, but the trade still needed a supply of cash grain. In February basis widened when farmers sold crops to pay taxes. In April and May tillage and planting took precedence over shipping grain, causing buyers to bid a premium (stronger basis) to keep grain flowing. In June as fertilizer and seed bills came due, farmers were motivated sellers; the grain trade "rewarded" this motivation by weakening basis.

Figure 7-11: Soybean Average Basis for Three Years, Monthly Cash Price Versus August Futures, Cincinnati and Wilmington, Ohio.

The Cincinnati basis reflects a large market area, sensitive to many market forces. The Wilmington market essentially feeds into Cincinnati with few other alternatives and consequently represents transportation costs and handling charges. Thus, the Cincinnati-Wilmington basis is fairly constant.

In this market, the historical average identifies December-January and May as profitable basis selling months. February and June are months to avoid. Each location will have its own characteristic pattern. A historical basis chart will reveal interesting conditions that exist at each location.

A historical average of basis provides a standard for evaluating basis sales opportunities. On a blizzard day in January, 1979, Cincinnati basis was 10 cents over the March corn contract for four hours. Compared to the historical average, a cash bid 10 cents over March futures was an exceptionally strong basis. The Great Lakes and Mississippi River above St. Louis were frozen, foul weather had slowed rail traffic and stopped trucks, but exporters in New Orleans needed grain to load waiting ships. They reflected those needs with a very strong bid until their needs were met by sellers around Cincinnati. The point is that commodity producers should know the typical basis patterns for their market and be ready to respond when the market signals unusually profitable opportunities. During periods of "abnormal" basis, a daily record flags the situation to help capture extra profit. Better still, a close working relationship with buyers, so they will call when an exceptional situation occurs, can mean extra dollars in the bank for a farmer–grain merchandiser. A historical basis chart is the first tool to identify those opportunities.

BASIS SIGNALS FOR CASH SALES AND PURCHASES

A strong or positive basis is an indication that, for some reason(s), buyers are willing to pay a higher than usual price (relative to futures prices) for a commodity for immediate delivery. These conditions may be confined to a limited area or prevail nationwide. They may last for a few hours or several days. An abnormally strong basis is a market signal for producers to deliver their production and make cash sales. Correspondingly, a weaker than normal basis signals producers to hold grain. It is saying that there is greater than normal discount relative to futures prices.

Basis signals dictate when a producer should make cash sales. But seldom do both a high price level and advantageous basis level exist simultaneously. In order to capture both, the producer must be prepared to employ the marketing tools described in Chapter 2 and reviewed at this point. High price level and weak basis are the signal to hedge. This permits a producer to lock in the high price with futures contracts but continue to speculate on basis. Because of convergence, there is a greater chance of significant basis improvement rather than basis deterioration. On the other hand, narrow basis occurs when farmers are reluctant to sell their production. They judge price level to be too low; they expect it to go higher, so they delay sales. The grain trade bids a basis premium to maintain flow of the commodity. Low price and strong basis call for executing a basis contract or a cash sale and "Buying the Board" to replace the cash speculative position. Both methods result in immediate delivery of the commodity in response to basis signals, but they delay fixing the price level until later, when the producer expects price level to be higher. They both capture a strong basis price but allow continued speculation on price level.

A combination of high price level and strong basis dictates a cash sale if the commodity is on hand, or forward contracting if it is still in the production stage. The cash sale results in immediate delivery of the commodity. Forward contracts for delivery at a later date anticipate a lack of supply availability at the later time and indicate that users are lining up supplies. However, it is seldom that elevator managers will bid away their basis profit months ahead of expected grain delivery.

Knowledge of normal basis patterns allows a producer to evaluate whether a forward contract offer is generous or not. Prior to harvest, a local elevator may offer $3.20 for corn delivered in March. Noting that March futures price is $3.25, (only −5 cent basis) and knowing that the usual March basis is 15 cents under, the producer can gain 10 cents above normal basis. If the price level is satisfactory, a producer can agree to the contract, knowing the decision has captured a narrow basis and a good price.

The general rule for farmer-sellers is to move their production with basis contract or cash sales when nobody else is selling and basis is strong. They should hold their grain when others are selling and basis is weak. Commodity buyers should take the opposite stance. They prefer an exceptionally weak basis. They should buy physical inputs when basis is weak. With basis discounts larger than normal, they can earn storage returns for their feeding operations. Those who do not have storage facilities can forward contract for future delivery at a fixed price. Cash contracts allow buyers to fix the price and basis but delay receipt of the grain. The cattle feeder or pork or poultry producer can reduce feed costs by locking in large discounts offered by wide basis. Correctly reading the basis market will signal when the situation exists.

Contract Month to Place Storage or Production Hedge

Commodities can be priced for future delivery in any one of several alternative option months. Price differences between various options are bids to grain owners for storage payments. The differences between option prices reflect the extent of carrying return being bid in the market. As such, they can help the commodity producer decide which month is best to use for hedging his or her production. The fact that futures price is higher for a July contract than in March does not necessarily mean July is the most profitable month in which to hedge. The calculation should be based upon the carry return offered by futures prices as compared to holding costs.

Whether the market is offering a return equal to costs of carry, excess of carry, or less than carry can be evaluated from relative futures prices. As a practical matter, futures prices for different delivery months seldom offer excess carry opportunities. Commodity traders capitalize on spreads that offer returns in excess of carrying costs and bid them out of the market. At the other extreme, futures prices can reflect a negative carry. A particularly strong demand for cash grain for immediate delivery can push the price of nearby contracts above distant months. It produces an inverted market, as in Figure 7-12. The market is penaliz-

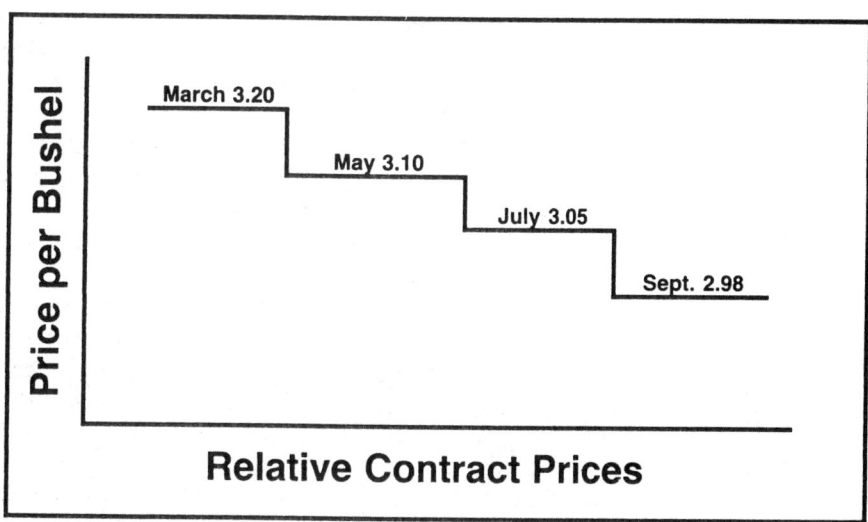

Figure 7-12: Inverted Futures Market. Prices Showing Negative Carry.

ing owners of grain for storing until later; it is signaling that the market wants the grain now rather than later. Immediate strong demand or limited current supplies produce inverted markets. The degree of inversion into future months depends on how desperately buyers want grain and how reluctantly owners are willing to part with it. An inverted market is most likely to occur in a year of short supply, that is, a small crop in relation to overall demand. Some users are willing to pay a higher price for cash grain they do not yet need — storage costs notwithstanding — in order to assure themselves an adequate supply and to avoid what they fear may be higher prices later in the season. But as cash prices rise significantly, other users find input prices too high to allow them to make a profit. They begin reducing the size of their poultry flocks, cattle herds on feed, or sow herds. High price is performing its function by rationing limited supply. Users typically overreact by cutting use too severely. In four or eight months after harvest, price level is lower than harvest because of the decline in demand. An inverted market tells commodity producers to sell immediately. It tells commodity users to wait until later to make purchases. All too often, both parties ignore these market signals, to their eventual regret.

It is easy to select the month for hedging in excess carry markets — set hedges as far out as possible. By contrast, an inverted market says "Sell Now" or use nearby contracts to hedge. The decision process for selecting the option to use for hedging is not so easy between these extremes. The decision depends on which contract month offers the greatest net return relative to costs of carrying grain in storage.

Suppose that during fall soybean harvest, the cash price offered for soybeans is $6.50. The January futures is $7.50, the following September option is $8.05, and the other contract months look like this:

Soybean Prices on October 15

Current Delivery $6.50	Futures					
	Jan.	Mar.	May	July	Aug.	Sept.
	7.50	7.60	7.75	7.95	8.00	8.05
Spreads						
Oct.-Jan.	1.00					
Jan.-Mar.		.10				
Mar.-May			.15			
May-July				.20		
July-Aug.					.05	
Aug.-Sept.						.05
Cumulative Returns Being Offered:	1.00	1.10	1.25	1.45	1.50	1.55

For simplicity, assume that the producer's local market usually has a zero basis in month of delivery.* The market is offering $8.05 ($6.50 + $1.55) for September delivery. Should the producer store and use the September contract to

---

*This assumption is invalid for most farmer markets. However, it does not invalidate the analysis. The important decision criterion is the change in basis, not its absolute value.

hedge in the $1.55 storage return? Not necessarily, as further analysis including storage costs will conclude. The objective should be to maximize return in excess of costs, not necessarily highest price.

Comparison of storage return compared to cost can be calculated in this fashion:

Calculation of Cumulative and Month-to-Month Net Return to Storage

Market Months	Basis Spreads		Added Cost of Carry	Net Return to Storage	
	Added	Cumulative		Added	Cumulative
	$/bu.	$/bu.	$/bu.	$/bu.	$/bu.
Harvest-Jan.	1.00	1.00	.44	.56	.56
Jan.-Mar.	.10	1.10	.16	(.06)	.50
Mar.-May	.15	1.25	.16	(.01)	.49
May-July	.20	1.45	.16	.04	.53
July-Aug.	.05	1.50	.08	(.03)	.50
Aug.-Sept.	.05	1.55	.08	(.03)	.47

In the calculation, the producer allocates all his or her fixed on-farm storage costs to the first month of storage, because placing the soybeans in the bins eliminates other alternative uses, and the producer insists on covering all fixed costs immediately. Monthly charges thereafter reflect 12 percent interest charges on capital tied up in inventory and 1.5 cents per month shrinkage and quality risk. Storing to January would yield 56 cents of profit, but holding across the next spread to March would reduce storage profit by 6 cents. The May-July spread offers profit but does not make up for the January-May loss. The analysis clearly defines January as the hedging month with intentions of moving the grain by then. Later returns do not cover monthly storage costs. Interest returns from money in the bank exceed interest returns on inventory, without the danger of spoilage loss.

Rolling the Hedge

Taking the analysis one step further, assume that in January, basis did not converge to zero as the producer expected. Instead, it is 20 cents under January. If this occurs, the hedger has two options: (1) sell in the cash market and buy back the futures contract, accepting the weak basis, which would produce a 36-cent return instead of the expected 56 cents, or (2) roll the hedge forward to another futures month. To roll a hedge forward, the hedger buys back the January contract and simultaneously sells another month to reestablish the position in the deferred futures. To determine whether to roll and what month to roll into, the hedger must again evaluate the spreads that exist. On January 10, cash and futures prices for soybeans look like this:

Prices on January 10

	Cash	\multicolumn{6}{c}{Futures}
		Jan.	Mar.	May	July	Aug.	Sept.
	6.45	6.65	6.75	7.00	7.20	7.25	7.20

Spreads

Jan.-Mar.	.10
Mar.-May	.25
May-July	.20
July-Aug.	.05
Aug.-Sept.	.05

Analysis of Spread Return Versus Cost of Carry

Months	Amount of Spread	Added Cost of Carry	Net Return to Storage Added	Cumulative
Jan.-Mar.	.10	.16	(.06)	(.06)
Mar.-May	.25	.16	.09	.03
May-July	.20	.16	.04	.07
July-Aug.	.05	.08	(.03)	.04
Aug.-Sept.	.05	.08	(.03)	.01

The analysis shows that it is profitable to roll into the July option and continue storing until early July.

Notice that by January 10, the futures price level had declined to $6.65 compared to $7.50 when the farmer established the initial hedge. The decline does not affect the farmer-warehouser. First, the original storage hedge protected the inventory against value decline; and second, the decision to sell cash grain or continue storage is made by analyzing contract spreads rather than price level or speculative urges.

The final results are evaluated in July. Suppose July cash and futures are trading at $7.00. Basis did reach zero as anticipated. An evaluation of spreads into the following months of August-September does not promise additional profit. The producer would sell the cash and lift the hedge with the following results:

Sell Soybeans for cash	$7.00
Futures gain on Jan. contract (7.50 − 6.65)	.85
Futures gain on July contract (7.20 − 7.00)	.20
Gross Price	$8.05
Cash value in October	− 6.50
Storage Return	$1.55
Storage costs (−0.44 − 0.16 − 0.16 − 0.16)	− .92
Profit from Storage	$ .63

Relationship of Futures Price Between Marketing Years

Each harvest begins a new marketing year with its own calculation of demand and supply arithmetic. Price levels may be higher or lower than the previous year. Since futures prices reflect both old crop and new crop conditions, the futures price in the last delivery month of one marketing year may differ sharply from futures price for the first delivery month of the next marketing year. In theory, if the same or similar supply-demand conditions were to exist in both years, December futures for a new crop year should trade at the same level as the previous December, not the end-of-year option. For example, if the remaining supply of old crop corn is expected to be adequate for the needs of feeders, processors, and exporters during the summer and early fall months, but a bumper new crop is in prospect, stairstepping corn futures prices might be as shown in Figure 7-13. The new crop would be priced at a significant discount to the old crop.

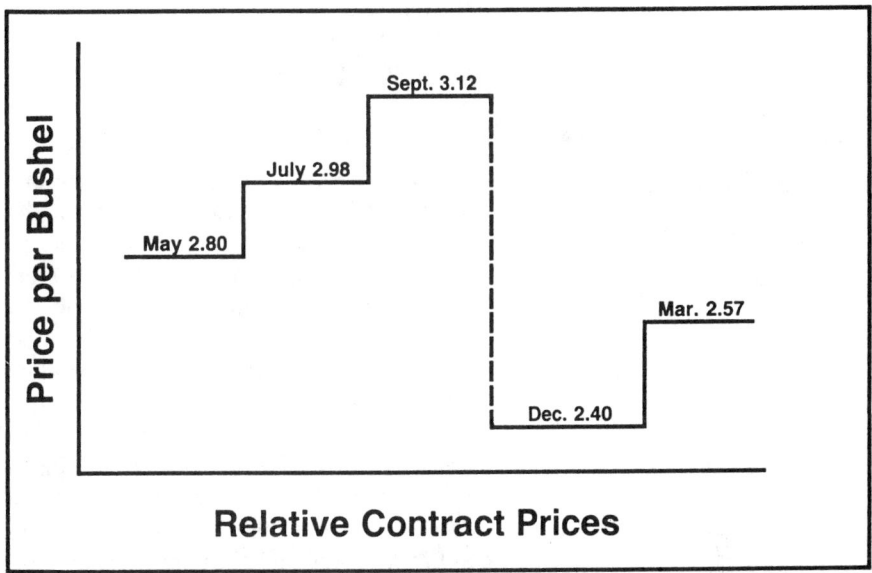

Figure 7-13: Relationship of Futures Contract Prices in Old Crop and New Crop Years.

The large down step in price does not say that old crop and new crop prices are totally unrelated. If there is an overabundance of old crop supplies as the marketing year draws to a close, and if a big new crop is in the offing, the large impending harvest will almost certainly depress end-of-year old crop prices. Conversely, if old crop supplies are barely adequate for the remainder of the year, and a small new crop is expected, the possible shortage will strengthen both old and new crop prices, and the new crop might trade at a premium to old crop.

Significant disparities between crop years can present profitable hedging opportunities for producers. Because speculators are accustomed to watching month-to-month spreads in old crop years, they sometimes forget that the new crop contract has its own set of price-making forces. Even though the old crop year might have been a short crop and produced high prices, next year is another planting season. When the most distant old crop contract and the first new crop contract trade at a premium to the nearby option, farmers can use the new crop contracts to set profitable hedges on growing crops or planned production. Since producers typically react to high early season prices by boosting production in the growing year, new crop prices decline as the season progresses. That situation existed in the spring seasons of 1981 and 1986. Drought had reduced the 1980 corn crop, causing higher price to ration the supply. Lack of rain continued throughout the winter and suggested the possibility of a short 1981 crop. New crop December futures traded at a 30-cent premium to old crop May during April. Farmers who expected to plant a big crop in response to strong price could have figured that December prices should have been similar to those in previous big harvests. They could have recognized how much drought possibilities had been bid into new crop contracts. The market offered an opportunity to place hedges at high prices before the new crop was planted. The year did in fact produce a bumper crop, and lower prices resulted. In fact, the highest prices of both 1981 and 1982 crop years occurred shortly after the 1981 harvest. When short old crops raise price levels, farmers should be prepared to cash sell the old crop and hedge the planned new crop. Production increases typically drive prices lower before the next harvest.

TO OWN OR NOT TO OWN GRAIN STORAGE

Commodity producers are constantly faced with decisions about whether to build storage facilities or, if facilities exist, how long to store grain. Storage is one of the market functions which must be performed by someone — farmers, elevator operators, processors, or consumers. Whoever performs the function incurs the costs and receives the return from it. A storage hedge can capture a storage return, but it does not guarantee a storage profit. There are costs of holding commodities in storage. They arise from owning the facilities and investing funds in the commodity. A storage profit is generated when basis gain (storage return) exceeds these costs.

Huge demand for space during harvest typically produces price discounts in excess of full carrying costs. When such discounts exceed annual costs of storage, farmers have an incentive to store their production at home or in commercial facilities. If elevators charge more for space than farmers would incur, then farmers can profit by building storage. On the other hand, if surplus storage space prevails relative to size of the locally produced crop, elevators will likely charge less than what on-farm storage costs would be. Elevators are filled and emptied several times each year, and therefore their per-unit storage cost is lower than most farmer storage facilities.

The farmer-producer has two incentives for building on-farm handling and storage facilities. The first is tied to production in that convenient on-farm facilities improve harvest efficiencies and reduce the risk of field loss from harvest delays. The timeliness benefit of farm storage means not having to wait in elevator lines at harvest. The second incentive converts the farmer-producer into a warehouser with expectations of profiting from basis improvement.

A worksheet in Table 7-4 outlines the process for calculating annual storage ownership and operating costs. A farmer-warehouser must recognize that costs exist beyond owning the facilities. He or she must incur additional costs of extra handling, shrinkage, quality control, and risks of loss with on-farm storage. The interest cost of holding inventory must be recognized. The farmer-warehouser could be earning interest on money invested in a savings instrument if he or she cashed in inventory rather than holding grain in storage.

The drying function should be considered part of the production process. Though closely associated with storage, it should be evaluated on its own profitability as compared to commercial rates. Nevertheless, efficient use of storage typically requires access to a grain dryer.

Table 7-4: Worksheet Procedure for Calculating Annual Ownership and Operating Costs of Grain Drying and Storage

INVESTMENT	Building	Equipment	Drying
1. Storage structure	$_____		
2. Handling equipment		$_____	
3. Drying equipment			$_____

ANNUAL COSTS OF OWNERSHIP AND OPERATION

FIXED COSTS

	Building	Equipment	Drying
4. Depreciation—investment divided by years of life	$_____	$_____	$_____
5. Insurance—mid-value times rate (.004)	_____	_____	_____
6. Interest—mid-value times rate (_____%)	_____	_____	_____
7. Taxes—assessed value times tax rate .	_____	_____	_____
8. Total annual costs	_____	_____	_____
9. Annual fixed costs per bushel (item 8 divided by bushel capacity) .	_____a	_____b	_____c

VARIABLE COSTS PER BUSHEL

Variable Drying Costs:

10. Drying fuel: LP gas _____¢/gal. x .0165 x _____ pts. moisture removed $_____
11. Electricity for drying _____¢/kwh x .01 x _____ pts. moisture removed .   _____
12. Total variable cost ....   _____
13. Total drying cost (items 9c & 12) ....   _____

Variable Storage Operating Costs:

14. Electricity for aeration _____¢/kwh x (.05 to .1) ....   _____
15. Loss from excess drying (3¢/bu. for each 1% under 15.5%) ....   _____
16. Dry matter shrink .5% x _____ price of corn ....   _____
17. Insurance on grain .00475 x _____ price of corn ....   _____
18. Rodent & insect control .002/bu. ....   _____
19. Property taxes (if applicable) ....   _____
20. Risk of grade loss (.01 to .10/bu.) ....   _____
21. Management cost (5% of value) ....   _____
22. Extra handling 0-5¢/bu. ....   _____
23. Total variable storage operating costs ....   _____
24. Total storage cost (items 9a, 9b, and 23) ....   _____
25. Interest on inventory per month: _____% interest rate divided by 12 x _____ price of grain times months stored ....   _____

26. TOTAL CARRYING COST ....   _____

The first task is to estimate annual fixed costs of ownership. These depend upon total investment in facilities, but costs are annualized. Investment should be broken down into drying and storage components. Annual fixed costs arise from the DIRTI-five: depreciation, interest on investment in facilities, repairs and maintenance, property taxes, and insurance on structures and equipment. Depreciation is calculated by dividing investment less salvage by years of life. Insurance and interest charges are calculated on midlife value of the investment. Taxes are charged according to assessed value times the local tax rate.

In addition to the procedure outlined in Table 7-4, a shortcut method to estimating annual costs is to multiply investment by an appropriate percentage rate, as shown in Table 7-5. The 14.1 and 21.4 percent rates reflect the DIRTI-five annual fixed costs. By multiplying the original investment by these rates, annual ownership costs can be estimated. Storage facilities requiring an investment of 50 cents per bushel would have annual fixed costs of about 7 cents per bushel (50

Table 7-5: Annual Cost of 20-Year Storage Structures and 10-Year Equipment, Calculated as a Percentage of Original Investment

Item	Structures (20-Yr. Life)	Equipment (10-Yr. Life)
	(% of Original Investment)	
Depreciation	5.0	10.0
Interest (half-rate)	7.0	8.0
Repairs and Maintenance	1.0	3.0
Taxes	.7	—
Insurance	.4	.4
Total	14.1%	21.4%

cents × .14); those costing $1.00 per bushel would cost 14.1 cents annually. This can be compared to the storage charge levied by commercial elevators in comparing alternatives for least cost.

The next step is to estimate variable costs incurred by placing grain in storage. The first item is the variable fuel and electrical costs for drying grain to grade standard of 15.5 percent and adding annual fixed drying costs to determine total drying costs. Drying should be estimated as a cost separate from storage. Though drying is necessary for safe storage, it could be done commercially. The on-farm drying cost can be compared with commercial rates to determine which is most economical.

To calculate variable storage costs, the farmer must add the costs of aeration to maintain quality, and extra drying to reduce moisture to safe levels for summer storage. Lowering moisture to 13 percent incurs fuel costs and causes weight loss; both must be included. In addition, grain handlers experience dry matter shrink during storage. The farmer-warehouser should include the costs of insurance on stored grain, costs of rodent and insect control, taxes on inventory where they are levied, and a charge to reflect the risk of maintaining quality. Extra management cost is also required for storage. On-farm storage requires charges to cover the costs of extra handling and shrinkage due to handling. Suggested coefficients for calculating these costs are included on the worksheet. The sum of variable costs are then added to per-bushel fixed costs to arrive at total storage cost.

Most of the above costs are incurred with the initial decision to store, and except for extra drying to lower moisture levels for safe summer storage, these costs do not change much during the storage period. The final item, interest on stored grain, is a function of time and increases day by day. Farmers have the option of having their money invested in the bank, earning interest, rather than having their investment in grain. And grain can be converted to cash any business day. Thus an interest charge should be assessed on grain inventory value. Adding interest charges to storage cost produces total carrying cost.

Examples of cost estimates from Illinois researchers for storing and carrying corn and soybeans in storage are presented in Tables 7-6 and 7-7.[3] Total on-farm storage costs for corn range from 24.6 cents for one month of storage to 58.4 cents for nine months. The cost on soybeans ranges from 28.9 cents for one month to 85.3 cents for nine months. As long as basis gain exceeds these costs, storage will produce extra profit.

Table 7-6: Farmer Costs of Storing and Carrying $2.80 Shelled Corn for Varying Periods of Storage

Item	Months Stored			
	1	3	6	9
	. . . . (¢/bu.)			
On-Farm Storage				
Storing				
Annual Bin Costs	9.8	9.8	9.8	9.8
Handling & Shrink	4.0	4.0	4.0	4.0
Other	8.0	8.0	8.0	8.0
Total Storing	21.8	21.8	21.8	21.8
Carrying Charges				
Interest (12%)	2.8	8.4	16.8	25.2
Extra Shrink (15.5%-13%)	—	—	9.4	9.4
Extra Drying	—	—	2.0	2.0
Total Carrying	2.8	8.4	28.2	36.6
Total Storing & Carrying	24.6	30.2	50.0	58.4
Rental Commercial Storage				
Minimum Charge (3 mos.)	15.0	15.0	15.0	15.0
Monthly Charge (1.5¢)	—	—	6.0	12.0
Shrink (15.5% to 13%)	9.4	9.4	9.4	9.4
Interest (12%)	2.8	8.4	16.8	25.2
Total Storing & Carrying	27.2	32.8	47.2	61.6

Source: R. B. Schwart and L. D. Hill, "Costs of Drying and Storing Shelled Corn on Illinois Farms."

Table 7-7: Farmer Costs of Storing and Carrying $7.00 Soybeans for Varying Periods of Storage

Item	Months Stored			
	1	3	6	9
	. . . . (¢/bu.)			
On-Farm Storage				
Storing				
Annual Bin Costs	9.8	9.8	9.8	9.8
Handling & Shrink	4.0	4.0	4.0	4.0
Misc. Items	8.0	8.0	8.0	8.0
Total Storing	21.8	21.8	21.8	21.8
Carrying Charges				
Interest (12%)	7.0	21.0	42.0	63.0
Insurance & Conditioning	.1	.3	.4	.5
Loss & Damage	?	?	?	?
Total Carrying	7.1	21.3	42.4	63.5
Total Storing & Carrying	28.9	43.1	64.2	85.3
Rental Commercial Storage				
Minimum Charge (3 mos.)	20.0	20.0	20.0	20.0
Monthly Charge (2¢/mo.)	—	—	12.0	18.0
Interest (12%)	7.0	21.0	42.0	63.0
Total Storing & Carrying	27.0	41.0	74.0	101.0

Source: R. B. Schwart and L. D. Hill, "Costs of Drying and Storing Shelled Corn on Illinois Farms."

The tendency for basis to narrow as the delivery month approaches provides farmer-warehousers with payments for storage. Figure 7-14 compares the theoretical basis pattern, reflecting carrying charges, to an actual historical average basis. It is the difference between basis gain, producing payments for storage, compared to carrying charge costs that yields storage profit to warehousers. As

Figure 7-14: Illustration of Storage Profit Due to Differences Between Carrying Charges and Basis Change.

long as basis gain exceeds storage cost (slope of basis line greater than slope of cost line), storage is profitable. When basis gain is less than storage cost, storage is not profitable (slope of basis line less than slope of cost line).

A farmer-warehouser makes the decision to build or not to build storage depending upon whether or not anticipated basis gains over the life of the bins will be greater than the total life cycle cost of the bins and interest cost of holding grain. In any one year after construction, however, fixed bin costs are irrelevant. The decision to store depends upon the variable cost of storage for that year compared to the expected basis gain. If a farmer erects a set of storage bins and basis patterns change drastically in two years, the decision of whether or not to dispose of the bins depends on only the variable cost of carrying grain and the annual opportunity cost calculated on the salvage value of the bins as compared to expected future basis gains.

LOCATION — WHERE TO SELL

Price variations between different locations cause a commodity to flow from areas of surplus production to consumption areas. The price difference is related to transportation cost, but not perfectly. Deciding where to sell depends on which location will yield the highest net return to the producer at the farm gate. Different markets have different prices and are worth shopping for. But the farmer must know what the transportation costs will be to move commodities to each market.

On the day identified by basis signals to make a cash sale, the seller should deduct transportation charges from the various price offers at different markets. Highest net price at the farm gate is the decision criterion. An example calculation of net returns on corn for a farmer located in southwest Ohio is shown in Table 7-8. The highest net price could be realized by delivering the grain to Cincinnati.

Basis variation from one market location to another is explained by several factors. The first is the actual transportation cost difference. Barges offer lower-cost shipping than trucks or railroads. Therefore, river markets usually bid higher prices than interior markets. Second, how much gross margin is the elevator or buyer taking? One may handle grain for 5 cents while another wants 10 cents per bushel. Finally, location differences depend upon the elevator operator's astuteness as a merchandiser. If he or she is adept at finding high sale price alternatives, that manager may bid a higher price to farmer-suppliers than a less active merchandiser. The reasons behind the differences are not important to a seller; the farmer's concern is finding the buyer who offers the highest net farm gate price.

Table 7-8: Calculation of Farm Gate Value to Determine Where to Sell in Various Markets with Various Transportation Costs

	Wilmington	Cincinnati	Chicago	New Orleans	Norfolk
Delivered Price/Bu.	$3.25	$3.43	$3.57	$3.53	$3.95
Transportation Costs	- .05	- .12	- .35	- .45	- .66
Net to Producer	$3.20	$3.31	$3.22	$3.08	$3.29

SUMMARY

This chapter emphasizes the difference between a price level sale and a basis level sale; they are two separate, but related, markets. Price level reflects worldwide demand and supply conditions; basis level reflects local demand for, and supply of, storage space. Market analysis of basis signals will tell the producer when and where to make cash sales.

We emphasized two features of the relationship between cash and futures prices — parallel movement and convergence. We reviewed their effects on types of hedges.

Our analysis looked at storage returns offered by futures contracts and costs of holding commodities in storage. As long as added returns exceed added costs, there is economic reason to store. Basis records and charts for individual producers guide their decisions.

Finally, we explained how to read basis market signals. Abnormally weak basis is the signal to withhold commodities from the market, delaying cash sales. Abnormally strong basis signals an immediate sale as the proper response. By separating basis decisions from price level decisions with the appropriate marketing tools, a farmer-producer is in a better position to capture advantageous prices in both markets.

Questions for Study and Discussion:

1. What two elements of marketing determine basis price levels?
2. How does the Law of One Price explain different prices that exist on a particular day for a commodity?
3. List and explain the two features which characterize the relationship between cash and futures price.
4. Where and when are cash and futures price the same? What assures this relationship?
5. How are cash and futures price related over time?
6. Why do grain futures contract prices normally stairstep upward?
7. What is the foundation of basis level negotiation in a market economy?
8. How does farmer reaction to price level affect basis?
9. What factors cause local basis to differ from Chicago basis?
10. Explain the two reasons for hedging price level.
11. Give examples of the five situations that produce narrowing of basis.
12. How does a warehouser with storage space earn a storage profit?
13. Why is it important that each producer-storer maintain his or her own basis record?
14. What do different basis charts show?
15. What are the basis signals for cash sales and purchases?
16. Explain the procedure for calculating which contract month to place hedges.
17. What is "rolling the hedge"? How is it done?
18. Why is the relationship of contract prices between marketing years important?
19. Explain the procedure for determining where to sell.

Basis Exercise:

Maintaining a basis record is an interesting exercise to help you learn more on the subject. Use a form such as the one suggested in this chapter to maintain a weekly record. Secure futures prices and local bids from newspaper or radio reports or from a local elevator on commodities you are producing. Local basis is computed by subtracting cash price from a selected futures option. The basis numbers can be plotted on a chart to visually illustrate its pattern over time. Three to five years of historical information can be used in identifying both average and extremes in a local market.

The end-of-year option (July for fall harvested crops and May for wheat) focuses on convergence of cash and futures over time. It is important in deciding whether to store or not and in estimating the amount of storage return to expect. It can also signal when to sell cash grain. Since most farmers are long-term storers, rather than daily basis traders, the end-of-year option serves them best in its simplicity.

Knowledge of basis and a feel for why it fluctuates will hone a producer's ability to make profitable basis transactions.

Homework Problems:

1. On October 25, you are harvesting corn and the elevator is bidding $2.10 for cash corn, and December futures are $2.40. Will a hedge in the December option return you a storage profit? Outline the procedure for making this analysis of profitability.

BASIS RECORD

Commodity _____

Month		Week 1	2	3	4	5
Jan	Futures ( )					
	Local Cash					
	Basis					
Feb	Futures ( )					
	Local Cash					
	Basis					
Mar	Futures ( )					
	Local Cash					
	Basis					

    2. On October 10, you are harvesting soybeans and you have empty storage you could use, or you could sell the beans to an elevator for $6.00 per bu. November futures are trading for $6.60. What facts do you need to determine whether you should sell at harvest or sell in January when you think January soybeans futures will reach $7.50?

    3. You were through harvesting corn on October 25 and have your 20,000 bu. stored in excellent condition in a new 20,000-bu. bin. On December 1 you note that cash corn is selling for $2.20 and March futures option is $2.30. What should you do if you think the price of March futures will reach $2.45 by March 1, when your elevator normally bids cash price 12 under March? How much more per bushel would you receive by your decision, assuming your guess about price level is correct?

    4. At planting time (April 15), you note the elevator is bidding $2.60 for new crop corn delivered at harvest while December option is $3.00. You have on-farm storage which would cost 1 cent per bu. per month (variable cost) plus 5 cents per bu. for the extra handling (in and out of storage). You have a loan on which you are paying 12 percent interest which could be reduced at harvest if you wanted, but which could be extended to May 1. May futures is $3.20 and your records show the elevator normally pays 10 under May during the last weeks of April. What should you do to maximize the storage income from your crop by April 15?

    5. It is mid-August. The summer drought scare is in full swing. Your crop looks good, and futures are: corn $3.95 Dec., $4.10 March; soybean $8.75 Nov., $9.50 March. You have 100,000 bu. storage and expect 100,000 bu. of corn production from your 800 acres. In addition, you anticipate harvesting 40,000 bu. of soybeans. Fall harvest bids on corn are $3.80. Historical corn basis is: harvest −70, Dec. −20, March −10. Interest cost is 6 cents/bu./mo. You decide price level is at its peak so you decide to price your crop. Normal soybean basis is: harvest −1.00, Nov. − 40, and March −10. How can you merchandise your bin space profitably? Even though price level may change, normal basis is ex-

pected to exist both Oct. and March. How much profit do you expect to earn on your storage space? What tool did you use to lock in the profit? Explain.

Footnotes for Chapter 7:

(1) Many of the ideas on basis are from Fred Bailey, *Understanding Basis—The Economics of WHERE and WHEN* (Chicago: Board of Trade, Clearing Corporation). Used by permission.

(2) Bailey, *Understanding Basis* 9.

(3) For further information on procedures and coefficients for calculating costs see R. B. Schwart and L. D. Hall, *Costs of Drying and Storing Shelled Corn on Illinois Farms*, Circular 1141 (Champaign: Cooperative Extension Service, University of Illinois, Aug., 1977).

Chapter 8

Financing a Hedging Plan

IMPORTANCE OF A MARKETING PLAN 268
THE LENDER'S ROLE 271
    Lender Policies for Hedge Loans 271
    Evaluating Hedge Loan Requests 272
    Structuring Operating Agreements 273
    Lender Documentation of Hedge Loans 274
    Monitoring Hedge Loans 275
SUMMARY 277

Financing a Hedging Plan

IMPORTANCE OF A MARKETING PLAN

Agricultural lenders who finance farm production join hands with farmers in their financial success or failure. By its nature, farming is a speculative process, subject to serious risk and uncertainty in both production and pricing. Consider, for example, an 800-acre corn farm producing an average yield of 130 bushels (Table 8-1). Yield fluctuations of 20 bushels per acre change gross receipts by $41,600. In comparison, market price swings of $1.20 per bushel change gross receipts $125,000. It takes yield variability of 60 bushels per acre to cause the same gross receipt variability as a $1.20 price swing. The farmer and lender must decide which is more likely to occur, a 60-bushel yield variation or a $1.20 price swing. A production plan, without a marketing plan, covers only part of the risk lenders share with farmer-borrowers.

Table 8-1: Change in Gross Farm Receipts Due to Change in Yield and Corn Price, Example Farm

Situation: 800 acres of corn; 130 bushels per acre base yield, $2.60 base price

Production base price $2.60		Marketing base yield 130 bu.	
Yield	Gross Value	Price	Gross Value
120	$249,600	$3.20	$332,800
130	$270,400	$2.60	$270,400
140	$291,200	$2.00	$208,000
Change high to low		Change high to low	
$41,600		$124,800	
17% change		60% change	

Lenders are accustomed to financing production plans without requiring marketing plans, thereby aiding and abetting cash market speculation. A Purdue survey of 37 agricultural banks showed that only 17 financed hedging accounts, with the average number of accounts per bank being six.[1] Most of the 20 banks not financing hedging cited a lack of demand from borrowers as the major reason for not financing such accounts. A second, but less cited, reason was lack of knowledgeable bank personnel available to supervise the accounts.

Farmers and ranchers are typically cash market speculators. They often double up on their price speculation with unpriced inventory stored in bins and an unpriced growing crop. Thus, it is not uncommon for farmers to speculate on price on two crops simultaneously.

Lenders and farmers need to form a team to improve farmer marketing. All too often a farmer's marketing strategy is dictated by "a payment is due" or "the banker says to sell." There is a better way. A planned strategy is better than a forced strategy. Knowing when and how to control cash market speculation can be a key to financial success for farmers and financial security for lenders.

Various marketing tools permit a borrower to protect against price and basis risk. The lender and borrower must mutually agree to use the appropriate marketing tool at the appropriate time. The degree of price speculation according to the marketing tool used to make pricing decisions is summarized in Table 8-2. A

Table 8-2: Summary of Degree of Price Speculation and Marketing Tools That Hedge Various Speculations Away

Degree of Speculation	Type of Price Speculation	Type of Contract (Tools)
Highest	Price Level and Basis	Unpriced Inventory (grain in bins & crop in ground)
	Price	Delayed Price or Basis Contract
	Basis	Futures Contract
	None	Forward Contract
Lowest	None	Cash Sale

cash sale or forward contract eliminates all price speculation. But it also eliminates the possibility for capturing more profitable opportunities if prices rise during the year. Basis contracting or hedging with futures removes respective parts of cash market speculation. The highest degree of speculation is practiced with unpriced inventory.

Before farmers and lenders begin using some marketing tools to reduce cash market speculation, they must recognize that adequate financing is a prerequisite to a sound marketing program. In particular, hedging in the futures market requires adequate liquidity. Without adequate financing, hedging increases financial risks. The producer who needs $25,000 to meet margin calls due to a price move against hedge positions, but has only $20,000, will be liquidated when margin calls are not answered. Invariably the market will turn the other way and move past the original hedged price. Thus, there is a double loss. Production will have to be sold at the lower price, and the futures loss will have to be deducted from that lower price. A hedging plan, without a financing plan, increases risk rather than reducing it.

Funds are required in a hedging account for initial margin and additional (maintenance) margin to cover short-run setbacks. Initial margins are "good faith money" deposits to insure the financial integrity of the futures position. Table 8-3 outlines the margin requirements for agricultural commodities as established on several exchanges by one firm. They are illustrative and subject to change. With a $600 initial margin per 5,000-bushel contract, a 100,000-bushel corn grower would require $12,000 as initial good faith money to hedge that production. The deposit must be made with the broker before a futures trade can be initiated.

Table 8-3: Hedger's Initial and Maintenance
Margin Requirements, 1989

Commodity	Initial Hedge	Maintenance (Min. Bal.)
Chicago Board of Trade		
Wheat	$ 700	$ 700
Corn	600	600
Oats	500	500
Soybeans	1,250	1,250
Soymeal	800	800
Chicago Mercantile Exchange		
Live Cattle	1,000	750
Live Hogs	700	500
Feeder Cattle	1,000	750
MidAmerica Commodity Exchange		
Wheat	140	140
Corn	100	100
Soybeans	250	250
Live Cattle	375	375
Live Hogs	250	250
Oats	100	100
New York Exchanges		
Cotton	1,500	1,500
Frozen Orange Juice	1,000	750
Sugar	1,000	750
Other Markets		
Minneapolis Wheat	800	800
Kansas City Wheat	800	800

Source: Stotler and Company.

Initial margin remains a cash asset for the farmer. Most brokerage houses allow the deposit to be held in Treasury Bills. The initial unit is $10,000, and subsequent $5,000 units can be purchased. With the initial margin invested in Treasuries, the farmer receives an interest return on the investment. Otherwise an idle balance held by the brokerage house does not earn interest for the farmer.

Since it is highly unlikely that any hedger will place a position at the extreme of a price move, additional liquidity for maintenance margin is required to hold a futures position once it has been established. Maintenance margin is the minimum balance that must be maintained in the brokerage account. Should the balance fall below maintenance levels, the broker will make margin calls for additional deposits to bring the balance back to minimum level. Due to volatile futures price fluctuations, margin calls can be large and required on short notice. A hedger of 40,000 bushels of soybeans who sets hedges $1.25 per bushel below the top of a price run up, or a hedger of 100,000 bushels of corn who sets hedges 50 cents too low, would each need $50,000 ready cash to meet margin calls.

The amount of liquidity required for hedging depends upon price volatility of the commodity being hedged and the type of pricing strategy that is practiced. Our experience with a particular system suggests that lenders should consider a **minimum** 60 cents per bushel of corn, $1.00 per bushel of soybeans, 75 cents per bushel for wheat, 20 cents per pound of cotton, and $10 per hundredweight for livestock. Different pricing strategies will have their own respective liquidity requirements. Those numbers mean that Midwest grain farmers may need $50 to $100 per acre liquidity commitments for a hedging program. This is in addition to the $150 per acre for seed, pesticides, fertilizer, fuel, and other out-of-pocket production costs. A total loan package represents a $250-per-acre liquidity requirement. How much of that liquidity is supplied by the farmer and how much by the lender is open to negotiation between the two.

The suggested liquidity figures assume that the pricing strategy will establish hedges near market extremes but still provide a margin of error. The amount required to meet margin calls is the difference between the hedge price and the extreme of any subsequent price move, if the farmer practices fixed hedging to hold the position until maturity. The soybean market, which went from $4.00 to $13.00 per bushel in 1973, would have required huge margin deposits had hedges been set before the top was reached. It is the inability to project total liquidity that might be required in a hedge that frightens many lenders.

It should be remembered, however, that margin money outstanding against a hedge position does not constitute a loss for a hedger.* The outflow of cash for maintenance margin when price moves against a position will be returned to a hedger when the cash sale is made. Because of parallel movement of cash and futures prices, the increased inventory value will be about as much as the deficit in the futures position.** But during the time period that maintenance margin is required, the change in inventory value produces a paper profit, while margin calls are cash transactions. A liquidity problem arises until the cash commodity is sold. It is this liquidity need that requires financing.

The true costs of hedging are limited to the interest charges on margin money while it is outstanding and to the commissions paid to the broker. The margin deposits are not losses. They are only lost pricing opportunity that all too often causes regret — regret that one did not wait for a better price. On the other hand, when a hedge position produces a surplus in the futures account, the surplus is not a profit. Inventory value of the cash commodity will have declined approximately equal to the gain in the futures position. The futures account accumulates cash while the inventory loss is still on paper, until the commodity is sold at a lower price. The lower cash price plus the futures gain raises the realized price to the level protected by the hedge. Lenders, as well as farmers, need to remember the offsetting gains and losses in both markets when hedges are set for price protection.

THE LENDER'S ROLE[2]

Lender Policies for Hedge Loans

Lenders who decide to finance marketing plans must recognize the importance of profitable pricing decisions and must consider the borrower's income and ability to repay. Many lenders have seen good producers suffer from poor marketing. They see improved pricing as a way to protect the quality of their loans. A sound hedging program should enhance a customer's borrowing capacity because it protects earning power. The lender who has a secured position in both the commodity and the hedging account should have a stronger loan.

Lending institutions will need to develop policies and practices to cover lines of credit for agricultural borrowers to use in futures market hedging. When the

---

*Review the parallelism discussion in Chapter 2 to understand offsetting gains and losses.

**This assumes the producer has full production to sell at the higher cash price. Often drought-reduced crops, leading to the price rise, leave the farmer with less than total production. The futures deficit thus exceeds increased inventory value.

lender decides to finance hedging, the first step is to formulate a statement of policy which governs the administration and control of all hedge loans. The statement might say: "The lender agrees to grant loans for the purpose of commodity hedging and will extend funds for financing initial margin, maintenance margin, and broker commissions that arise out of trading contracts in the futures markets."[3] In addition to the formal statement of policy, a lending institution should make clear its position on the basic elements of a hedge loan and develop criteria for loan officers to follow.

Efforts to educate loan officers on hedging practices should be an early step toward a sound and effective loan program. Loan officers must become familiar with alternative marketing tools and the mechanics of hedging to properly evaluate loan requests and monitor accounts.

The institution should define what percentage of initial and maintenance margins will be lent to qualified borrowers. Some lenders will advance total financing of margin requirements; others will advance a percentage of each margin. Another variation is to require borrowers to provide initial margin while the lender agrees to provide liquidity for maintenance requirements. Some lenders may finance the total margin requirement on commodities in storage, but only a portion on anticipated production since production risks still exist.

Once the hedge loan is granted, a lending institution must always be able to provide all the funds required to follow through with the plan. If the lender cannot provide all the liquidity the market requires, the borrower's position in the futures market will be liquidated, and funds deposited to meet prior margin calls are lost.

It is impossible to predict accurately the maximum commitment needed to meet margin calls, because no one knows the extremes the market will attain once hedges are set. If the lender anticipates that legal loan limits may be approached, he or she should secure a commitment for an overline loan from another bank to meet full requirements.

Lenders must keep in mind that loan quality is not damaged by margin deposits, however large they may be, as long as the producer has commodities as collateral. A hedge that is liquidated prematurely because of inadequate liquidity can be disastrous for both borrower and lender.

In developing the mechanics for establishing hedging loans, lending institutions must:

1. Develop procedures to evaluate loan requests,
2. Structure operating agreements,
3. Define documentation required to secure hedge loans, and
4. Establish methods to monitor the hedge account.

Evaluating Hedge Loan Requests

Policy guidelines should define who is a qualified borrower for a hedge loan. Normal credit standards and criteria should be used. Furthermore, the lender must be certain the borrower understands different techniques and implications of using marketing tools to set pricing decisions. The lender should be convinced the borrower has a clearly defined marketing plan and pricing strategy and can manage the emotions of a hedge account. The lender should also determine that the borrower's production or production capacity meets the specifications of futures contracts with regard to quantity and quality.

A balance sheet statement and a profit and loss projection determine a borrower's financial condition. Analysis of the first shows the borrower's ability to withstand adverse production or pricing risk. The second spells out cash flow requirements and price levels required to produce profit for repayment.

The borrower's pricing strategy must be evaluated against the financial strength on the balance sheet in formulating a marketing plan consistent with risk-bearing ability. Weighing the degree of acceptable cash market speculation against risk-bearing ability of the borrower is the first step in evaluating the marketing plan that a lender agrees to finance. A high-equity operator can afford to be a cash market speculator much longer than a low-equity operator. A highly leveraged farmer might be held to the least speculative alternative with early cash contracting.

Which marketing tool to use to confirm pricing decisions, and how each is used, depends upon experience and knowledge of the farmer-hedger. An experienced hedger is more likely to have the knowledge and discipline to move in and out of cash market speculation when it is financially advantageous to do so. The lender and borrower must decide if fixed hedging, selective hedging, or flexible hedging best fits the farmer's knowledge, experience, and financial resources. No one plan fits all situations, but a specific, preagreed plan and pricing strategy solves potential problems. It also provides a way to do a historical analysis on how well the plan would have performed in past market situations.

Structuring Operating Agreements

Due to the speculative element present in farmer use of futures markets, a special working relationship is required between borrower and lender. Lenders abhor the idea of financing speculation. (But they do so each time they finance a production program based on cash markets without a marketing plan.) Forward planning and stated procedures formalized into a written operating agreement prevent misunderstandings. The agreement provides the details of the understanding reached by both parties.

The borrower's production estimates, marketing plan, and pricing strategy should become parts of the agreement. They can be reviewed from time to time and used as a standard for evaluating progress. How much production will be committed to the plan? Will the producer set pricing decisions on a calendar schedule, determine price objectives from technical analysis, or employ a trend following strategy? It should specify whether true hedging or selective hedging will be followed. The plan should be specific about setting and lifting hedges and about what criteria will be used to make the decisions.

The agreement should describe how the flow of funds will occur. In some cases funds are transferred directly from the bank to the brokerage house on a call from either the broker or borrower. In other cases they move from the bank to the farmer, and then to the broker. Wire transfers are quicker and save mailing details, though they are more costly. Excess brokerage account funds often go back to the bank directly to keep the outstanding balance as low as possible to reduce interest charges, or to earn interest on surplus balances. The borrower and lender should negotiate beforehand what interest rates will be paid on credit and debit balances in the hedge account.

Periodically, the hedging account should be settled with payments to the borrower or against the loan balance. The agreement should spell out time of settlement. The logical settlement time is when cash transactions are made, but it need not be so limited. For crops, settlement could occur at the end of the crop year. For livestock, settlement could be made as each contracted group of animals is sold. It might take a week's, a month's, or two months' production to equal the contract amount. Even though the producer sold $45 hogs in the cash market, that producer could take home a $56 effective hedged price as he or she settled up on the way past the bank. When cash price was higher than the futures contract, sale of cash animals would be the time to pay the deficit balance on the futures position to the lender.

The agreement should also state that the customer will not speculate in commodities that are not produced. Borrowers should agree to use a reputable broker who understands the farmer's objectives and marketing plan. The loan could be made callable to protect against any terms of the agreement being violated.

Lender Documentation of Hedge Loans

In order to secure a lien on funds used for trading futures contracts, lenders need to have three documents: (1) the note, (2) the security agreement, and (3) the hedge agreement and assignment of hedge account, known as a tri-party agreement.

1) The note should be written for the amount of the maximum borrowing anticipated by the hedging plan, the interest rate, and the maturity date. It can be either a time or demand note. Since agricultural products may become marketable at times different from original plans and expectations, a demand note may be more advisable. It may eliminate the need for drawing up a new note, whereas a time note might mature before the commodity becomes marketable.

2) The security agreement provides the lender with a lien on the assets used as collateral on the loan. It further spells out the lender's interest in receipts from sale of the inventories used as collateral. Whatever collateral is required to secure a production loan, the amount should be increased to secure the hedging loan.

3) The third document is a hedge agreement and assignment of hedge account. It is used by the lender to secure funds in the brokerage account as part of collateral. With this provision, the effects of lower prices on the farmer's products are protected by surplus in the brokerage account. Conversely, a deficit in the brokerage account is offset by higher inventory value of the farm products.

In addition to the borrower and lender, the tri-party agreement further involves the broker in a three-way business arrangement. The document authorizes the lender to expand the loan to meet margin calls. The broker also has authority to request and receive additional margin as required, directly from the lender. Operating in the other direction, it allows the lender to withdraw excess margin funds from the brokerage account to reduce the borrower's loan. The lender can make an "excess margin call" on the broker. This document further authorizes the brokerage firm to furnish copies of the hedge transactions to the lender. These make monitoring the hedge much easier for the lender.

Several provisions should be included in the tri-party agreement. These provisions should be listed after naming the parties concerned and after making the statement of intentions of the loan to finance a hedging account. The following provisions may be included:[4]

1. "The lender agrees to advance funds to the borrower for hedging. The borrower agrees to deposit the funds with a broker and to use the funds only for hedging as defined in the marketing plan.
2. "As security for the funds, the borrower grants a security interest in the hedge account, an assignment of the funds presently in the account, and those to be accumulated in the future.

3. "The broker is authorized and directed to pay funds from the hedge account to the lender upon request. Further, the borrower agrees not to withdraw funds from the account without the consent of the lender.
4. "The lender will advance funds for margin requirements directly to the broker.
5. "The borrower gives the lender power of attorney to operate the hedge account if necessary.
6. "The borrower may execute transactions in the account, unless prohibited by the lender.
7. "The lender is authorized to liquidate the account when the lender deems it necessary for his own protection.
8. "Any funds paid by the broker to the lender will be applied to the borrower's loan. The interest rate charged on outstanding balances will be _____%; the rate paid on surpluses will be _____%.
9. "The lender is authorized to receive from the broker, and the broker is authorized to deliver to the lender, copies of all futures transactions made in the account.
10. "The agreement will remain in full force until canceled in writing.
11. "The borrower warrants to the lender that the account has not been alienated or assigned.
12. "The agreement is signed by the borrower and the lender.
13. "The agreement acknowledgment is made by the broker or brokerage firm."

After the agreement has been completed and signed, the original should be filed and recorded in accordance with local, state, and federal laws. The lender and borrower each should retain an acknowledged copy.

Monitoring Hedge Loans

The tri-party hedge agreement between farmer, lender, and broker provides measures of control over activities in the account. For the most part, the borrower should manage his or her own sales decisions or employ a market consulting service. Lenders are in the business of making loans and providing advice as requested, but not trading commodities. Understandably, most lenders are reluctant to help manage a borrower's marketing decisions. However, in order to safeguard funds, a lender may insist on being informed of market positions and may reserve the right to intervene to prevent losses for the borrower and ultimately the bank.

The broker provides copies of all brokerage statements to both borrower and lender. These statements serve as a report to the lender of market activity and status of the account. The statements can be used as a check to determine whether a borrower is following the marketing plan.

Brokerage firms provide customers with several account statements, as shown in Figure 8-1.

1) The "confirmation of sale" itemizes trades that are made. It reports the position taken, number of contracts, commodity traded, and price information. Confirmations of sale are issued within 24 hours of the transaction and mailed to appropriate parties.

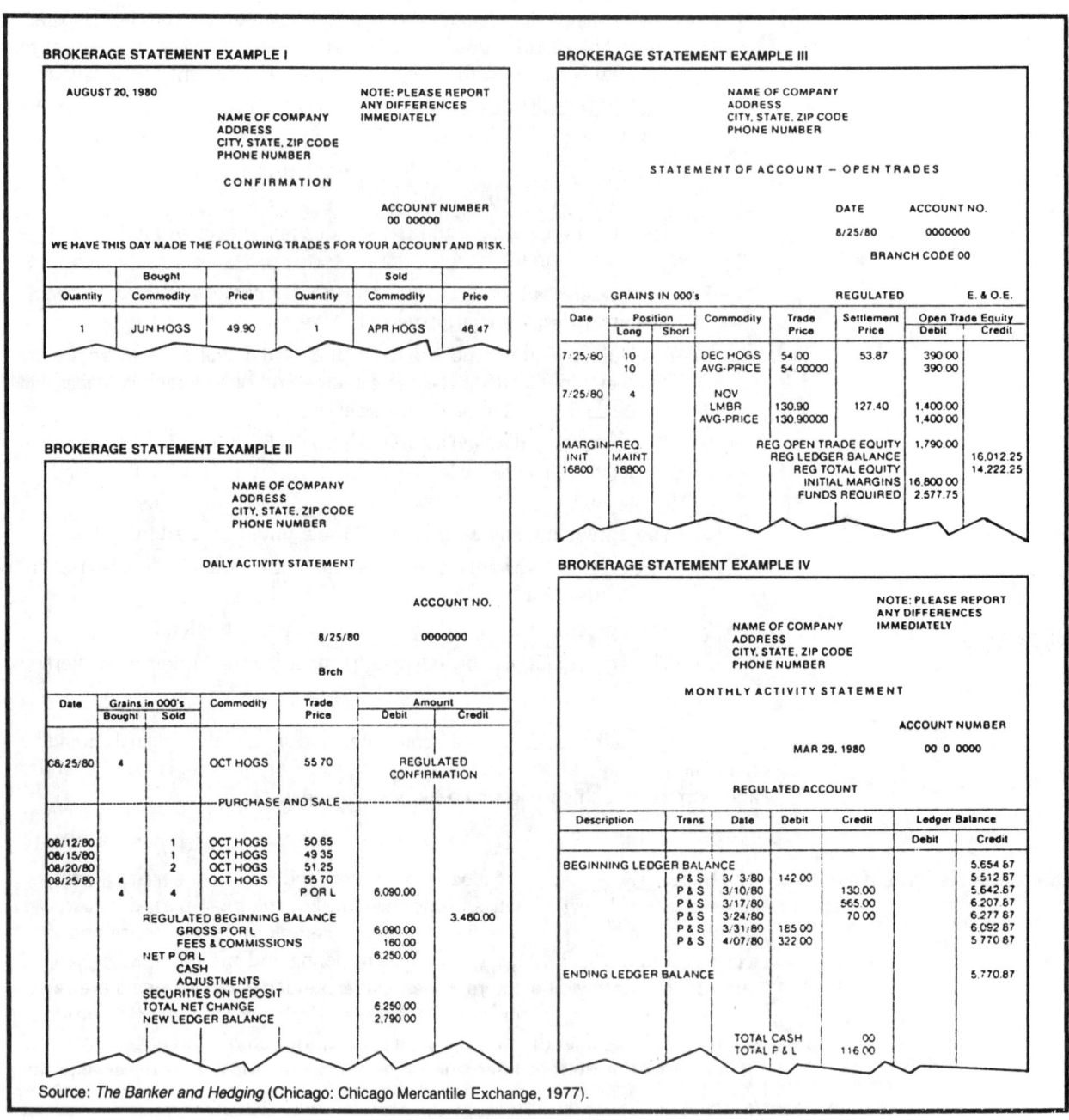

Figure 8-1: Examples of Brokerage Firm Account Statements.

2) The "statement of purchase and sale" (Daily Activity Statement) reports each time a position in an account is offset and lists date of transactions, prices, profit or loss for each trade, commissions paid, exchange fees charged, and a customer's ledger balance.

3) The "statement of open trades" shows the status of an account at the close of a trading day. This report notes the status of open trade equity for positions not

yet closed out. Equity is the difference between initiation price and the settlement price on close of the report. It is included automatically with each monthly activity statement. Daily statements of open trades are available upon request.

4) The "monthly activity statement" is similar to a bank statement, as it lists all credits or charges to an account for a 30-day period. It reports the current bottom-line balance.

Additional loan monitoring requires a ledger of cash flows and inventory records on the physical commodity. A separate ledger for the commodity account should record the amount of margin funds advanced or excess margins received and repaid on the loan. This ledger can be balanced with the month-end statement issued by the brokerage firm. The borrower may be required to furnish a listing of inventory on hedged commodities, especially when advances are requested. In addition, a collateral inspection should be made occasionally by a loan officer or third party to verify inventory and reconcile the listings provided by the borrower.

With proper understanding of operating procedures, documentation, and monitoring, loans for hedging can be adequately secured without undue time demands on lenders. Improved earnings for the producer, gained from executing an effective marketing plan, improve the quality of an operating loan. The alternative is for the lender to finance cash market speculation.

SUMMARY

Lenders usually are called upon to provide financing for hedging as well as production. A hedging plan must have adequate liquidity to carry it out. Without adequate liquidity, financial risk is increased rather than minimized by hedging. Earning power of loans can be enhanced by a sound marketing plan and pricing strategy.

Lending institutions have to develop policy guidelines and operating procedures to properly structure and administer lines of credit for hedging. Adequate documentation and monitoring of hedge loans properly secures them for the lender.

The alternative to financing production and hedged marketing is to finance cash market speculation.

Questions for Study and Discussion:

1. Why is adequate financing an important part of a hedging program?
2. Calculate a minimum line of credit needed to finance hedging on production from a 1,000-acre crop-livestock farm producing 80,000 bushels of corn, 20,000 bushels of soybeans, 1,500 head of market hogs, and 200 head of fat cattle, using the initial margins and suggested liquidity requirements in the chapter.
3. Why is money to finance maintenance margins a liquidity problem and not an expense?
4. What items should be reviewed by lenders in evaluating hedge loan requests?
5. What items should be included in an operating agreement governing hedge loans?
6. What documents are required by lenders to secure a hedge loan?
7. What steps should a lender take to monitor a hedge loan?

Footnotes for Chapter 8:

(1) Freddie L. Barnard and J. H. Atkinson, "The Current Financial Situation in Agriculture As Viewed by Agricultural Bankers," *Purdue Farm Management Report* (West Lafayette, Ind.: Purdue University, April, 1982) 7.

(2) Information in this section was gleaned from *Risk Management Guide for Ag Lenders* (Chicago: Chicago Mercantile Exchange, n.d.). Used by permission.

(3) *Risk Management Guide for Ag Lenders* 6.

(4) *Risk Management Guide for Ag Lenders* 16.

Chapter 9

Livestock Marketing

CHARACTERISTICS OF PERISHABLE COMMODITIES.................280
FUNDAMENTAL PRICE FORECASTING IN LIVESTOCK 281
USING HEDGING TO FIX CATTLE PRICES..........................282
      Basis Problems and Opportunities 284
BEHAVIOR OF NON-STORABLE FUTURES CONTRACT PRICES 286
MAKING LIVESTOCK HEDGING WORK: BEAR SPREAD HEDGING 288
TIMING CASH SALES 291
AGRICULTURAL PRICE CYCLES...................................294
   The Cattle Cycle...295
      Ranching and Cow-Calf Production........................295
      Decision Making by Cow-Calf Producers 297
SUMMARY...299

Livestock Marketing

CHARACTERISTICS OF PERISHABLE COMMODITIES

Perishable agricultural commodities such as live cattle, hogs, eggs, broilers, and feeder cattle have an entirely different relationship to futures markets than storable commodities. The differences change the role of futures pricing; they eliminate storage decisions from pricing, and they generate price cycles which must become part of a production and pricing plan. Thus, pricing decisions must be made differently.

Corn, wheat, and cotton have relatively fixed planting, production, and harvest dates. Timing of the production cycle is fixed by environmental conditions. Production from a short harvest period must be rationed into use over a longer consumption period. Storable commodities can be held for several months, a year, two years, or longer, if properly stored. They can be drawn out of storage and into consumption by high current prices relative to lower price offers for future delivery.

Livestock products, in contrast, do not have the same storability characteristics nor marketing flexibility. Production is more nearly continuous. Once the breeding decision is made, production will be forthcoming in due time. When livestock is market-ready, it cannot be withheld from the market, nor can the products be preserved and stored economically for long periods. Prior to being market ready, production cannot be moved into consumption ahead of schedule. Producers do not have the flexibility of withholding production from the market in response to low prices, nor can they bring immature production to the market in response to high prices. Non-storable commodities conform to a rigid production schedule that dictates that they move into consumption when, and only when, market ready, give or take a few days or weeks. The price the product will bring depends upon what consumers will pay at that time. An unlaid egg can be neither poached nor fried, only done without; surplus eggs sink in value because of limited alternative uses.

Individual futures option prices under such conditions are considered "true" forecasts of the equilibrium price that will prevail in the future time at maturity of each contract. The market participants' assessment of supply and demand conditions that will exist at the contract maturity date determines the equilibrium price they bid into futures prices. Various options are not tied one to the other by carrying charge as are grain contracts.

However, in real life, futures prices have not been accurate forecasts of cash prices when the future becomes the present. Based upon early data, Leuthold concluded that futures prices for live cattle and live hogs tend to be biased and inefficient forecasts of cash prices.[1][2] Garcia and Martin studied later data and introduced methods for including cyclical, seasonal, and feed grain cost factors into their analysis. They concluded:

In a forecasting sense, cattle futures appear to add little information beyond that available in lagged cash prices. Live hog futures appear to have performed the forecasting function well, relative to both live cattle futures and lagged cash prices, except during the period when feed grain prices were highly variable. However, when forecasting was most difficult, hog futures failed to predict wide price swings.(3)

Even though futures prices on a particular day are thought to be the best estimates of value at the time, these studies concluded that futures prices are poor predictors of realized cash prices, because of supply and demand adjustments made in the interim.

What is the farmer or rancher-producer to do if he or she cannot rely on futures as accurate price predictors? How can the farmer determine whether to produce or not, or when to make pricing decisions? First, from a fundamental perspective, the farmer must develop skills as a price forecaster to gauge when he or she feels the futures market is wrong. If futures are higher than he or she expects cash price to be, the farmer can hedge production; if lower, the farmer can remain unhedged. Second, the farmer can calculate costs of production and make decisions to hedge whenever he or she is satisfied with profits the futures are offering. It is a matter of saying, "I don't know what price will be, but I will be satisfied with the profit the market is offering in order to reduce my risks." Finally, the farmer can forget fundamentals, develop a technical trading strategy, and follow its signals to price production.

FUNDAMENTAL PRICE FORECASTING IN LIVESTOCK(4)

Hog price forecasting, based upon fundamentals, is tied to measures of demand and pork supply projections. Supply projections can be based on quarterly pig crop reports. Upon issuance of each report, hog price forecasts for the next 12 months can be calculated from the supply projections. These forecasts should be compared to futures pricing opportunities and plans developed for the sale of each quarter's production. This procedure would involve modifying previous plans for the next three quarters and new plans for only the farthest-out quarter. The producer would be making pricing plans at the same time that breeding plans are made.

The above decision pattern applies when no change in production is anticipated by the producer. However, increasing or decreasing production should always be considered an option. When feed supplies are small and higher input prices drive up cost of production, or hog numbers are so large as to drive prices down, production will be unprofitable. The market is dictating reduced production. On the other hand, low feed costs or reduced pork supplies signal producers to increase production to capture higher profits. The forecasting reference to the pork producer for changing production level is the hog cycle.

The time period to consider production level changes is late summer, when feed supplies are known and intended hog production information is available. At that time, producers can use the futures market to lock in the cost of inputs for 12 months. They can hedge-buy corn and soybean meal supplies and hedge-sell planned production if relative prices offer profits. This idea was introduced earlier in terms of locking in profitable cost-price spreads before initiating production increases.

Supply estimates for beef production can be developed from semi-annual cattle inventories and quarterly cattle on-feed reports. Demand for beef can be estimated from disposable personal income and consumers' willingness to spend money for beef. The other factor in predicting price is the supply of competing products. Other meats such as pork, lamb, veal, broilers, and turkeys are most important. Other foods such as eggs, dairy products, and vegetables can also affect demand for beef, but only slightly.

The resulting price forecast, compared to cost of production, guides producers to decide whether to fill feedlots or not. If a profit level is not bid into the market by relative production costs and sale price, the producer may decide to leave the lots empty. Whether or not to leave feeding facilities idle can be a month-to-month decision. Breeders do not have as much flexibility as feeders because breeding herds are more difficult to liquidate and reestablish.

A producer who intends to use futures contracts to set pricing decisions should start planning cattle sales about six months before cattle are to be put on feed. During that time, futures price relationships between feed, feeder livestock, and fat cattle will hopefully adjust to equal or exceed a profit objective. When to establish hedges to capture margins offered by futures price relationships depends upon price forecasts and the willingness to accept the risk of lesser profit in exchange for the opportunity for higher profit.

If the feeding enterprise is entered into without forward pricing fat cattle, prices must be constantly monitored in order to decide whether to cover the sale via hedging or not. The price should be fixed when the probability of further gain is less than the probability of loss from the existing price level, or when further loss from lower prices would endanger economic survival of the enterprise. It is better to hedge in a small loss rather than suffer large losses due to a further price decline.

While the above reasoning appears logical, analysis of various cattle-feeding hedging strategies to implement that reasoning have not produced encouraging results.

USING HEDGING TO FIX CATTLE PRICES

A Kansas study reports the results on cattle feeding profits of several hedging strategies over a 10-year period.[5] The operation was patterned on a 20,000-head (one-time capacity) feedlot with approximately 400 head of feeder cattle scheduled in every week and 400 finished cattle sold every week. Costs of inputs and prices of finished cattle were based on current quotations throughout the period. Weekly, annual, and 10-year average feeding profits were calculated on a per-head basis. The feeding period was 147 days, so 505 lots of cattle were involved in the simulated analysis.

Seven program options were considered:

1. All lots fed out unhedged (the unhedged program was a benchmark or control for the other alternatives),
2. All lots routinely hedged,
3. Hedge only when the futures hedge price equaled or exceeded the calculated break-even price,
4. Hedge only lots when the futures hedge price equaled or exceeded the cash price for finished cattle when the feeder cattle went into the feedlot,
5. Hedge only lots when the futures hedge price equaled or exceeded *both* the break-even price and the current cash cattle price,

6. Hedge only lots finished during the September to December period, and
7. Sell finished cattle by cash forward contract the day feeder cattle went into the feedlot, at the contract price offered for finished cattle that day.

Results of the study are summarized in Table 9-1.

Table 9-1: Average Profits, per Head, from Seven Alternative Cattle Hedging and Contracting Programs, May, 1965–December, 1974

Alternative	1965	1966	1967	1968	1969	1970	1971	1972	1973	1974	Average 10-Year Profits	Variance	Lots Hedged
						($)							
I. Unhedged	36.30	13.68	2.97	18.67	26.77	-1.46	21.52	29.85	8.55	-53.11	9.55	1,079.737	0
II. Routine Hedge	15.51	16.12	14.01	6.23	2.23	-6.98	-7.36	-1.54	-29.06	-2.67	0.18[1]	417.243[1]	505
III. Futures ≥ Break-even	19.05	17.22	14.01	16.71	18.98	-5.32	21.45	27.28	2.05	-11.31	11.81	980.095	218
IV. Futures ≥ Cash	29.89	23.48	14.01	19.36	26.10	-6.23	10.48	29.85	-10.89	-0.23	13.08[2]	732.439[1]	204
V. Futures ≥ Break-even & ≥ Cash	29.89	23.33	14.01	19.36	26.10	-4.76	21.32	29.85	2.05	-12.28	14.43[3]	1,060.335	145
VI. Seasonal Hedged (Fall)	31.65	17.22	6.66	14.39	28.29	0.96	10.61	24.24	20.80	-44.48	10.38	907.302[3]	174
VII. Contracted	19.69	17.43	-1.33	13.72	21.42	-0.97	2.06	21.43	-29.98	-34.38	2.41[1]	199.556[1]	0

[1]Indicates that difference as compared to unhedged value is statistically significant at the 1 percent level.
[2]Indicates that difference as compared to unhedged value is statistically significant at the 10 percent level.
[3]Indicates that difference as compared to unhedged value is statistically significant at the 5 percent level.

Option 1 — Average profits on completely unhedged operations for the 10 years were $9.95 per head. The trend in profits was typified by instability. Under this system, 349 lots (69 percent) returned profits, while 156 lots experienced losses.

Option 2 — Routine hedging produced average profits of 18 cents per head; significantly below Option 1. In fact, routine hedging was the worst alternative in 7 out of 10 years. Routine hedging of all lots on the day they entered the feedlot improved profit on 183 lots but reduced profit on 322 lots.

The major reason routine hedging and other price-fixing techniques did not improve profits in this study was that cattle prices were in a general up trend for most of the period studied, as illustrated by Figure 9-1. Hedging nullified windfall profits that otherwise would have accrued during general rise in price level.

Option 3 — This is one of several selective hedging programs. Cattle were hedged only when the effective futures hedge price (futures adjusted for location basis) equaled or exceeded the break-even price. This strategy generated an average profit of $11.81 per head. This program assumed that when prices did not meet the hedge criterion, the cattle were fed out unhedged. Fifty-seven percent (287 lots) did not meet the criterion for hedging. Profits increased on 112 of the 218 hedged lots, while profits decreased on 106 lots.

If cattle had not been placed on feed when the effective futures price was less than break-even, average profits would have equaled $10.36 per head. This variation of Option 3 would have left the lots empty part of the time. Profits on the ani-

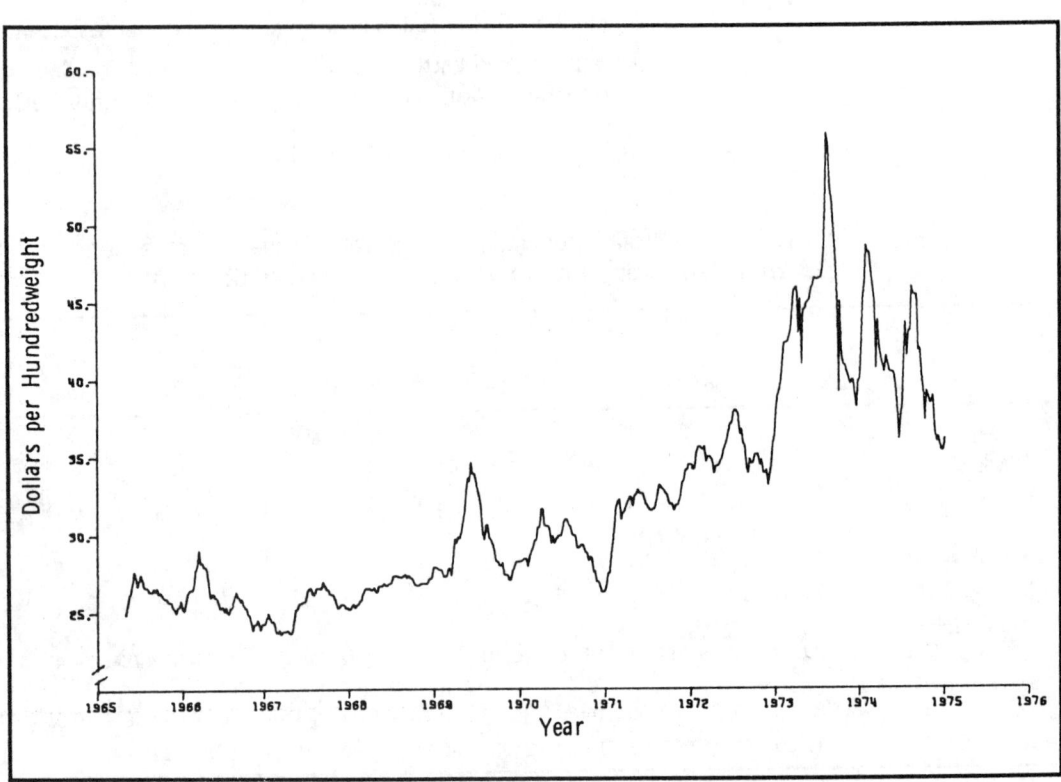

Figure 9-1: Prices of Choice 900-1,100# Slaughter Steers, Kansas City, May, 1965–Dec., 1974. Source: John H. McCoy, *Cattle Hedging Strategies*.

mals actually fed were less than profits on unhedged cattle. Thus, there was less return to total costs during the period studied.

Option 4 — Hedging was done when futures were equal to or greater than current cash price. Profits were $13.08 per head. Forty percent (204) of the 505 lots were hedged. Leaving the lots empty 301 times when the criterion was not met dropped profits to $3.21 on those actually fed out.

Option 5 — This option showed the highest profit of all strategies tested. The combination specified that cattle would be hedged only when the effective futures price equaled or exceeded both the break-even price and the current cash price. Profits were $14.43 per head. But only 29 percent of the lots were hedged. Leaving lots empty when the two criteria were not met dropped average profits to $7.10 per head.

Option 6 — A seasonal program of hedging produced average profits of $10.38. Only cattle finished for market during September through December were hedged; all other lots were fed out unhedged.

Option 7 — The final option did not use the futures market but assumed that the day they went on feed, cattle were contracted for delivery when finished. Profits were $2.41 per head, largely because that system precluded windfall profits during rising price trends.

Basis Problems and Opportunities

In the Kansas study, McCoy and Price noted particular problems in their analysis because of basis variation.[6] For satisfactory hedging, the basis relationship should be relatively constant, or at least predictable. The record shows substantial instability in the Kansas City basis for live cattle (Figure 9-2). Before 1973, the Kansas City basis averaged about minus $1.00, ranging from approximately minus $2.50 to plus $2.00. After 1973, however, the basis ranged from minus $7.70 to plus $4.40, with an average of approximately minus $2.00. The 10-year study used a basis of only minus $1.00 in deciding when to hedge. However, the realized basis was often much worse at maturity. A hedge sale that originated with an expected minus $1.00 basis, but ended with minus $3.00 basis, produced less than the expected price protection. Thus, several lots in the study that were hedged to cover break-even price finished with unexpected losses.

Basis decisions with non-storable commodities are meaningless. Unlike storable commodities, there is no way to capture strong basis levels, nor avoid poor basis levels. Live cattle contracts are traded on a bi-monthly schedule, but finished animals must move into slaughter and consumption when and only when ready. Cattle finished in mid-May cannot be delivered against an expiring April contract, nor profitably held into mid-June to deliver against the June contract. A weak basis during May cannot be avoided. Likewise, a favorable basis in April cannot be captured because the animals are not ready for market. Hedging non-storable commodities with futures contracts, therefore, does not provide total protection against price change because of basis variation. The basis variation is beyond the control of producers, and it is impossible to capture either strong basis or avoid weak basis at the time cash sale is made. The cash sale must be made when animals are market-ready.

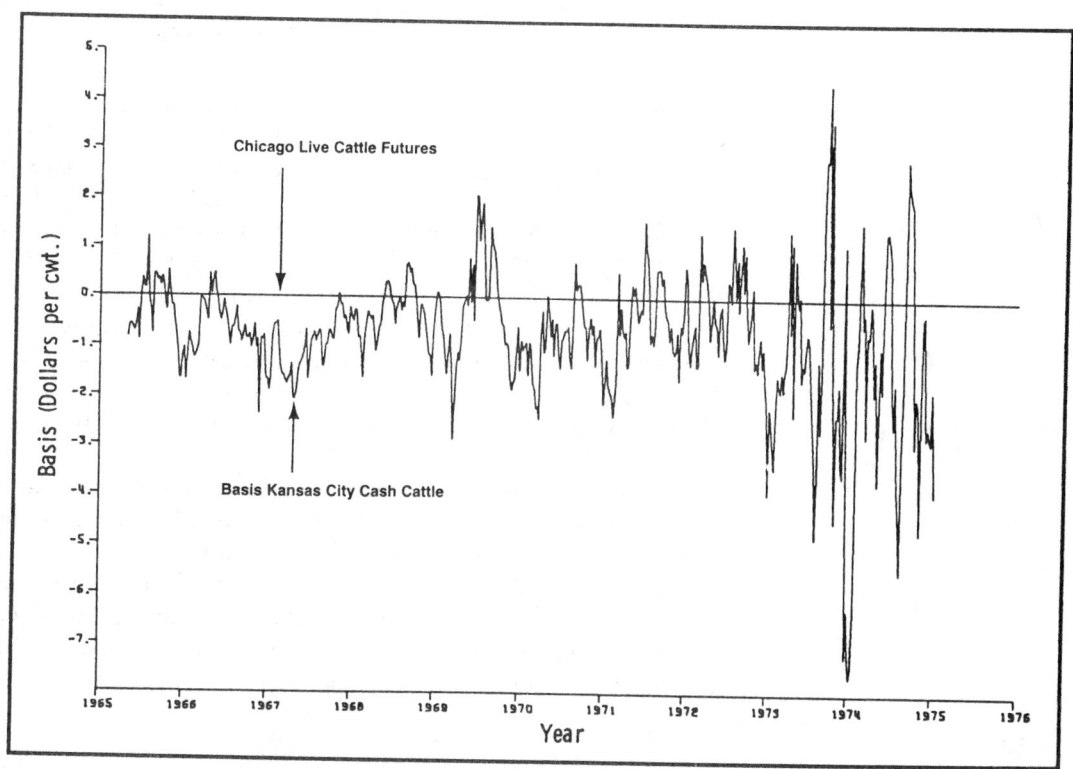

Figure 9-2: Kansas City–Chicago Live Cattle Basis, May, 1965–Dec., 1974. Source: John H. McCoy, *Cattle Hedging Strategies*.

Only during delivery months are cash price and futures price related by transportation and transaction costs of delivering against contracts. The only way a hedger can avoid a bad basis situation is to deliver production to satisfy the futures contract. Basis projections could be more meaningful and predictable if futures contracts on non-storable commodities were traded every month and delivery problems were minimized. Cash settlement of contracts, rather than physical delivery of livestock, might be beneficial in stabilizing basis. Until procedural changes occur, hedging livestock with futures contracts poses a serious hazard to a pricing plan because of basis variation.

Basis variation can eliminate hedge profit. The feeder who buys feed and feeders at a basis premium to the hedged prices effectively pays more for the inputs than the hedged costs would suggest. Similarly, if the finished product is sold at basis discounts, gross receipts are reduced. The feeder may have a futures profit of $50.00 per head locked into a pen of cattle, but if he or she pays $14.00 per head more for feeders (2 cents above hedged price on 700 pounds) and $13.50 more for corn (54 bushels per head times 25-cent basis premium), and receives $22.50 less on finished cattle (2 cents below hedged price on 1,100-pound cattle), the hedge profit margin is eliminated (a total of $49.50 higher costs and reduced returns). Profitable hedging does not guarantee profitable production, because of basis variation.

Viewed from the opposite perspective, basis variation on hedged feed purchases, hedged feeder cattle purchases, and hedged live cattle sales can be sources of additional profits. Futures transactions could lock in a generally acceptable profit level, but astute cash buying and selling to improve basis could yield significant additional returns. The feeder who can buy feed ingredients at basis discounts in excess of carrying costs will lower his or her cost of feed inputs below the hedged price. Likewise, if able to make cash purchases of feeder cattle below the hedged price, the buyer will have lowered the effective purchase price of feeders. On the product selling side, the feeder who can make cash sales of finished animals at higher than hedge prices will improve profits above the hedged selling price. Used as a marketing tool, with both a basis and price perspective, hedging removes price level speculation while focusing attention on basis level trading. Astute basis trading is as important to profitability in livestock feeding as is fixing price level decisions.

BEHAVIOR OF NON-STORABLE FUTURES CONTRACT PRICES

Hieronymus studied cash and live cattle futures over the period of 1973-1975.[7] His observations can be helpful in formulating a pricing strategy for hedging price level. The years of study covered a period of unusual price variability. Demand for meat was strong, as incomes rose. During 1973, there were production cutbacks on meat, and price ceilings were imposed by presidential decree. During 1974, poor crops led to herd reductions, forcing large supplies to market. Prices fell significantly in early 1975 but rebounded later as available supplies declined.

The futures price gyrations of that time (reported in Table 9-2) led Hieronymus to the following observations about futures pricing of live cattle:

1. "All prices rise and fall together.
2. "Variation in cash price and nearby deliveries is greater than with more distant deliveries.

Table 9-2: Cash Cattle and Cattle Futures Prices, Chicago Mercantile Exchange, Monthly Averages of Daily Closing Prices, Dollars per Cwt.

Month	Choice Steers Omaha	Feb. 1973	Apr. 1973	June 1973	Aug. 1973	Oct. 1973	Dec. 1973	Feb. 1974	Apr. 1974	June 1974	Aug. 1974	Oct. 1974
1973												
Jan.	40.31	42.84	42.71	42.70	41.42	41.28						
Feb.	43.07	44.47	43.84	44.09	43.43	42.62	42.50	42.34				
Mar.	45.28		45.92	45.74	44.91	43.80	43.32	43.37	43.28			
Apr.	44.76		44.38	43.25	43.11	42.65	42.81	43.06	42.99			
May	45.71			45.91	45.52	45.15	45.25	45.35	45.03			
June	46.65			47.63	47.24	47.14	47.21	47.35	47.02	46.83		
July	48.11				50.93	51.75	52.41	52.59	51.87	51.68	50.94	
Aug.	53.39				56.72	55.64	56.50	57.07	56.67	56.72	55.92	
Sept.	44.79					43.48	45.40	48.11	48.31	48.42	47.91	47.42
Oct.	40.65					41.71	44.91	48.30	48.94	49.16	48.59	47.89
Nov.	37.47						40.56	45.28	46.27	46.38	45.71	45.02
Dec.	38.87						38.64	46.06	48.00	48.13	47.60	46.86

Month		Feb. 1974	Apr. 1974	June 1974	Aug. 1974	Oct. 1974	Dec. 1974	Feb. 1975	Apr. 1975	June 1975	Aug. 1975	Oct. 1975
1974												
Jan.	47.22	50.95	53.80	54.30	52.93	51.81	51.60					
Feb.	46.22	46.57	48.61	51.50	52.26	52.07	52.14					
Mar.	42.42		43.17	46.23	48.32	48.83	49.36					
Apr.	40.86		42.20	45.65	47.64	49.91	47.02	46.96				
May	40.04			41.04	41.59	39.69	39.26	39.13	37.06			
June	37.50			36.47	37.21	37.49	37.42	37.12	36.73			
July	43.60				46.02	46.30	45.73	45.51	45.41	45.65		
Aug.	46.52				48.46	46.70	45.19	44.81	45.38	46.10	46.33	
Sept.	41.24					39.41	38.11	41.21	41.77	43.02	43.30	
Oct.	39.72					41.09	42.96	44.06	44.03	45.29	45.02	
Nov.	37.77						38.53	40.57	41.28	42.46	42.24	
Dec.	37.13						38.12	41.14	41.90	42.60	42.68	42.43

Month		Feb. 1975	Apr. 1975	June 1975	Aug. 1975	Oct. 1975	Dec. 1975	Feb. 1976	Apr. 1976	June 1976	Aug. 1976	Oct. 1976
1975												
Jan.	36.17	37.10	38.00	38.70	38.82	38.69	38.83					
Feb.	34.68	35.08	36.72	37.54	37.22	36.81	36.93					
Mar.	36.14		37.96	38.59	37.65	36.77	36.90	37.02				
Apr.	42.68		42.50	43.29	41.08	39.27	38.92	38.74				
May	48.35			48.47	43.86	40.33	39.39	38.81				
June	52.21			53.42	48.63	43.51	41.74	40.67	40.17			
July	50.41				46.77	41.91	40.84	40.37	40.05	40.61		
Aug.	47.84				47.27	42.07	42.04	42.77	42.98	43.65		
Sept.	49.64					47.79	47.21	45.29	44.45	44.92	44.77	
Oct.	48.27					47.49	44.59	41.69	40.81	41.72	41.53	
Nov.	45.49						45.83	43.48	41.90	42.79	42.16	42.05
Dec.	45.62						47.03	43.70	41.71	42.81	42.69	42.16

Note: Read across the table to discern the relative cash and various futures option prices that existed in any month. Read down the columns to see how the price of cash or a particular option changed over time.

Source: Hieronymus 168.

3. "There is substantial independence of price change in the more distant deliveries and the cash.

4. "The dominant influence on cattle futures is the cash cattle price.

5. "The dominance of the cash price suggests that the market seems to think cash price is as it should be and will not change.

6. "Traders are very skeptical about the validity of recent price change. They view a change as an aberration which will not last, and they assume that prices will return to their earlier level.
7. "And finally, markets do look ahead, anticipate supplies to come to market at a later time, and adjust prices in line with expectations."

These generalities are important in understanding some of the differences between non-storable and storable futures. Non-storable commodities do not have a reason to reflect a market carrying charge. Therefore, price in each contract month should be the same as every other month, save for changes caused by anticipated supply or demand change in each respective contract month. This explains the dominance of cash price in setting nearby futures price level and the unresponsiveness of distant contract months to fluctuations in nearby contracts. Since real changes in future demand or supply cannot be accurately predicted, the present situation dominates the price-making forces. Unpredictable events that occur over time cause futures prices to fail miserably as an accurate forecast of cash prices. As with grain contracts, futures prices are a daily pricing offer to accept or reject; they are not accurate price forecasts because price-making forces change.

MAKING LIVESTOCK HEDGING WORK: BEAR SPREAD HEDGING

Textbook, fixed hedging in distant months when production is to be market ready seldom works to capture a high price level in livestock. To gain pricing rewards from the futures markets, livestock producers need to employ a different strategy.

Livestock hedgers are frustrated trying to set hedges with futures contracts because of the time lag between nearby high prices and lower prices on distant options that coincide with when they will have market-ready animals. A reduced supply of immediately available market-ready animals causes strength in cash prices as packers bid for animals. Strong cash price leads nearby futures contract price higher. But more distant contract months typically do not rise as strongly as nearby months, if at all. Breeders and feeders cannot capitalize on price protection for expected production when current cash prices are high, and distant futures offer significantly lower pricing opportunities.

To illustrate the point, suppose the cash hog market and futures rise to $60 per hundredweight and then decline to $40 over a 12-month period at a steady rate. A producer selling equal numbers each week realizes the annual average price of $50.

A fixed hedging strategy would sell contracts in each option month, hold those contracts until the live hogs are sold, and liquidate the contracts upon sale of live animals. Since not all calendar months have future options, it is necessary to place two contracts in the next option to cover non-traded months. The prices in Table 9-3 are futures quotes for respective contracts that traded on September 15, 1986. A farmer producing 1,800 head of hogs annually would market 150 head each month, or about one contract. The price locked in by executing hedges on September 15 averages $52.28. After subtracting a $2.00 basis (applicable for Ohio) and 20 cents commission, the average net hedged price is $50.08. Hedging results are almost exactly the same as doing nothing and speculating in the cash market!

However, suppose that cash hog price rose to $70.00 before starting the 12-month decline to $40.00. In this instance, the cash market speculator would average $55.00 per cwt. The fixed hedger, locked into $50.08, would be worse off.

Table 9-3: Live Hog Futures Prices of Various Contract Months, as Traded September 15, 1986

Option	Price/Cwt.	
Dec.	$56.10	
Jan. sales on Feb. contr.	55.17	
Feb.	55.17	
Mar. sales on Apr. contr.	49.70	
Apr.	49.70	
May sales on June contr.	51.00	
June	51.00	
July	50.40	
average	52.28	
less	2.00	Ohio basis
	50.28	
less	.20	Commissions
	$50.08	

With option-to-option price discounts such as on September 15, fixed hedging will match the results of not hedging only when hedgers execute their positions at the extreme top of futures market prices. When distant options are discounted less, hedgers might gain. But if distant options are discounted more, hedgers lock in lower returns than cash marketing.

The secret to capturing returns in livestock hedging is to execute futures contracts equal to a full year's worth of production in the nearby contract (trading at a premium to other months), hold the position until the futures market breaks, and then lift all of them at once whenever the bottom of the trend is signaled. When the break occurs, the nearby option typically falls more sharply than the deferred contracts. Since the break may not occur while holding the nearby option, the contracts must be rolled forward into the next nearby month, waiting for the break. Three or four rolls may be required.

The 1982-1983 hog price situation serves to illustrate the procedure. The weekly plot of the various contract months in Figure 9-3 shows the strength in nearby contracts when strong cash markets led the August, 1982, futures contract to historic highs. A pork producer looking at cash prices in late August and planning to expand production would not have extra animals market-ready until about June, 1983. The August contract traded up to $66.08, but the June contract was trading at $50.81 per hundredweight at the same time. This meant that June futures were offering $33.60 ($15.27 × 220 lbs.) less per head for the hogs. Farmers usually expect current prices to hold a constant level into the future. Each farmer thinks he or she is the only person who sees strong price as a market signal to expand, and does not believe other producers will increase production. Commodity traders feel differently, however, by discounting distant options significantly.

It is psychologically difficult for farmers to hedge when distant contract prices lag significantly below current cash prices. They may have been stung by hedging too soon in the market rise and do not want to risk another drubbing. Figure 9-3 shows that prices did fall in response to increased production as the June

Figure 9-3: Relationship of Various Futures Contract Months, 1982-1983, Live Hog Contracts.

contract dropped to $46.00. However, a June contract sold at $51.00, when the August contract peaked, would have offered little protection. A June hedge set at the high in January of $56.00 would have provided more price protection, but picking tops is virtually impossible.

A solution to the livestock hedging problem is bear spread hedging. It permits capturing the strong nearby futures price level and relies on anticipated producer response to present high prices as they expand production. It further depends upon futures prices eventually flattening out month to month in characteristic fashion, because they have no carrying charges to reflect.

Bear spread hedging is initiated by using some method to decide that futures price of the nearby contract has reached its high, or is sufficiently high to be acceptable. At this time, hedges on 12 months of production are executed by using the nearby contract, for example, selling the October contract in early September near the top depicted in Figure 9-3. There is no way that the producer can speed up production to have animals available for delivery against the nearby contract, as a true hedge would dictate. But as the nearby contract reaches the time of maturity the price protection is rolled into a more distant contract month, such as December. This is done by simultaneously lifting the nearby (with a buy) and resetting the hedge in another contract month (with a sell). As the price level declines, the second contract continues to provide the price protection sought. As the second contract approaches maturity, the protection is again rolled into another contract month, such as June, by the same procedure as before.

To exit the hedge, and lift the contracts, producers can follow one of two methods. As production equal to one contract is delivered to the cash market, one contract is lifted and not replaced. This procedure satisfies the definition of hedging, because contracted amounts are kept in balance with production amounts. Eventually, all the production will be sold and all the contracts lifted. An alternative procedure is to hold all the contracts and lift all of them when the market analysis signals that the market has bottomed. By lifting one at a time, the hedger may run out of positions before the market finishes sinking. Lifting them all at once, near the market bottom, allows the producer to capture more of the total move on the full year's worth of production.

Bear spread hedging permits a producer to set hedges at higher price levels than deferred contracts offer. But it does not guarantee that the hedges will capture all of the price move that eventually occurs. Rolling contracts month to month has its hazards and challenges.

Figure 9-3 shows the October option dropping to the level of the February option. Rolling from October to February could have been done at the same price. But during February, the roll to June might have been done at a $4.00 discount (February $59.00 versus June $55.00). Such slippage cannot be avoided. There is no way to capture the option-to-option price discounts; one can only try to trade them at minimum amounts. But slippage is not the real issue; the real issue in January is whether hog price will be $56.00 or $46.00 in June. The proverbial optimist anticipates $56.00 forevermore; the realist sells a year's worth of production in January by rolling into the June contract.

In due time, the producer will have sold the planned production and lifted the hedge. The producer should have achieved a greater degree of price protection than if he or she had initially established hedges in lower-priced distant contract months. If the producer is lucky, he or she might even set contracts on 12 months of production near the top of the market and capture most of the price move.

TIMING CASH SALES

Within a restricted time frame, a producer can read the market for signals on when to sell cash livestock. He or she can shorten or lengthen the feeding period slightly in order to alter selling weight in response to changing price and changing production cost to maximize feeding profits. Hogs can be sold as light as 190 pounds or fed longer and sold at 260 pounds. Cattle can be sold early or late by adjusting feeding period and selling weight. Present prices, cost of gain, weight discounts, and the probability of price change can be combined into a decision framework to determine when to move livestock to market. Proper evaluation of costs and returns can give the producer some measure of control in selecting relatively more profitable selling weights and market dates.

At heavier weights, livestock are less efficient in converting feed into weight gain. Cost of gain goes up at an increasing rate as animals put on additional weight. Table 9-4 notes the amount of feed required for each 30 pounds of additional gain and the added cost of gain at various corn prices. As long as market price per hundredweight exceeds the cost of adding a pound of gain, it is profitable to feed to heavier weights. For example, if market hogs were selling for $40.00 per hundredweight and corn price was $3.30 per bushel, the added cost of feeding hogs from 220 to 250 pounds would be profitable because the added cost would be $39.49 per hundredweight. The profit-maximizing goal is to equate *added* cost of each pound of production to the *added* returns from that extra production. As long as added returns exceed added costs, net returns continue to increase. The principle applies to all livestock.

Table 9-4: Cost of Adding 30 Pounds of Gain on Hogs Weighing 190 and 220 Pounds at Various Corn Prices

	Total Cost of Gain[1]			
	190-Lb. Hog Fed to 220 Lb.		220-Lb. Hog Fed to 250 Lb.	
Price of Corn per Bushel	Cost of 30 Lb.	Cost per 100 Lb.	Cost of 30 Lb.	Cost per 100 Lb.
	.... ($)			
1.50	5.22	17.40	5.44	18.13
1.60	5.57	18.56	5.80	19.33
1.70	5.92	19.72	6.16	20.54
1.80	6.26	20.88	6.53	21.75
1.90	6.61	22.04	6.89	22.96
2.00	6.96	23.20	7.25	24.17
2.10	7.31	24.36	7.61	25.38
2.20	7.66	25.52	7.97	26.58
2.30	8.00	26.67	8.34	27.79
2.40	8.35	27.83	8.70	29.00
2.50	8.70	29.00	9.06	30.21
2.60	9.05	30.17	9.43	31.42
3.00	10.34	34.45	10.62	35.39
3.30	11.37	37.90	11.85	39.49
3.60	12.40	41.35	12.92	43.07

Source: Derived from data in John T. Larsen and Robert L. Rizek, *The Costs of Feeding to Heavier Weights*, ERS-371 (Washington, D.C.: USDA, February, 1968), 27-28.

[1] Calculated on the basis that 2.4 bushels of corn or its equivalent is required to raise a hog from 190 to 220 pounds, and 2.5 bushels from 220 to 250 pounds. An extra 45 percent is added to the cost of feed as the approximate costs other than feed.

During periods of either rising or declining price trends, the analysis must be modified in response to changing price. With a changing price trend, total value of the animal is changing, not just the additional pounds. Rising price increases the value of the first 220 pounds on a market hog, not just the value of the 30 pounds added by going to 250 pounds. Conversely, a declining price means that value loss occurs on all pounds, not just the added weight.

To decide when to sell in either up trending or down trending markets, the producer must calculate the change in total value of the animal against the cost of adding the additional weight.* If price is expected to increase to $42.00 per hundredweight over the next two weeks, the value of the animal increases by $17.00. The original 220 pounds increases in value by 2 cents per pound and the extra 30 pounds are worth 42 cents each.

Depending upon added cost and extent of price increase expected, additional weight increments might be profitable in rising markets. One risk is price discounts for over-weight hogs that reduce the selling price. Assume that by increasing selling weight from 250 to 260 pounds, a $1.00 per hundredweight price discount is applied (from $42.00 to $41.00). The original 250 pounds would be discounted $2.50. The extra 10 pounds would be worth $4.10. Added cost of feed would be $4.00. Thus, the net result would be a reduction in profit of $2.40 per head ($4.10 value of weight gain, minus $4.00 added cost, minus $2.50 discount on original 250 pounds).

This method of analysis also pinpoints the profitable sale weight in a declining market. If price is expected to decline $2.00 (from $42.00 to $40.00) per hundredweight in the time required to grow from 220 to 250 pounds, the total value of the animal will increase only $7.60 (220# × 42¢ = $92.40 versus 250# × 40¢ = $100.00). As costs increase by $11.85, the result is a $4.25 reduction in profit ($7.60 − $11.85).

The conclusion of this analysis is to feed to heavier weights in up trending markets until overweight discounts are incurred, and to sell at lighter weight in down trending markets.

Detection of market price up trends or down trends is critically important in making cash sale decisions. Price charts of both futures and cash prices identify the general trend of market price for the analysis. The producer cannot predict exactly how much price will change over any time period, but if the producer has a feel for the trend, he or she can capture more profit than by just being an "every Monday" seller. A cattle feeder who watches price trends and reacts to them will certainly be better off than neighbors holding 1,600-pound steers, hoping for a price recovery.

Selection of "the" selling day can be further refined by watching the trend on a daily cash price chart and selling on up-price days. Buyers often raise their bids on days of light runs. The day to sell is when nobody else is selling.

If all producers utilized an added value versus added cost analysis to select the weight for marketing cash livestock, price declines in down trending markets would be halted earlier by reductions in the total tonnage that the market would otherwise have to absorb. In up trending markets, the price rise would be stopped short of the extreme by increased tonnage from heavier animals. Some short-term price variability would be removed from the market as producers reduced tonnage in weak markets, or increased tonnage in strong markets. In practice, feeders tend to hold heavyweight animals, hoping for price recovery. The practice is self-defeating, as increased total tonnage actually depresses price further when it eventually does come to market.

---

*This assumes that the cost of weight gain is the only variable. This assumption is invalid in limited-space operations where overcrowding results from delayed marketing. Space and time costs are thus also important in calculating the optimum selling weight but are ignored here in order to simplify the analysis and explain the procedure.

AGRICULTURAL PRICE CYCLES

Long-term agricultural price cycles refer to fluctuating price levels over time caused by periodic expansion and contraction of the supply of agricultural products. Cycles reflect farmer production decisions and the subsequent market response to move an adjusted supply into consumption. High price entices increased production which must be sold at low price when the increase is market-ready. Low price discourages production, and the reduced supply fetches a higher price at a later date. Price changes are much larger than production changes due to inelastic demand for farm products. Examples of production cycles and the resulting price cycles are shown in Figure 9-4.

Price cycles are based upon producers' expectation that current price levels will continue to hold into the future. Producers tend to base production plans on current profits rather than on future expectations. They expand production in response to current high prices, thinking that no other producer will do the same. When the aggregate of increased supply becomes market-ready, prices decline to move the added volume into consumption. The cattle cycle is a good reference for understanding price cycles.

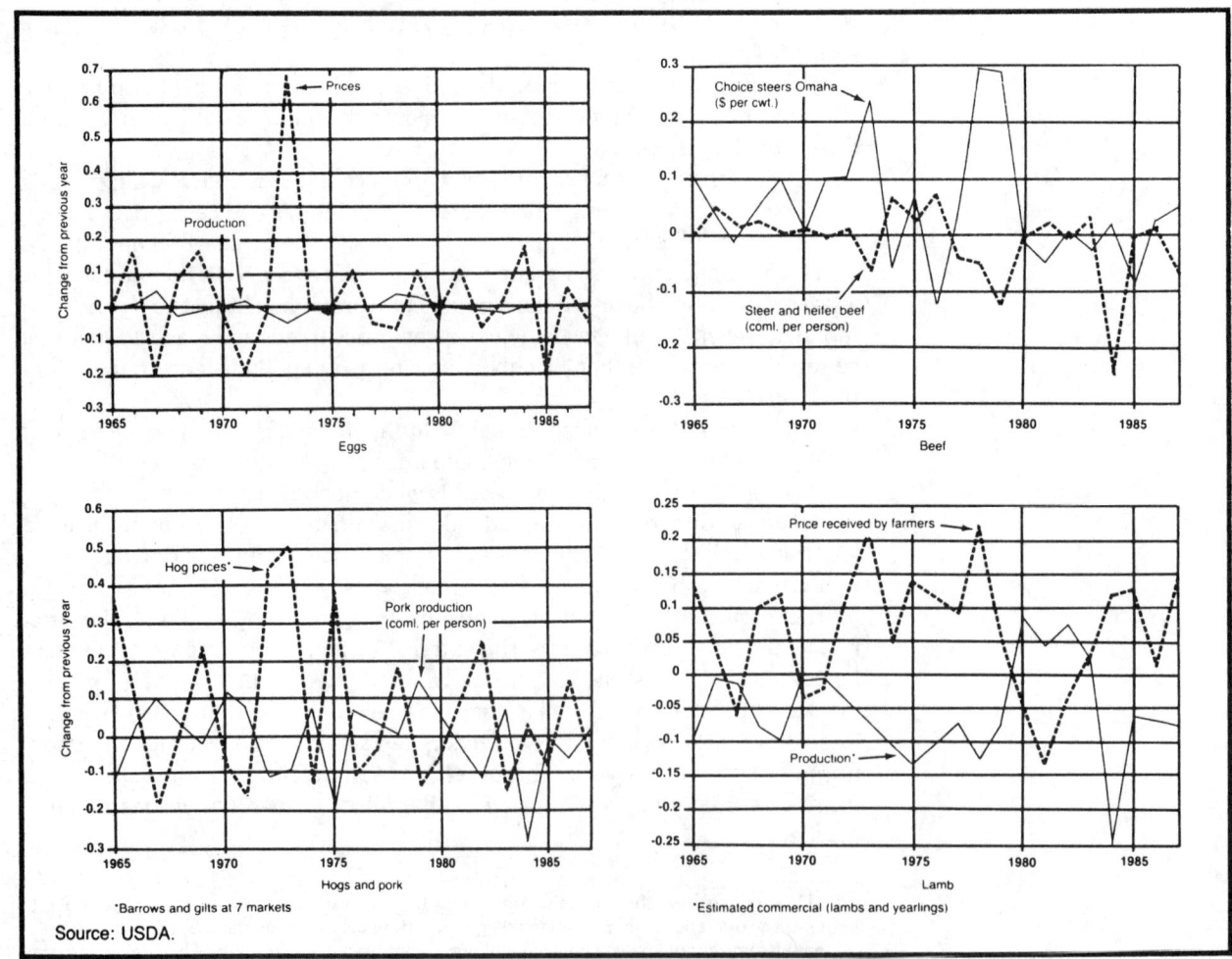

Figure 9-4: Price and Production Cycles for Four Commodities.

The Cattle Cycle

Ranching and Cow-Calf Production.

> The cattle cycle is an economic disease. We spend millions on animal diseases. How much do we spend on the real income killer, the cattle cycle?... We view the cattle cycle like death and taxes, as inevitable. Cycles are predictable and their causes are known. Through research and education, they could essentially be eliminated.
>
> Dr. Willard Williams

Characteristics of the beef enterprise create a predictable supply pattern. All the product that is produced will be consumed, at some price. Supply of livestock on the market thus becomes the dominant force in setting price. Outside factors exert relatively minor influences. Supply is controlled by thousands of independent cow-calf producers who tend to react in unison to high prices or low prices. The result is a predictable cattle cycle supply curve and its concomitant resulting prices.

Reduced beef supplies initiate the expansion phase of the cattle cycle. At the meat counter, prices rise to ration available supplies. High retail beef prices are reflected through the channels of trade to affect prices of slaughter cattle, feeders, and breeding stock. At relatively high price levels, heifers that normally would have been marketed as feeder calves are held back by ranchers and added to breeding herd. Slaughter numbers are further reduced, and beef supplies become more limited. Retail beef prices increase further to intensify the production signal to producers.

In response to high prices, producers ultimately expand production. In due time, expanded production means larger meat supply in the retail markets. Larger supplies in the market lead to reduced retail meat prices, lower fat cattle prices, reduced feeder prices, and lower inventory values on breeding stock. As producers reduce inventory to maintain cash flow or eliminate unprofitable production, the market must move an abnormally large volume of production to liquidate inventories. Prices drop severely because inelastic demand causes a much larger percentage drop in prices than the percentage change in volume. After a period of liquidation, marketings decline and prices improve for cattle producers. Then the expansion phase of the cycles takes over and the process is repeated.

A key factor in the cattle cycle is the biological time lag in the production process. Quick reactions to market signals are impossible because so much time is required in cattle reproduction (Figure 9-5). Heifers are bred for their first calf when they are 14 to 18 months old. Their gestation period is 9 months. The calf in turn will reach mature slaughter weight in 17 to 19 months, depending on its individual rate of gain and the feeding program. If a heifer is retained to expand the herd rather than sent to slaughter, it could take about 5½ years from the time she is born until her offspring reach slaughter weight to increase the available beef supply. On the consumption end, however, slaughter animals move through the packing plant, arrive at food stores, and are usually consumed within two weeks after leaving the feedlot.

The biological time lag is further complicated by producer response. Outside forces such as inflation or deflation, weather, feed supplies, and feed prices influence producer decisions to expand or contract production. The psychological lag in producer response to price signals causes production cycles to extend beyond the length of time involved in the biological lag. The combined effect has caused cattle cycles to run about 10 years in length from peak to peak in numbers during the past three decades. Figure 9-6 shows the production cycles since 1930.

The cattle price roller coaster was particularly severe during the 1970's and 1980's. The early part of this period was characterized by general inflation, rising consumer income, and increasing demand for meat. During 1973, per-capita supply of beef decreased for the first time in many years, putting increased up-

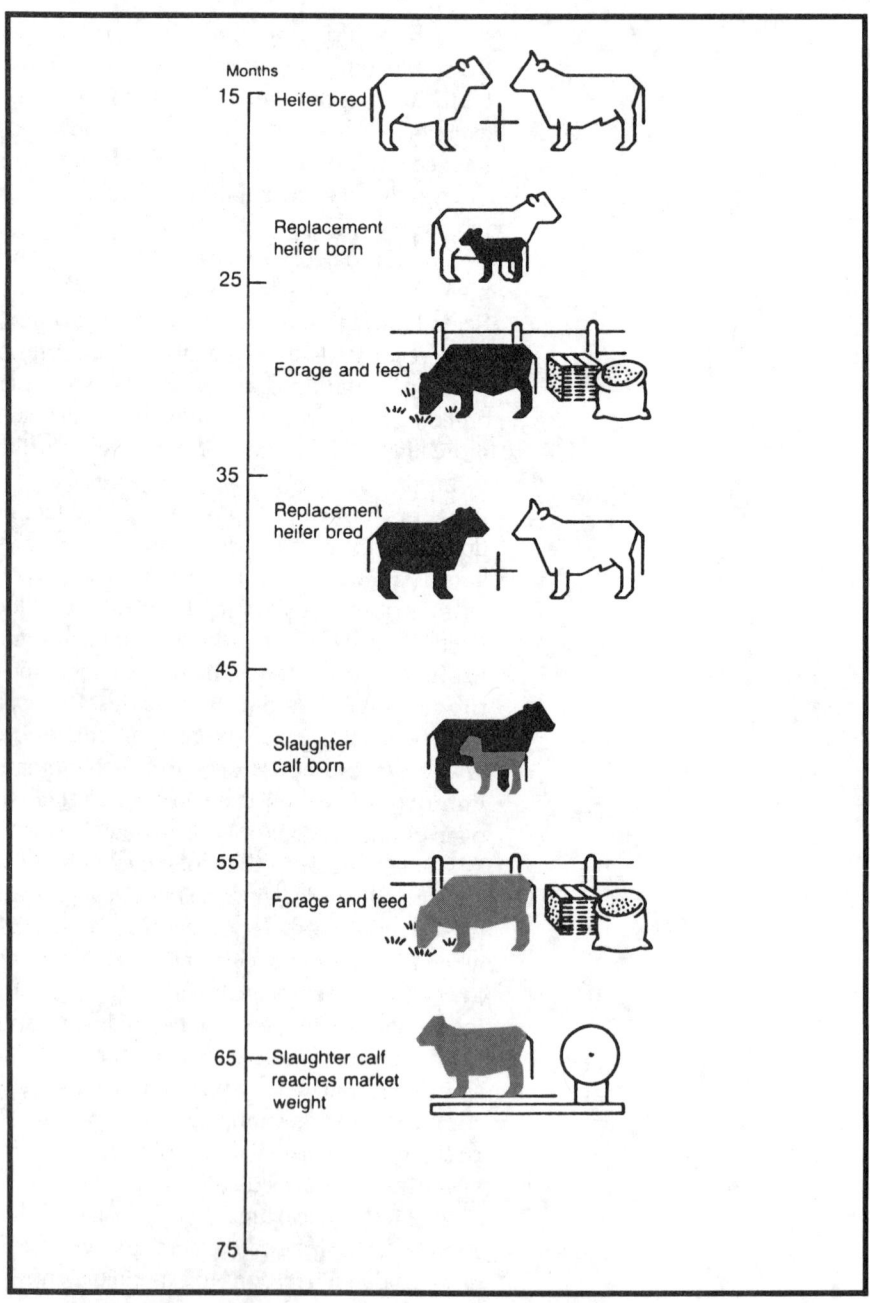

Figure 9-5: Expansion Plans Are Subject to Biological Timetable.

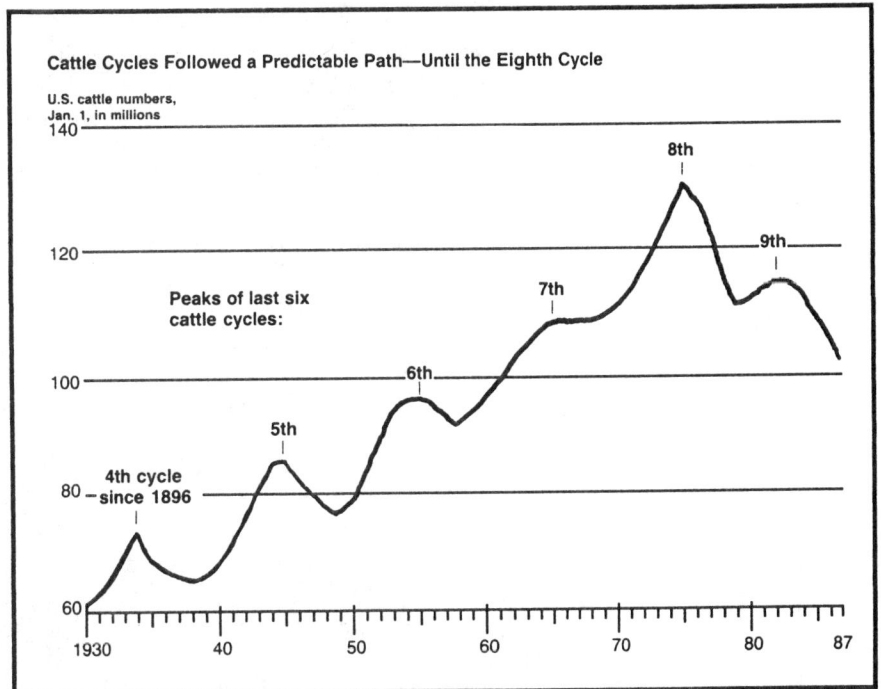

Figure 9-6: Herd Rebuilding Signals the Start of a New Cattle Cycle. Source: *Farmline*, "Why the Cattle Cycle Went Awry," May, 1987.

ward pressure on prices. Meat boycotts, price ceilings, and panic buying distorted already high prices. Producers reacted to high prices by diverting heifers from slaughter to breeding to increase the size of their cattle herds. Producers expanded the U.S. total cattle herd to a record 132 million head by 1975. A severe drought in 1975 reduced feed and forage production. Production adjustments caused high slaughter rates and a severe price decline. In response, cattle numbers declined 16 percent (21 million head) over the next four years from 1975's record high. Expansion was then encouraged by excellent grazing conditions and record high prices in the spring of 1979. By January, 1981, the U.S. inventory had increased by about 4 percent to 115 million head and then stalled. From pressures of the faltering agricultural economy, ranchers sent heifers to feedlots and cows to slaughter. On January 1, 1987, the U.S. cattle herd slumped to 102 million head, the lowest number since 1962.

Prices of beef calves, as one measure of changing prices, gyrated in response to the changing supply-demand conditions. They rose to about $70.00 per hundredweight in 1973, only to plunge into the $20's and $30's late in 1974 and 1975. After gradual improvement, feeder calf prices skyrocketed to over $100 per hundredweight in the spring of 1979. They slipped to $65 in 1986 and began recovering to $76 in 1987.

Decision Making by Cow-Calf Producers. Fluctuations in supplies and prices could be reduced considerably if production were geared to consumer demand.

One way for a rancher to gear production to demand is to calculate the ratio of the number of cattle (cows only) per 100 people in the U.S. population.[8] This ratio provides an indication of the supply-demand balance. It averaged 24.5 cows per 100 people from 1959 to 1975 (Figure 9-7). The ratio was higher before 1959 because of more dairy animals on farms. In the 20 years from 1955 to 1975, a ratio greater than 24.5 suggested an over-supply relative to demand and therefore lower prices; a lower ratio signaled an under-supply and higher prices. In 1975, the number of cows per 100 people rose to 26, a 20-year high, and the price of feeder cattle in constant (1967) dollars fell to the lowest level in the 40-year period from 1940 to 1980. The cow herd was then liquidated, and the number of cows per 100 people declined to 22 by the start of 1980. In response, feeder calf prices rose to the level of the early 1970's. Since 1982, cow population continued a dramatic and unprecedented decline, but price level also declined to pre-1940 levels in constant dollars. Something changed.

Prior to the 1980's, it was thought that multiplying the average ratio by the population would give some idea of growth in herd size that would need to occur in future years.[9] The U.S. population was forecasted to rise to 241 million by 1990. Using the measuring stick of 24.5 cows per 100 indicated the need for a cow herd of about 60 million head by 1990, assuming no change in per-capita consumption. This herd size would provide a per-capita supply of 89 pounds of retail-weight beef. Of the total cow herd, about 10 million head would be dairy cows. From a cow herd of 50 million cows in 1982, growth to 60 million did not seem unreasonable. But the 44.5 million herd in 1987 presented a different story; there was no way the herd size could rebound by 1990 to 60 million head. The predicted path of previous cycles changed in 1982 because the assumption of per-capita consumption proved invalid as consumers changed eating habits. Ap-

Figure 9-7: Cow-People Ratio and Feeder Animal Prices, 1925-1987.

parently due to health concerns, they switched from beef to poultry consumption, and beef consumption declined.

In order for producers to benefit from cattle cycles — and avoid economic squeezes — it is necessary to recognize potential over- or under-supplies, before they are reflected in the market place. By monitoring herd size and the rate at which replacement heifers are entering the herd, producers can anticipate excessive expansion or contraction. To develop a cyclical marketing strategy, a rancher must take appropriate action to create a counter-cyclical reaction to what others are doing. As others are building inventory, the individual rancher should be reducing numbers and shortening the length of time he or she holds animals. The counter-cyclical rancher would cull fast, sell early, and sell light. Cows should be culled in the early stages of a price decline; they will be worth less in salvage value later on. The individual should sell calves rather than hold them as yearlings. A stocker operator could buy in the spring at lower price than the previous fall and also not incur the expense of winter feed. Purebred cattle producers should sell bull calves rather than holding them until they are breeding age. Contracted sales of cattle should be made early in the season rather than waiting until heavy fall runs materialize at weaning time. Hedging feeder sales with futures could be used to protect price level.

When the market turns, the reverse strategy applies. The time to build inventory and hold animals longer is when others are liquidating at lower prices. Calves should be kept to sell as yearlings. Heifer calves can be held at low price levels for sale as bred heifers or cows when the market for females is more active. Stocker operators should buy in the fall and overwinter the stock for the following grass season. About halfway in the build-up phase of the cycle, lots of capital tends to flow into the cattle market. It creates a particularly active market for breeding stock. Purebred producers should hold young stock for a two-year-old bull and bred heifer sale. Contracting or futures hedging should be avoided. In the up trending phase of the cycle, the objective is to own stock as long as possible to profit from the rising price trend.

Such a counter-cyclical production strategy creates uneven cash flows, but it could improve long-run profitability. If a small number of producers practiced the counter-cyclical strategy, there would be moderation of the extremes of the cycle. If every producer followed the strategy, the cycle would be changed. In reality, since most cattle producers have reacted the same way in the past, they can be depended upon to do so in the future. Cattle production cycles, with boom and bust prices, will continue in the future.

SUMMARY

This chapter draws attention to differences in livestock product pricing compared to storable grains. Because of these differences, futures pricing and hedging strategies also have to be different. The study cited on cattle hedging showed that no pricing strategy panacea exists. But bear spread hedging and timing cash marketings should improve profits.

The problems associated with basis complicate hedging as a means to protect prices. Basis variability does not permit accurate localizing of futures prices to project the results of hedging. Because of the biological schedule for selling livestock, the producer does not have the flexibility that grain sellers have. The delivery process and handling of live animals make delivery against contracts more complicated. Until the futures markets solve the basis problem, hedging is limited in its ability to reduce price risk.

Finally, the chapter reviewed agricultural price cycles in the context of cow-calf production. The problems and opportunities posed by cattle cycles can be extended into other livestock products. The important thing is to recognize their existence and formulate comprehensive strategies for reacting to changing production cycles.

Questions for Study and Discussion:

1. What particular feature makes pricing non-storable commodities different from pricing storable commodities?
2. Explain the key fundamental factors you need to analyze in forecasting hog and live cattle prices.
3. What pricing strategies have been used for making futures pricing decisions in live cattle?
4. How do basis considerations complicate futures pricing decisions?
5. What general conclusions have been reached about the behavior of non-storable futures contract prices?
6. Explain bear spread hedging.
7. What causes agricultural price cycles?
8. Why do agricultural prices change relatively more than agricultural production?
9. Explain the biological and psychological lags in the cattle cycle.
10. Explain the system suggested as one way for farmers and ranchers to predict impending over-supply and under-supply of cattle.

Footnotes for Chapter 9:

(1) Raymond M. Leuthold, "The Price Performance on the Futures Markets of a Non-storable Commodity: Live Beef Cattle," *American Journal of Agricultural Economics* 55(1974): 271-279.

(2) Raymond M. Leuthold and Peter A. Hartman, "A Semi-strong Farm Evaluation of the Efficiency of the Hog Futures Market," *American Journal of Agricultural Economics*, Vol. 61, No. 3 (1979): 482-489.

(3) Philip Garcia and Larry Martin, *Futures Markets for Selected Non-storable Commodities: Price Forecasting Revisited*, Bulletin No. 80E-148 (Champaign: University of Illinois, 1980).

(4) Refer to Chapter 4 for a more detailed discussion of fundamental pricing, and to Good, Hieronymus, and Hinton, *Price Forecasting and Sales Management*, for an in-depth analysis.

(5) John H. McCoy and Robert V. Price, *Cattle Hedging Strategies*, Bulletin 591 (Manhattan: Agricultural Experiment Station, Kansas State University, 1975).

(6) McCoy and Price, *Cattle Hedging Strategies*.

(7) Thomas A. Hieronymus, *Economics of Futures Trading* (New York: Commodity Research Bureau, 1971) 167-169.

(8) USDA, "New Cattle Cycle: Can Producers Take Charge?" *Farmline* (Washington, D.C.: Economic Research Service, July, 1980) 14.

(9) USDA, "New Cattle Cycle."

Chapter 10

730-Day Crop Marketing and Profit-Maximizing Livestock Hedging

THE MARKETING CHALLENGE 302
ONE-TIME SELLING OR CONTINUOUS MARKETING? 303
    One-Time Selling 303
    Pricing Strategies 303
AN INTEGRATED PLAN OF CONTINUOUS 730-DAY MARKETING 304
    Time Framework: Marketing for 730 Days 304
    Selective Hedging/Selective Speculating 304
    Storage Pricing for 12 Months 304
    Flexible Hedging 305
    Basis Sales: Capturing Storage Profits 307
        Storage Decisions for 730-Day Marketers 309
PROFIT-MAXIMIZING LIVESTOCK HEDGING 310
    Selective Hedging of Finished Animals 310
    Hedging Feeder Purchases 310
    Profit Margin Hedging 311
    Basis Signals for the Feed Buyer 312
    Futures Contract Months 313
SUMMARY: THE 730-DAY CROP MARKETING PLAN AND PROFIT-
MAXIMIZING LIVESTOCK HEDGING 313
    730-Day Marketing for the Crop Farmer 314
    Profit-Maximizing Hedging for the Livestock Farmer 315

730-Day Crop Marketing and Profit-Maximizing Livestock Hedging

Commodity marketing is not a simple process; it consists of several pieces. Each of the pieces has its unique characteristics, uses, and limitations, and each forms a part of the overall picture. The individual pieces can be combined in a number of ways. An intelligent marketer understands the pieces and fits them together into an integrated plan to serve his or her particular situation.

THE MARKETING CHALLENGE

In the chapters that preceded, we have considered a number of concepts as background for formulating a marketing plan. Let us review the pieces to see how they fit together in an integrated marketing plan.

1. Price volatility offers the opportunity to capture high profits and the challenge to avoid disastrous losses created by rising and falling commodity prices.
2. Many marketing tools are available. Each has its own characteristics and limitations. All uncouple pricing decisions from production decisions.
3. Futures markets dominate the pricing aspect of marketing. They function within a fixed business structure and play several significant roles in the marketing process.
4. Supply and demand fundamentals ultimately determine cash price in a market economy. But unknown and changing physical, economic, political, social, and psychological conditions create uncertainty about future price realization.
5. Technical market analysis attempts to gauge the effect of human judgment and emotions in price discovery, especially in the short run. Tools of technical analysis identify price objectives and provide insights into reading market action.
6. A farmer's market strategy evolves as a method to react to market signals and to cope with risk. A strategy must be developed consistent with personal philosophy, attitudes, and financial condition.
7. Pricing decisions are required in two separate but related markets — futures price level and local basis level.
8. Adequate financing is vital in a marketing plan that utilizes futures contracts. Risk is increased, rather than reduced, by inadequate liquidity.

9. Storable and non-storable commodities have different pricing conditions because of the nature of the products. Pricing decisions must be different.

These individual pieces must now become parts of an integrated marketing plan that addresses the issue of profitable marketing.

ONE-TIME SELLING OR CONTINUOUS MARKETING?

To integrate all the pieces into a successful marketing plan, let us consider a fundamental question: Is a farmer wise to be a speculative, one-time marketer, or is he or she wiser to become a continuous marketer?

One-Time Selling

On July 1, a person farming 1,000 acres could have 100,000 bushels of grain stored in bins and 100,000 bushels growing in the field. Tons of money dance through this speculator-producer's mind, as one contemplates the booty reaped from selling both the stored crop and the growing crop for $3.80 per bushel. This farmer speculates on 200,000 bushels from both crops, hoping that another part of the country will have a drought, or that the Soviets will make a record purchase. Either or both events will reward the speculation.

Because the farmer intends to maximize profit by catching the TOP cash price of the year, he or she stands beside the market, day by day, delaying the pricing decision. One speculates that price will rise above today's level and provide an even higher price opportunity in the future. One knows that once grain is hauled to town, or a forward contract signed, that irreversible, one-time decision locks one into a profit level for the year. But when one speculates on higher price, one risks having to take lower price.

Lenders join in the speculative game by financing both the production of a new crop and storage of the old crop. They too have visions of the rewards from $3.80 corn. Two corn crops (200,000 bushels) would gross $760,000 for their borrower.

Unfortunately, the market often fails to cooperate with these visions. Widespread rains in July and August can produce a huge crop. Hoped-for international buyers may fail to place orders. Market price typically drops to move a bountiful crop. In a $2.00 corn market, 200,000 bushels gross only $400,000, less $45,000 carrying charge on the old crop to net only $355,000. From high price expectations to low price reality, with $400,000 swings in gross receipts, a farmer's net income all too often moves from riches to rags when speculating on two crops.

Cash market speculation is *one-time, passive marketing*. Emotional selling based on fear, hope, and greed influences most passive marketing decisions. Farmers don't sell at bonanza prices because they firmly believe prices will go higher, but they do sell at the bottom because they fear lower prices. Others submit to "wishful thinking" — $10 soybeans, $4 corn, and $6 wheat — and wait and wait. Both types typically sell when loans are due, before or well after the price has peaked.

Pricing Strategies

Confronted by the challenge to capture the high of market price and in the face of an uncertain future, farmers have tried many pricing strategies to cope with passive marketing. Some farmers sell by the calendar: equal amounts every Wednesday; or on the 15th of every month; or a third contracted in the growing year, a third from storage in December-January, and a third from storage during the June-August "summer rally." At best they capture the average market price, minus holding costs.

Then there are the calculators who try to improve their one-time pricing decision. Some base sales decisions on costs of production and attempt to scale up. Others analyze cycles, seasonals, fundamentals, or charts. With luck they sometimes make high-price sales on part of their production. But usually they look back with 20/20 hindsight and say, "That's when I should have sold both crops."

AN INTEGRATED PLAN OF CONTINUOUS 730-DAY MARKETING

There is a better way. It is an entirely different marketing concept, not passive marketing, with its irreversible one-time decisions, but *active farmer pricing* through continuous 730-DAY MARKETING — a 730-day "marketing year." This approach converts a farmer from a one-time passive seller into a continuous active crop merchandiser.

Time Framework: Marketing for 730 Days

In building an integrated marketing plan, crop producers must keep in mind the fact that pricing decisions on *a single crop span a two-year period*: the growing year and the storage year. The first stage of a crop "marketing year" begins in November as production plans are being made for the new crop and continues throughout the growing season until the end of harvest. During the second stage of the "marketing year," pricing of the harvested, or old, crop begins at the end of the growing 12 months and continues for the next 12 storage months. Thus, pricing of a single crop spans 730 days — the growing year plus the storage year.

Selective Hedging/Selective Speculating

The practice of 730-Day Marketing relies on selective hedging/selective speculating to reduce, not increase speculation. A producer establishes short futures positions equal to planned or growing production when futures prices trend down. Price protection from short futures contracts offsets value erosion in the growing crop. When market price stops declining and turns into an up trend, the short futures positions are liquidated. The value of a growing crop is allowed to appreciate, unfettered by a futures position, as market price rises.

Selective hedging solves some of the problems of fixed hedging. With fixed hedging, the producer sets hedges in down trending markets but maintains the position until the product is sold in the cash market. A corn producer who sets a profitable pre-harvest hedge at, say, $3.00 per bushel in April enjoys price protection as prices decline to $2.70 in June. But if the market turns around in July and rises back to $3.00 before harvest, the hedger will have seen price protection accumulate in the futures account, then disappear as the market rebounds. With selective hedging, one would lift the hedge position in July to let the value of the growing crop rise along with the market, rather than return the futures account surplus to satisfy the futures contract.

Storage Pricing for 12 Months

In up trending markets during the storage year, the crop producer benefits by owning the old crop. A rising price level increases the value of stored commodities. There are two methods for extending pricing of the old crop across the entire 12 months of the storage year. One may either own the physicals or replace the physicals with paper ownership through "Buying the Board." Once cash sales are made, it is impractical to repurchase cash grain to refill bins for further price appreciation; it is much cheaper and more convenient to replace the cash position with a paper position (long) in the futures market. Thus, the way of retaining ownership throughout the storage year is by "Buying the Board." A producer, understanding that holding unpriced grain in anticipation of rising price is pure speculation, should see little difference in holding long futures contracts to accomplish the same objective. To maximize the value of farm production, the 730-Day Marketing marketer utilizes long speculative positions in futures contracts during part of the marketing year to replace a cash position with a paper one.

The practice of reowning inventory has a major feature: It separates the basis pricing decision from the price level decision. It allows the commodity producer to capture BOTH a strong basis and a high price level. A high price level is typically accompanied by weak basis. Profitable basis sale opportunities are usually eliminated, therefore, in cash market speculation. The farmer-storer who is willing to make a cash sale at low price levels in response to strong basis knows full well that he or she can participate in the market price rise when it occurs by "Buying the Board." One does not have to hold cash commodities to speculate. One thus has a market philosophy that permits one to react to two sets of market signals and capture advantageous prices in both. Replacing a cash position with a paper one also eliminates the physical storage risks and carrying costs of storage speculation.

In practice, when old crop is stored after harvest and before cash sale, it is stored unhedged in price up trends. Its value is rising as the price level rises. Should the basis market dictate a cash sale of old crop, ownership of the physicals is replaced by buying futures contracts to continue the benefits of the price rise. When the up trend runs its course and a down trend sets in, the long futures position is liquidated. The 730-Day Marketer re-buys the old crop grain in each subsequent up trend throughout the storage year to ride its price upward on each move. One practices selective speculating to "Buy the Board" when appropriate. Thus, one is able to add price level increases from grain held on paper throughout the year, while having sold the physicals at the most advantageous time to do so.

Any stored old crop grain still waiting for a basis signal to sell is protected by short futures positions during price down trends. Corn stored at harvest when price is $3.00 per bushel loses value day by day as price declines during the storage period. Elevator operators who buy cash grain from farmers at harvest do not risk such value loss; why should farmers storing grain on the farm not protect the value of inventory? When the down trend has run its course, contrary to grain warehouse practice, farmer-storers should lift their short futures position and become selective speculators by allowing the inventory value to rise along with the market.

Across the entire 730-Day "marketing year" of pricing one crop (the growing year and the storage year) as a selective hedger, a farmer hedges in down trends but speculates in up trends. Our analysis shows that such a practice cuts cash market speculation in half. The farmer-producer is hedged about half the time. This is in direct contrast to most farmers who never hedge but speculate 100 percent of the time, all too often on 2 years' worth of production.

Flexible Hedging

Flexible hedging is required in order to execute 730-Day Marketing in a 12-month calendar year. A grain producer with 100,000 bushels of old crop and 100,000 bushels of growing new crop must make separate decisions on pricing of each crop. As price trends up and down during the calendar year, one seeks to protect the growing crop against declining markets with hedges, and one seeks to capture price appreciation on old crop production in rising markets by owning inventory or "Buying the Board."

The flexible marketer deals with one crop at a time in the futures market, depending on direction of the price trend. In order to profitably market both old crop and new crop in a calendar year, flexible hedging reverses futures market positions as trends change. When market down trends change to up trends, new crop hedges are lifted and old crop is repurchased with long futures positions. When up trends change into down trends, long positions on repurchased old crop are liquidated and new crop hedges are reinstituted. The flexible hedger holds a market position on one crop or the other at all times, but not both, as cash market speculators do. The two market positions are summarized in Figure 10-1.

Figure 10-1: 730-Day Marketing Plan: Own Old Crop in Rising Markets, Protect New Crop in Declining Markets.

Flexible hedging differs considerably from pure hedging. Textbook hedging requires owning both the physicals and an offsetting futures contract at the same time. Farmers who hedge anticipated production do not have the commodities on hand. Furthermore, when the old crop is replaced with a long position in the futures market, the "opposite" requirement of the textbook definition is again not met. These differences emphasize the peculiar objectives of farmer use of the futures market in contrast to the grain trade. They further emphasize the difference between pure and flexible hedging.

Flexible hedging has one additional side benefit — it breaks the link between the biological clock and the market clock. Farmers who wait for the crop to be "knee high" or wait for it to tassel often find that the most profitable pricing opportunities have vanished. It is not uncommon for grain prices to drop 30 cents per bushel between December and May — long before the crop tassels. Flexible hedging dictates sales based on the market clock — not the biological clock.

Producers can rely on various methods for determining when market reversals occur. Our research and analysis has shown a trend-following procedure to be effective. Discussion in Chapter 5 outlined a method for calculating 10- and 40-day moving averages. The crossover definitions provided the signals for trend changes. A discussion in Chapter 6 reviewed advantages and disadvantages of moving average analysis. A commercially available source of moving averages may be obtained from the Commodity Research Bureau.[1] In addition, there are a number of private consulting firms that provide comparable services.

Basis Sales: Capturing Storage Profits

Decisions to make cash crop sales are determined by basis signals from the market, totally separated from price level decisions. The spread between local cash price and a futures price signals when to store and when to sell cash grain. Weak basis offers a storage return and suggests that a storage commitment be made. Strong basis does not offer a storage return and suggests an immediate cash sale.

Whether farmer-storers decide to sell at harvest because a strong basis offers the prospect of low storage returns, or if they decide to store and speculate the basis, they should remember that the timing of the cash sale should be determined by the prospects for basis gain, *not* by hoped-for price level gain. Selling cash grain as dictated by basis signals, and "Buying the Board" to speculate on a price rise is preferable to holding cash grain to speculate. Holding grain on paper is cheaper, so that net profit of a correct decision is greater, and net loss of a wrong decision is less.

Storage, or immediate sale, depends upon the storage return offered by the market, relative to the costs incurred by storage. Storage profit is the difference between the basis gain and the costs of carrying the commodity forward.* A picture of the relationship between changing basis and changing cost of carry is presented in Figure 10-2. It shows that as long as the *slope* of the line representing basis change *exceeds the slope* of the line of cost change, storage is profitable. Storage is unprofitable when the *slope* of the basis change is *less than the slope* of the cost change. Ideally, to capture all basis gain, a farmer should fill bins at A, empty them at B, refill them C, and empty them at D. Bins should be empty from B to C.

To express the concept, presented as a picture in Figure 10-2, in numbers, rely on the methods of calculating basis gain and carrying charges in the earlier chapter. As long as the amount of basis gain day to day in the market exceeds the increase in carrying cost day to day, continued storage is profitable. Whenever basis gain drops below the daily increase in cost, storage is no longer profitable and the cash sale should be made.

An analysis of daily basis change is required to execute this concept. Gather daily cash prices from a local market and an important terminal market. Subtract both from the end of market year futures option (July corn and soybeans, May wheat). The daily change in basis is the gross return to storage. It may be 1 cent, 2 cents, 5 cents, zero, or a negative amount per day. Against the daily basis change, compare the daily change in costs of storing the grain. From the earlier analysis, we calculated variable storage cost to be about 4 cents per month for corn and wheat, and 8 cents per month for soybeans, or 1 cent and 2 cents per week, respectively. Only variable costs (interest, shrinkage, and extra drying) are important, because fixed cost of bin and facilities ownership is incurred whether the bins are full or empty. In practice, since the daily basis change is small (and

---

*At this point, readers should review Chapter 7 to refresh their understanding of cash-futures price convergence and calculation of carrying charges.

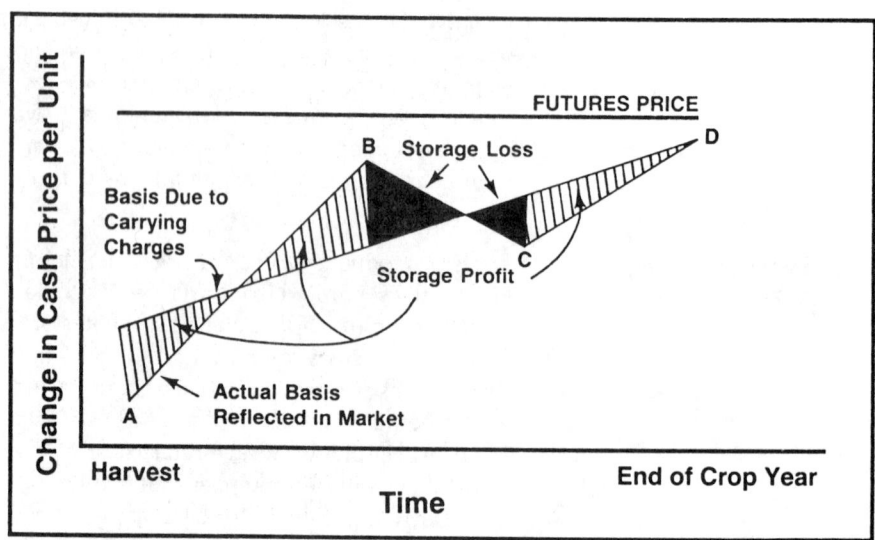

Figure 10-2: Basis Decisions for Cash Sale, Using Rate of Change in Basis Versus Change in Carrying Costs.

variable) and the weekly cost change is small, we watch for a distinct change or a deteriorating basis to confirm the signal for a cash sale. A two-day basis deterioration confirms the signal to sell cash grain. This deterioration may occur in the week after harvest is completed, or it may occur after January 1 as tax sellers begin emptying bins. A deteriorating basis says that farmers are more willing sellers than the grain trade is a willing buyer.

The daily decision as to when to sell out of storage does not answer the first question of whether to store or sell directly out of the field. In those years when harvest-time basis is weaker than the theoretical basis based upon carrying charges, the decision is easy — store. The market is saying that it does not need the grain for immediate use because the cash price being offered is less than the cost of carrying the grain forward in time. That pattern existed in most Midwest corn markets for many years. However, recent years have witnessed a dramatic shift in basis patterns. The original storage decision has become more complex and requires speculation on *expected* basis change.

The experience of the Cincinnati corn market in Figure 10-3 illustrates the shift. Over the seven years of historical basis from 1974 to 1981, harvest basis averaged 52 cents under the July futures, and it converged to zero by July. The market thus offered an average 52-cent gross storage return. That return exceeded the carrying charge of 28 cents by 24 cents. In 1983 through 1987 the historical basis pattern changed significantly (as reported in Table 7-3 in Chapter 7). Over the latter period, the October 15 basis averaged only 18 cents under July; March basis averaged 7 cents over July; and June basis averaged about 14 cents over July. Basis gain from October to June averaged 32 cents, barely enough to cover total carrying costs. The significant changes in basis pattern were the narrow harvest-time basis and the positive premium basis that was bid into the market, typically by January. Grain surpluses and storage programs changed the picture. Grain sellers (Ohio and Indiana farmers) had the alternative of placing grain under government loan rather than having to sell it at low cash prices, but a hungry chicken in Georgia required cash corn to maintain production. Futures prices reflected worldwide supply and demand, with huge surpluses dragging the general price level downward, but the cash market traded at a basis premium to entice cash grain into the market. Thus the basis picture changed.

The shift in basis is not the important issue. Although it makes the original storage decision more speculative, storage profits are not ruled out by the change. As long as one expects the new basis pattern to parallel the old basis pattern, the two have the same rate of change, or slope of the basis line. Under those circumstances, the decision whether or not to store is the same — store. It is only when the slope of the basis line, or **gain** in basis, is reduced that the decision is affected. With a storage cost of 28 cents per bushel, when a harvest basis of −20 might be expected to increase to +8 or more by July 1 (gain of 28 cents or more), corn storage should be initiated. Whenever there is the expectation of less than 28 cents basis gain — sell out of the field.

The decision criterion to store or not thus reduces itself to the same as the day-to-day decision: Store and continue to hold grain as long as the change in cost of storing is less than the expected change in basis over the same time period. As soon as added costs exceed added return, sell.

Storage Decisions for 730-Day Marketers. When the 730-Day Marketer decides to store, an analysis of the trend in price level of the commodity must be made. It is at this point that price level change comes back into the storage decision in deciding whether to hedge stored grain or store it unhedged. If futures

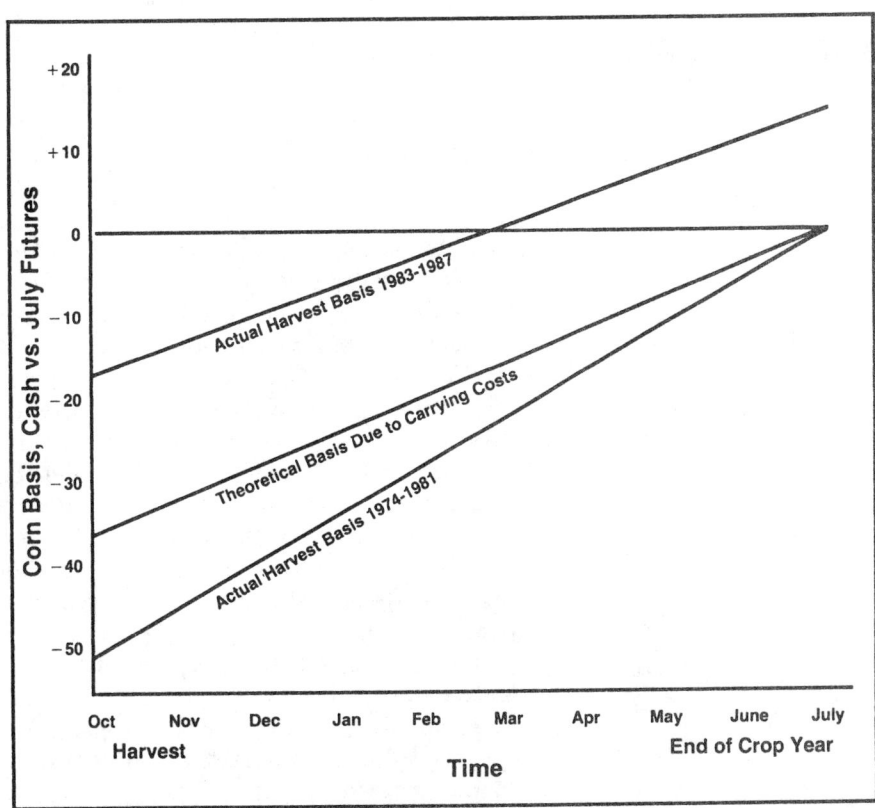

Figure 10-3: Cincinnati Corn Basis Patterns, 1974-1981 and 1983-1987, as Compared to Theoretical Basis Due to Carrying Costs.

prices are in an up trend, one should delay hedging as inventory value is increasing. If the market is in a down trend when the crop is committed to storage, the inventory should be hedged immediately. If a pre-harvest production hedge had been set earlier in the season to protect the growing crop, it should be maintained during the down trend in order to protect the inventory value of the harvested crop in storage. A storage hedge, set in down trends, by selling futures contracts equal to the amount stored, protects inventory value while allowing prices to continue to converge (basis to strengthen).

Once grain is in storage, the 730-Day Marketer analyzes the basis market every day to decide when to cease storage. One does not have to wait until maturity of the futures contract to sell cash grain and lift the hedge. The market may offer a more profitable opportunity prior to contract maturity. The decision of whether to make a cash sale and lift the hedge or continue storing is made daily by comparing changing basis to changing costs.

PROFIT-MAXIMIZING LIVESTOCK HEDGING

Selective Hedging of Finished Animals

The livestock feeder can employ the selective hedging procedure to protect the value of animals on feed during the production period. The livestock marketing plan is executed in response to changes in price trend on the commodity one is producing. As a selective hedger, one establishes short hedges in down trending markets to prevent value erosion on growing animals. Short positions are liquidated in up trends to realize inventory appreciation, unfettered by brokerage account deficits.

Hedging Feeder Purchases

Livestock feeders who buy feeder calves or feeder pigs can use futures contracts to lock in low-cost replacement animals. Whenever the feeder cattle market enters an up trend, a hedger would buy contracts equal to his or her planned purchases to protect against the risks of having to pay higher prices at the time feeders are to be laid in. At a market turnaround, one would liquidate the long feeder hedges.

The price relationship between fat cattle and feeder cattle essentially provides the livestock producer with a two-sided, flexible hedge in live cattle contracts. Rising live cattle prices tend to pull up feeder cattle prices; declining live cattle prices depress feeder prices. In the futures market, spread traders who execute feeder spreads keep the two futures prices aligned (see the section in Chapter 6 entitled "Profit Rather Than Price" for definition of feeder spread and reverse feeder spread trading). Prices of feeder cattle contracts rise and fall with rising and falling live cattle contracts. The spread difference between fats and feeders gets wider as corn price declines, and it gets narrower as corn price rises.

A cattle feeder can use price fluctuations in the futures markets to capture the best of both rising and falling prices. One sell-hedges live cattle production in price down trends, and one buy-hedges feeder animals in up trends. By using flexible hedging to deal with two markets, one effectively has a short position in down trending markets and a long position in up trending markets to protect price on both the finished product and input of feeders. The number of contracts is adjusted to reflect relative weight of feeders bought and live cattle sold.

Feeder pig buyers could rely on the live hog contract as a surrogate method for hedging feeder pig purchases, since no feeder pig futures contracts exist. High live hog prices create strong feeder pig prices; low hog prices drive down the value of feeder pigs. Because that economic relationship is bid into cash markets, a feeder can use flexible hedging to hedge-sell finished hogs in high-priced markets and hedge-buy feeders in low markets.* At market turns the hedger liquidates one of the positions and institutes the other.

Converting feeder pigs into live hog contracts can be done by three methods. First, the hedger could buy and sell the same number of head as finished. A 1,500-head producer would sell 12 contracts to hedge 360,000 pounds of market-weight hogs. At a market turn one could buy 12 contracts to hedge in feeders. The 360,000 pounds of volume is too great for feeder needs, however. As a second alternative, the feeder could buy only 2 live hog contracts, reasoning that he or she was purchasing only 60,000 pounds (1,500 head at 40-pound purchase weight). A 12-2 combination based on weight appears out of proportion in terms of value. The third method is to hedge equivalent values. A budget analysis to calculate value of feeder pigs relative to hog prices typically shows feeder value per head to be roughly equal to the hundredweight price of hogs. Consequently, a long feeder pig hedge matched on value appears most appropriate. The procedure is to buy half as many long futures positions to hedge feeder pigs as the farmer would sell as short positions to hedge the live hog production. The example flexible hedger thus sets 12 short hog positions in down trends and 6 long feeder pig positions in up trends.

Profit Margin Hedging

Cattle, hog, and poultry producers should also be concerned about hedging in low costs on their feed needs. Livestock producers should seek to lock in profit as the maximum difference between cost of input items and value of production by hedging both.

In a livestock operation that buys all its feed, hedge-buying feed ingredients and hedge-selling production tends to maximize a profit margin. In up trending feed ingredient (corn and soymeal) markets, long hedges are placed to protect against cost increases. When the market turns down, the hedges are liquidated as in a selective hedging program. They are reinstituted on the next market upturn. Each time a feed ingredient price rises, the producer protects against the rise by hedge-buying.

The need to lock in low feed prices on a farm that produces its own feed needs is less obvious. But the same procedures should be used to create an integrated marketing plan. To understand profit-maximizing hedging on an integrated farm, the enterprises should be viewed as two divisions — crop division and livestock division. The crop division should be concerned with maximizing profit by realizing as high a price as possible for the crops grown. The livestock division endeavors to maximize profit by buying feed inputs as cheaply as possible. The fact that one manager, rather than two vice presidents, makes pricing decisions for both divisions does not alter the basic comprehensive marketing plan. To effect the plan, the manager places short hedges in down trending markets to protect against value erosion of the growing crop and stored feed grain. In up trending markets, one lifts the crop hedge and initiates a long hedge to lock in low feed costs. The manager changes hats as price trend changes. In down trend markets, one wears the crop producer hat; in up trend markets, one wears the livestock feeder hat. The flexible hedging strategy of the crop farmer/livestock feeder is illustrated in Figure 10-4.

---

*Basis risk is very great in this kind of hedge.

By separating the decisions as to when to sell futures to hedge livestock production, when to buy futures to hedge in feeder animal costs, when to hedge-buy feed needs, and when to hedge-sell crop enterprise production, the integrated crop/livestock farmer can use market price fluctuation to widen the profit margin. It requires daily market analysis to detect price trend changes. But with an integrated hedging plan that takes different market positions at different times during the year, the integrated hedger can improve profitability.

Basis Signals for the Feed Buyer

The same technique for basis decisions used by crop producers is applicable for feed buyers in deciding when to make cash feed purchases or whether to build feed storage facilities. Feeders, however, react in a manner opposite of crop producers. When basis is weak, feeders make cash purchases; when basis is strong, they delay purchase until feed is required by the livestock.

Figure 10-4: Flexible Hedging of Feed Needs and Crop Production by Farmer-Feeder.

Livestock producers who erect feed storage facilities incur ownership costs just like the grain producer with storage. Buying feed ahead of needs incurs interest costs on inventory purchases and the ownership and operating cost of facilities. Their total costs of storage will be similar to those of any warehouser. They must capture basis advantages in order to pay ownership costs, if their decision is to be profitable.

Basis discounts in the cash market greater than carrying costs dictate that feeders should own storage and purchase feed ahead of needs. Their purchase of cash grain at greater discounts than storage costs would produce savings for them. On the other hand, strong basis would not offer sufficient discount to cover holding costs. In that case they would be better off buying futures contracts as price protection rather than holding cash feed.

Futures Contract Months

Nearby futures contract options typically have the largest trading volume and trade on more information. Distant months, especially those at greatest distance, have relatively little activity. The increased volume of nearby contracts produces market liquidity that allows quicker and easier execution of trade orders at smaller discounts. Current information trades more actively in nearby options. The nearby contracts are much more responsive to price-making factors as a consequence. Thus, the 730-Day Marketer focuses trading in nearby contract months to gain the benefits of volume trading and market sensitivity. The usual trading months for new crop production would be November soybeans, December corn and cotton, and July wheat. As the first day of each of these months arrives, contracts are rolled into next-high-volume months such as March, then July for fall harvested crops; and December, then May for wheat. Not every nearby month has to be used; those with sufficient volume will do. Use of fewer months requires fewer contract rolls as positions have to be moved forward.

Full hedging of the entire production should be established on each trade. Less than full hedging leaves the untraded portion subject to cash market speculation. Trading futures contracts in excess of production and inventory quantities creates "Texas hedges" that double up the effects of price changes. This too should be avoided. A hedger deals only in the commodities produced and in quantities consistent with production and inventory.

Livestock hedging should also be practiced in nearby contracts to improve responsiveness to fundamental price-making forces and to assure filling trade orders easily. The lead months move to wider extremes as they react to cash market forces. The hedger has the opportunity to benefit from wider fluctuation in futures as a consequence. Bull markets that create significant spreads between nearby and distant months can offer particularly profitable opportunities.* It should make no difference that the production is yet unborn for it to be profitably priced. Plans for production and intention to produce create a pricing requirement. It is better to set price when profitable opportunities present themselves than to have finished animals on hand that must be sold for whatever they will bring. Forward pricing of anticipated livestock production is no different from pricing anticipated crop production. Pricing ahead one year's production, or even two years', is not beyond the realm of possibilities. As long as contract obligations will be met at some time with actual production, the hedge definition of equal quantity is satisfied.

SUMMARY: THE 730-DAY CROP MARKETING PLAN AND PROFIT-MAXIMIZING LIVESTOCK HEDGING

Farmer-producers must understand markets and marketing. They must understand themselves and their psychology. They must understand the structure of the futures market. They must design a marketing plan that fits their situation. They also need to develop a financing plan, usually with their lender. They need to employ a broker who understands the plan and can effect the futures market transactions required in it. Some busy farmers may add a fourth person to their marketing teams, an advisor. The advisor becomes the market watcher and analyzes the market signals. The advisor becomes the farmer's market watchdog.

A marketing program is not a simple process if it is to be rewarding. 730-DAY MARKETING AND PROFIT-MAXIMIZING LIVESTOCK HEDGING are offered as integrated, profitable hedging programs.

---

*See "bear spread hedging" in the previous chapter.

730-Day Marketing for the Crop Farmer

730-Day Marketing for the grain producer deals with pricing decisions in three stages: the growing crop, the cash sale, and the stored crop. The marketing period for a crop starts during the winter planning stages of the growing year and continues for 730 days until the end of the storage year.

Throughout the planning, planting, and growing season, the growing crop is protected against price erosion by hedging with short futures positions.

At harvest time, the decision to sell out of the field or to store is determined by basis market analysis. A strong basis dictates sale from the field; a weak basis dictates storage. When storage is chosen, timing of the cash sale decision depends upon the improvement in basis as compared to the added cost of holding grain. 730-Day Marketers make cash sales to capture a strong basis when the market dictates the move. They sell to capture basis profits and do not store to speculate on price level.

Throughout the second year, the storage year, farmers own grain when market price is going up. But farmers DO NOT HAVE TO OWN GRAIN STORED IN BINS TO SPECULATE ON A PRICE RISE. They can replace their cash position with a futures position and speculate on the Board instead of speculating with grain in bins. And it is less expensive to speculate on paper (commission is 1 cent per bushel) than to pay interest and storage to speculate with stored cash grain.

When using the 730-Day Marketing concept, a farmer hedges the growing-year crop against price erosion, owns the stored crop in bins or on paper for price appreciation in the storage year, and sells grain on a strong basis to capture storage profits.

Squeezed into a calendar year period, 730-Day Marketing constantly deals with one or the other of two crops. In price down trends, the farmer hedges the growing crop. At a market turnaround, he or she lifts the short hedge and simultaneously holds the old crop unhedged in storage, or buys it back on paper to "refill" empty bins. During price up trends the old crop is owned as a long futures position after the cash sale has been made.

When the price up trend turns into a down trend, the long position is liquidated and a short hedge on growing production reinstated. The 730-Day Marketing plan goes a step beyond selective hedging. It employs FLEXIBLE HEDGING. At changes in price trends, market positions are reversed. The producer is almost always in an active marketing position.

Finally, the cash sale is made after an analysis of the basis market. Whenever prospective gains in basis no longer exceed added holding costs, the cash sale is made. This practice converts farmers into grain merchandisers, intent on capturing storage profits, as opposed to storage speculators.

730-Day Marketing incorporates two innovative procedures not found in other marketing strategies.

First, the farmer can speculate without holding grain in bins. This allows one to capture BOTH a strong basis AND a high price level. One seeks to capture high prices in *both* markets.

Second, pricing decisions are not an irreversible one-time decision, but a constant process. The farmer holds a futures market position almost all the time, either short or long. One is not standing outside the market as a cash market speculator wondering when to fix a one-time pricing decision. With continuous pricing, a person is always in the market, letting the market work for, not against, him or her. In practice, continuous marketing reduces cash speculation significantly. Production is hedged about half the time. Speculation is elected the other half of the time only for reasons that move the market (drought, strong demand) and is not carried out as the standard accepted practice of passive marketing.

With continuous 730-Day Marketing, farmers can capture marketing profits in all three stages:

1. price protection from growing crop hedges,
2. storage profit from basis sales, and
3. speculative profit from owning cash or paper-stored grain.

With these added together, 730-Day Marketing offers greater potential marketing profits than most other marketing plans.*

Profit-Maximizing Hedging for the Livestock Farmer

The concepts practiced in flexible hedging serve the livestock producer with an integrated PROFIT-MAXIMIZING HEDGING plan as well as 730-DAY MARKETING serves the grain farmer. Profit-Maximizing Hedging deals with pricing decisions in three important activities:

1. the sale of finished animals,
2. the purchase of feeder stock, and
3. the purchase of feed ingredients.

The objective of the plan is to maximize returns over costs of major input items. A full year's production and requirements are committed to the program. Throughout the year, value of production is protected from price erosion during market price down trends. At price trend change into up trends, selling hedges are lifted to enjoy inventory value appreciation. Correspondingly, at market turns, feeder animals are hedge-bought to protect against rising costs, since feeders and live cattle tend to trend upward together. At the end of the up trend, feeder hedges are liquidated and the production hedges reestablished. Whether the price level is going up or down, the livestock feeder enjoys the results, protecting either high price on product or low cost on feeders.

Feed purchases of corn and soybean meal are added into the hedging program to hold these major cost items at low levels. When the ingredient markets turn into up trends, corn and soymeal are hedge-bought for cost protection. In down trends, feed hedges are liquidated and buying of cash commodities proceeds hand to mouth, taking advantage of dropping cash market prices.

Farmer-feeders producing on-farm feed inputs (corn, oats, barley, or sorghum**) carry trading activities one step further. They switch to their seed corn cap in taking short positions to hedge growing crops when detection of down trend signals says to lift the feed corn hedge. They hedge feed in market up trends and grain in down trends. They let the market work at capturing both $4.00 corn "sales" and $2.00 feed cost. There is nothing illegal or immoral about making both the corn enterprise AND the hog enterprise profitable!

---

*The 730-DAY MARKETING concept has been successfully employed by AG COM, INC., a commodity marketing consulting firm for farmer-hedgers. Ag Com, Inc., has defined the tools for making 730-Day Marketing an effective marketing program.

**Feed commodities produced but not traded on exchanges may be converted into corn equivalent as a surrogate and traded in the corn pits, as long as various feed items trade almost lock step to corn.

Profit-Maximizing Hedging works for the farmer-feeder to capture, in effect, wide corn/hog or corn/steer ratios, irrespective of price levels in the individual markets. A farmer seeks price protection in all three segments of the operation:

1. Price protection on fed livestock,
2. Purchase price of feeders,
3. Major cost feed ingredients.

When these are added together, integrated Profit-Maximizing Hedging offers a greater profit potential than hit-and-miss programs.

The name of the game in farming today is PROFIT. Much of production is a necessary exercise; the bottom line depends on marketing. A logical, integrated plan makes profit possible — we wish you the best of 730-Day Marketing years.

Questions for Study and Discussion:

1. Over what time period should a producer plan to price his or her production?
2. How does the 730-Day Marketing Plan operate to protect price level on new and old crop production?
3. Why should basis decisions be separated from price level decisions?
4. What is the key feature of making profitable basis decisions?
5. Outline a method for calculating basis profit.
6. How is "Buying the Board" speculation similar to storing cash grain?
7. Explain how to lock in production profits in livestock.
8. What are the basis signals that the input buyer should watch for in the market?
9. Explain why nearby futures contracts offer advantages over distant contracts in placing futures hedges.

Footnote for Chapter 10:

(1) Commodity Research Bureau, "CRB Futures Chart Service" (Chicago: Knight-Ridder Business Information Service, 30 S. Wacker Dr., Suite 1820, 60606).

Glossary of Terms

ACREAGE ALLOTMENT: A voluntary limitation on the number of acres of a given crop farmers plant as established under a federal farm program.

ACREAGE RESERVE: An arrangement under which farmers agree to withdraw a stated acreage of cropland from production for a specified number of years. This is a conservation measure under the Conservation Reserve Program, which provides for annual compensation for any loss of income.

ACTUALS: Physical commodities, as distinguished from futures contracts.

ASK: An offer to sell a specific quantity of a commodity at a stated price.

AT THE MARKET: An order to buy or sell at the best price obtainable when the order reaches the trading floor.

AT THE MONEY: An option with a strike price equal to the current price of the futures contract the option is based on.

BAR CHARTS: Charts of futures prices drawn with vertical lines, showing the daily high, low, and closing prices.

BASIS: The difference between a cash price at a specific location and the price of a particular futures contract.

BASIS QUOTE: An offer or sale of a cash commodity in terms of the difference above or below a futures price.

BEAR: One who believes prices will move lower.

BEAR MARKET: One in which prices are declining.

BEAR SPREAD: The simultaneous purchase and sale of two futures contracts in the same or related commodities with the intention of profiting from a decline in prices; the sale of a nearby futures contract (short) while being long the distant cash.

BID: An offer to buy a specific quantity of a commodity at a stated price.

BREAK: A rapid, significant decline in price, frequently falling below a commonly recognized "support" level.

BREAKOUT: A movement of prices to outside a well defined trading range.

BROKER: (1) A person paid a fee or commission for acting as an agent in making contracts or sales; (2) **floor broker:** in commodities futures trading, a person who actually executes orders on the trading floor of an exchange; (3) **account executive:** the person who deals with customers in commission house offices.

BULL: One who expects prices to rise.

BULL MARKET: One in which prices are rising.

BULL SPREAD: The simultaneous purchase and sale of two futures contracts in the same or related commodities with the intention of profiting from a rise in prices; being long the nearby futures and short the distant futures.

BUYING THE BOARD: Buying a futures contract to replace a long cash position (grain in a bin that has been sold) with a long futures position to speculate on a price rise.

CALL OPTION: A contract that gives the buyer/taker the right but not the obligation to buy a fixed quantity of a commodity at a stipulated striking price at any time up to the expiration of the option.

CARRYING CHARGES: Those costs incurred in owning a physical commodity, generally including interest, insurance, and storage; may also include transportation, grading, sampling, and so forth.

CARRYOVER: That portion of current supplies of a commodity comprised of excess stocks from previous production/marketing seasons.

CASH COMMODITY: Actual stocks of a commodity as distinguished from futures contracts; goods available for immediate delivery or delivery within a specified period following sale; or a commodity bought or sold with an agreement for delivery at a specified future date.

CASH FORWARD SALE: The sale of a cash commodity for delivery at a later date. Price may be fixed at the time of the agreement, or there may be an agreement to determine the price at the time of delivery on the basis of prevailing local cash price or on some future price. Also known as "deferred delivery" or "forward sale."

CASH MARKET: (1) The market for the physical commodity; (2) an organized, self-regulated central market, such as a commodity exchange; (3) a decentralized over-the-counter market; (4) a local organization, such as a grain elevator or meat processor, which provides a market for a small region.

CASH SETTLEMENT: Payment for physical commodities received.

CHARTER: A technical trader who reacts to signals read from graphs of price movement.

CHARTING: The use of graphs to view market patterns in order to plot trends and patterns of price movements, volume, and open interest, in the expectation that such graphs and charts will help one to anticipate and profit from future price movements. An aspect of technical analysis as contrasted with **fundamental analysis.**

CLEARINGHOUSE: An agency connected with commodity exchanges which provides the mechanism by which all futures contracts are made, offset, or fulfilled through delivery of the commodity and through which financial settlement is made.

CLOSING OUT: Liquidating an existing long or short futures position with an equal and opposite transaction, also known as "exiting."

CLOSING PRICE: The price for a commodity futures contract generated by trading through open outcry during the closing period of a trading session.

COMMISSION: The charge which a commodity broker must make to the client when the broker buys or sells contracts. Traditionally this charge is levied at the time of liquidation; however, there is a shift toward charging one-half when the position is entered.

COMMODITY CREDIT CORPORATION (CCC): A government-owned corporation established in 1933 to assist U.S. agriculture by providing price-support programs in which the CCC purchases excess supplies of commodities and by assisting in the export of agriculture commodities.

COMMODITY FUTURES TRADING COMMISSION (CFTC): A federal regulatory agency empowered under the Commodity Futures Trading Commission Act of 1974 with regulation of all commodities trading on all domestic contract markets. The CFTC consists of five commissioners, one of whom is chairperson. All are appointed by the President subject to Senate confirmation. The CFTC replaced and assumed all powers of the Commodity Exchange Authority.

CONGESTION: In technical analysis, a price range within which buying power and selling pressure are about equal, resulting in a sideways movement of prices.

CONSOLIDATION: A price pattern represented by a prolonged period of narrowing price movement.

CONTRACT: An agreement to buy or sell a specified amount of a particular commodity. The contract details the amount and grade of the product and the date on which the contract will mature and become deliverable, if it is not liquidated earlier.

CONTRACT MONTH: The month in which a given contract becomes deliverable, if not liquidated or offset before the date specified.

CONTRACT UNIT: The specific amount of a commodity represented by the futures contract.

CONTRARIAN: A trader who takes positions opposite to the market direction anticipated by the majority of traders.

CONVERGENCE: The tendency for prices for physical and futures to approach one another, usually during the delivery month. Also called a "narrowing of the basis."

CORNER: The securing of significant control of a commodity, enabling price manipulation, often called a "squeeze"; in its extreme, the control of more contracts requiring delivery than the available supply.

COVERING: Offsetting a previously established futures short position by a PURCHASE. Also "evening up" or "liquidating."

CROP YEAR: The period beginning with the harvesting of a crop to the corresponding time the following year.

CYCLE: The tendency for an entity to move through phases of high and low with sufficient regularity to be at least partially predictable, e.g., the hog cycle.

DAY TRADE: A trade which is entered and liquidated on the same trading day, as different from an **overnight trade.**

DEFERRED DELIVERY: The more distant months in which futures trading is taking place, as distinguished from the nearby futures delivery months. See also **Cash forward sale.**

DELIVERY: Tender of a commodity by (1) issuing a warehouse receipt or a bill of lading, (2) issuing a shipping certificate for some commodities, or (3) delivering actuals against a futures delivery contract.

DELIVERY DATE: Date on which the commodity must be delivered to fulfill the terms of the contract.

DELIVERY MONTH: The calendar month in which the futures contract matures and within which delivery of the physical commodity can be made.

DELIVERY NOTICE: Written notice from a seller through the clearinghouse, stating intention to make delivery on a short futures position on a particular date. This notice also specifies quantity, grade, and place of delivery. A notice precedes and is distinct from a warehouse receipt or shipping certificate, which are instruments representing transfer of ownership.

DELIVERY POINTS: The places specified by a commodity exchange to which delivery of a physical commodity can be made in fulfillment of a contract.

DELIVERY PRICE: The price fixed by the clearinghouse where futures deliveries are invoiced. Also, the price at which a commodities futures contract is settled when deliveries are made.

DELTA: The amount of change in the option premium compared with the change in price of its underlying futures contract.

DEMAND: The quantity of a commodity that consumers (users) are willing to buy at various market prices.

EFFICIENT MARKET: A market in which new information is immediately and costlessly available to all investors and potential investors; a market in which all information is instantaneously assimilated and therefore has no distortions.

ELASTICITY: The percentage change in quantity relative to the percentage change in price. This concept applies to both supply and demand.

ELEVATOR: A commercial facility for grain storage.

EQUILIBRIUM: A state of rest from which there is no tendency to change. Where the quantity of a commodity supplied equals the quantity of that good demanded at a price that is acceptable to all successful buyers and sellers.

EXCESS: The amount of money that can be withdrawn from a futures account without liquidating the positions; the amount by which equity exceeds margin requirements.

EXCHANGE (COMMODITY): An organized market place where particular commodities are bought and sold.

FEED/VALUE RATIO: The relationship of the cost of feed to market-weight sales prices of livestock, expressed as a ratio, such as the *hog/corn ratio*. These ratios serve as indicators of the profit margin or lack of profit in feeding animals to market weight.

FIRST NOTICE DAY: The first day on which notices of intention to deliver actual or physical commodities against a short futures contract can be presented by sellers through the clearinghouse.

FLOOR BROKER: A member of the commodity exchange or representative of a member firm of the commodity market who is "on the floor" (i.e., in the market, ready to do business) at all times when the market is open.

FORWARD CONTRACT: A cash market transaction in which two parties agree to the purchase and sale of a commodity at some future time under such conditions as the two agree—in contrast to **futures contracts**.

FORWARD SALE (OR PURCHASE): See **Forward contract**.

FUNDAMENTAL ANALYSIS: The forecasting of price movement traditionally using factors such as income, acreage, inventory, yields, government programs, and techniques of econometrics, as contrasted with **technical analysis** or **charting**.

FUTURES: Contracts for the purchase and sale of commodities for delivery some time in the future on an organized exchange and subject to all terms and conditions included in the rules of that exchange.

FUTURES COMMISSION MERCHANT (FCM): An individual, association, partnership, or corporation engaging in soliciting or in accepting orders for the purchase or sale of any commodity for future delivery (on and subject to the rules of any contract market) and registered with and regulated by the CFTC.

FUTURES CONTRACT: An agreement to make or take delivery of a standardized amount of a commodity, of standardized quality grades, during a specific month, under terms and conditions established by an authorized exchange.

FUTURES EXCHANGE: A membership organization whose activity involves trading sessions between members in the specific commodities listed for trading on the exchange.

GOOD TILL CANCELED ORDER (GTC): An order which is valid at any time during market hours until executed or canceled by the client. Also known as an "open order."

GRADES: Standards set for the quality of a commodity.

HEDGER: One who sells futures contracts in anticipation of future sales of cash commodities as a protection against possible price declines, and who purchases futures contracts in anticipation of future purchases of cash commodities as a protection against the possibility of increasing costs.

HEDGING: The establishment of an opposite position in the futures market from that held in the physicals.

HISTORICAL VOLATILITY: The actual amount that prices increased or decreased in the past. Usually expressed as a percentage.

IMPLIED VOLATILITY: The amount prices are expected to go up or down in the future. Usually expressed as a percentage.

IN THE MONEY: A term used to describe an option that has intrinsic value. A call at $4.00 on wheat trading at $4.10 is in the money 10 cents.

INTRINSIC VALUE: A measure of the value of an option if immediately exercised. Amount by which the option is "in the money," i.e., has real value.

INVERTED MARKET: A futures market in which distant-month contracts are selling below near-month contracts.

LAST TRADING DAY: The day on which trading ceases for a particular delivery month. All contracts that have not been liquidated by an offsetting purchase or sale by the end of trading on that day must therefore be settled by delivery of the actual physical commodity.

LEVERAGE: Buying or selling with deposits, collateral, or margin less than the face value of the item being traded. Commodity futures markets usually require about 10 percent of the face value of the contract, giving 90 percent leverage.

LICENSED WAREHOUSE: A warehouse designated for delivery by the exchange on which a commodity is traded. Only such designated warehouses may be used to store a commodity for delivery.

LIFE OF CONTRACT: The period of time from the first trading day in a futures month through the last trading day.

LIMIT (UP OR DOWN): The maximum price advance (or decline) that the rules of the exchange permit a commodity to trade at from the previous day's trading session.

LIMIT ORDER: An order that has some restriction of execution, such as price or time.

LIMIT PRICE: The highest or lowest price that a commodity may trade at during a trading session, as fixed by the contract market's rules. See also **Limit (up or down)**.

LIQUIDATION: The closing out of a long position. The term is sometimes used to denote closing out a short position, but this is more often referred to as "covering."

LIQUID MARKET, LIQUIDITY: A market where selling and buying can be accomplished with ease, because of the presence of a large number of interested buyers and sellers willing and able to trade substantial quantities at small price differences. Often measured by volume and open interest.

LOAN PRICES: Prices at which the U.S. government will lend producers (farmers) money for their crops placed in storage and under loan.

LOAN PROGRAM: The U.S. government agricultural price support operation. A program under which farmers commit their crops to the government with the assurance that they will receive a certain minimum loan price. If the price of the commodity rises above the loan price, the farmers may sell their crops in the open market and repay the loan. Even if the price of the commodity falls below the low level, the *non-recourse* character of the loan makes it possible for farmers to deliver their crops to the CCC (Commodity Credit Corporation), discharging their obligation in full.

LOCALS: Members of the U.S. exchange who trade for their own account and/or fill orders for customers and whose activities provide market liquidity.

LONG: One who has bought a cash commodity or a commodity futures contract, in contrast to a **short,** who has sold a cash commodity or futures contract.

LONG HEDGE: Purchase of futures to lock in a low (favorable) price in anticipation of acquiring the resource (feed) at some future date.

LONG LIQUIDATION: Closing of long positions.

LONG THE BASIS: A hedge consisting of ownership of actuals against short futures in a particular commodity. See also **Hedging.**

MARGIN: (1) An amount of money deposited by both buyers and sellers of futures contracts to insure performance against the contract, i.e., to deliver or take delivery of the commodity (not an equity or down payment for the goods represented by the futures contract); (2) **profit margin:** the difference between the price which one pays for goods and the price at which the goods or products of them are resold.

MARGIN CALL: A call from a brokerage firm to a customer to bring margin deposits back up to minimum levels required by exchange regulations; similarly, a request by the clearinghouse to a member firm to make additional deposits to bring clearing margins back to minimum levels required by clearinghouse rules.

MARKET: A region or place where buyers and sellers come together to transact business.

MARKET IF TOUCHED (MIT) ORDER: An order to buy or sell when the market reaches a specified price. An order to buy becomes a market order when the commodity sells (or is offered) at or below the order price; an order to sell becomes a market order when the commodity sells (or is bid) at or above the order price.

MARKETING: A process whereby buyers and sellers transact business in the purchase and sale of commodities.

MARKET ORDER: An order to buy or sell at the best price available and as soon as possible after the order reaches the trading floor of the exchange.

MAXIMUM PRICE FLUCTUATION: The limit of fluctuation in the price of a futures contract during any one trading session, as established by the exchange.

MINIMUM PRICE FLUCTUATION: The minimum amount by which the price of a commodity can fluctuate during one trading session, as established by the exchange.

MOMENTUM: In technical analysis, the relative change in price over a specific time interval.

MOVING AVERAGE: In technical analysis, a method of averaging prices of the most recent days in an effort to determine and/or forecast the direction of movement.

NEARBY DELIVERY (MONTH): The futures contract closest to maturity. Also called "spot month."

NON-MEMBER FCM: A brokerage firm (FCM) that does not own its own seats on futures exchanges; a firm that executes its orders through a member FCM.

NOTICE OF INTENTION TO DELIVER: A notice that must be presented by the seller to the clearinghouse. The clearinghouse then assigns the notice and the subsequent delivery instrument to the longest-standing buyer on record.

OFFER: A price at which a trader is willing to sell; the asking price.

OFFSET: The liquidation of a purchase of futures (longs) through the sale of an equal number of contracts of the same delivery month, or the liquidation of a short sale of futures contracts through the purchase of an equal number of contracts of the same delivery month. Either action transfers to other persons the obligation to make or take delivery of the actual commodity.

OPEN CONTRACTS: Contracts bought or sold and not offset by an opposite trade.

OPENING RANGE: The range of prices at which transactions took place during the period designated as "the opening" of trading by an exchange.

OPEN INTEREST: The total number of futures contracts of a given commodity which have not yet been offset by opposite futures transactions nor fulfilled by delivery of the actual commodity; the total number of open transactions, with each transaction having a buyer and a seller.

OPEN OUTCRY: A method of public auction for making bids and offers in the trading pits of a commodity exchange.

OPEN POSITION: A market position which has not been closed out.

OPTION: An agreement which gives the taker (purchaser) the right, but not the obligation, to buy from or sell to the grantor (seller) of the option at any time before its expiration a specified quantity of the commodity concerned at an agreed price.

OPTION BUYER: The person who buys calls, puts, or any combination of calls and puts.

OPTION SELLER: Also, **option writer.** The person who originates an option contract by promising to perform a certain obligation in return for the price of the option.

ORIGINAL MARGIN: See **Margin.**

OUT OF THE MONEY: A term used to describe an option that has no intrinsic value. A call at $4.00 on wheat trading at $3.90 is out of the money 10 cents. The right to sell $4.00 wheat in a $3.90 market has no real value.

OVER-SOLD: In technical analysis, an opinion that the market price has declined too steeply and too fast in relation to underlying fundamental factors.

PARALLEL PRICES: The constant relationship between cash and futures markets, the difference between the two being the basis.

PARITY: A theoretically equal relationship between farm product prices and all other prices. In farm program legislation, parity is defined in such a manner that the purchasing power of a unit of an agricultural commodity is maintained at its level during an earlier historical base period.

PHYSICAL: See **Actuals.**

PIT: The place where futures are traded on the floor of a commodity exchange.

POINT: A unit expressing minimum price change. Synonymous with "minimum price fluctuation."

POINT-AND-FIGURE: A method of charting which uses prices to form a pattern of movement without regard to time. It defines a price trend as continued movement in one direction until a reversal of a predetermined criterion is met.

POSITION: A market commitment. A buyer of futures contracts is said to have a long position, and conversely, a seller of futures contracts is said to have a short position.

PREMIUM: (1) The amount by which a cash commodity sells over a futures price or another cash commodity price; (2) the excess of one futures contract price over another; (3) in options, the price paid by the purchaser of the option to the grantor (seller).

PRICE LIMIT: The maximum price fluctuation on a futures contract during one trading session, as determined by the exchange. Also known as "limit."

PRICE TREND: See **Trend.**

PRICING: The act of fixing a price.

PUBLIC ELEVATORS: Storage facilities for grain, licensed and regulated by federal and state agencies. Grain of the same grade but owned by different persons is usually commingled, or mixed together. Some elevators are designated as depositories for delivery by exchanges dealing in commodities stored at these elevators.

PUT OPTION: A contract that gives the purchaser the right, but not the obligation, to sell a fixed quantity of a commodity at a stipulated strike price at any time during the life of the option.

RALLY: An upward movement of price after a decline.

RANGE: The difference between the highest and lowest prices recorded during a given trading session, week, month, life of contract, or any given period.

RESISTANCE: In technical trading, a price area where new selling will emerge to dampen a continued rise. See also **Support**.

RETRACEMENT: A reversal within a major price trend.

REVERSAL: A change of direction in prices.

ROLL, ROLLING FORWARD: See **Switching**.

SCALPER: A speculator on an exchange floor who trades in and out of the market on very small price fluctuations. The scalper, trading in this manner, provides market liquidity but seldom carries a position overnight.

SELLER: A risk taker who takes a short future position or grants (sells) a commodity option.

SELLER'S OPTION: The right of the seller to select commodity quality, time, and place of delivery within the limits prescribed by the exchange upon which futures contracts are traded in making physical delivery.

SETTLEMENT PRICE: The price on the close of the day, or a price determined to be in line with actual closing values. Settlement prices are used to set the next trading day's price fluctuation limits and are used as the basis for adjusting margins and determining equity by the clearinghouse for contracts traded on each exchange.

730-DAY MARKETING PLAN: To hold a position in the futures market at virtually all times during the 365-day growing year and the subsequent 365-day storage year. A short futures position is held in down trending markets to protect the value of the growing crop and the value of grain in the bin. Cash grain is sold when basis calculations signal that the added cost of holding grain longer exceeds the added return from basis improvement. Cash grain is replaced by a long futures position when the market is in an up trend to participate in price appreciation.

SHORT: Starting a transaction by the sale of a futures contract. An open position on a futures market which was initiated by a sale.

SHORT COVERING: See **Covering**.

SHORT HEDGES: The sale of futures against holdings of the physical commodity.

SHORT THE BASIS: The purchase of futures as a hedge against a commitment to sell in the cash or spot market.

SPECULATOR: A person using the futures market for a purpose other than hedging and willing to assume varying degrees of risk in exchange for potential profit.

SPOT: (1) The characteristic of being available for immediate (or nearly immediate) delivery; usually refers to a cash market price for the physical commodity available for immediate delivery; (2) sometimes used in reference to the futures contract of the current month, in which delivery is possible at any time.

SPOT COMMODITY: An actual or physical commodity.

SPOT MARKET: The market in which cash or spot transactions occur.

SPOT MONTH: The first deliverable month for which a quotation is available on the futures market.

SPOT PRICE: The commodity cash sale price, as opposed to a futures price.

SPREAD (STRADDLE): The purchase of one futures delivery month against the sale of another futures delivery month of the same commodity; the purchase of one delivery month of one commodity against the sale of that same delivery month of a different commodity; or the purchase of one commodity in one market against the sale of that commodity in another market. The purpose of any one of these transactions is to take advantage of distortions in normal price relationships. The term "spread" is also used to refer to the difference between the price of one futures month and the price of another month of the same commodity. There are four basic types of spreads: (a) **Interdelivery spread:** The purchase and sale of the same commodity, in the same market, in different delivery months. Also called an **intramural spread.** (b) **Intermarket spread:** The purchase and sale of the same commodity, in the same or different delivery months, in two different markets. (c) **Intercommodity spread:** The purchase and sale of different but related commodities, in the same or different markets, in the same or different delivery months. (d) **Commodity product spread:** The purchase of futures raw material and the sale of the derived processed products futures, or vice versa.

STOP ORDER, STOP LOSS ORDER: An order which becomes a market order to buy only if the market advances to a specified level, or to sell only if the market declines to a specified level. As soon as this specified level is traded, the order is executed for the client at the next obtainable price. There is no guarantee the order will be executed at the price specified. A stop loss order is used to prevent or minimize losses in either a short or long position.

STRIKE PRICE: The price at which an option can be exercised irrespective of current market conditions.

SUPPLY: The quantity of a commodity that producers are willing to provide at various market prices.

SUPPORT: In technical analysis, a price area where new buying is likely to come in and stem any decline.

SWITCHING: Liquidation of a position in one delivery month of a commodity and simultaneous initiation of a similar position in another delivery month of the same commodity. When used by hedgers, this tactic is referred to as "rolling forward the hedge."

TECHNICAL ANALYSIS: An approach to forecasting commodity prices based on the study of price movement itself without regard to underlying fundamental market factors. Contrasted with **fundamental analysis.**

TECHNICAL RALLY: Price variations arising from factors other than those affecting supply and demand for a commodity.

TECHNICAL TRADING: Systematic trading based on charts or technical analysis that indicate buy and sell signals. See also **Technical analysis.**

TERMINAL ELEVATOR: In the movement of grains, a storage facility located at a point of accumulation or distribution.

TIME VALUE: The part of an option's value that cannot be realized by immediately exercising the option.

TRADER: A person who takes positions in the futures market, usually without the intention of making or taking delivery.

TREND: The general direction of price movement, within a specified time interval.

TRENDLINE: In charting, a line drawn across the bottom or top of a price chart indicating the direction or trend of price movement. If up, the trendline is called **bullish;** if down, it is called **bearish.**

UNDERLYING FUTURES CONTRACT: The futures contract that a particular option gives a person the right to buy or sell.

VARIABLE PRICE LIMIT: A price limit schedule, determined by an exchange, that permits variations other than the normally allowable price movements for any one trading day.

VOLUME: The quantity of business or transactions done.

WAREHOUSE RECEIPT: A document guaranteeing the existence and availability of a given quantity and quality of a commodity in storage.

Index

A

Account executive (*See* Broker)
Account surplus, 49, 70, 131, 273
Acreage allotment, 130
Acreage reserve, 130, 214
Actuals, 232
Advisor, 89, 275, 313, 315
Ag Com, 315
Aggregate demand, 104, 129
Agriculture price cycles (*See* Price cycles)
Analysis of grain pricing strategies, 188–191, 215–217
Arbitrage, 44, 236
Assessing a futures price, 62
At the market, 70
At the money, 218
Average price of year, 188, 215–217

B

Bar chart, 152–153
Bar chart rules, 168, 170
Basis, basis level, 26–27, 183, 190, 213, 232–266, 305, 314
    components of, 233
    defined, 233
    and hedging, 236–240, 284–286
    livestock, 284–286
    location, 262
    records, 246
    signals, 252
Basis contract, 34–36
Basis negotiation, 242–246
Basis records and charts, 246–252
Basis signals, 252–256, 307–310, 312
Basis speculation, 240, 307–310
Bear, bear market, 62
Bear spread, 288–291
Bid, 65–66
Biological time lag, 135, 295
Black market, 131
Bonded warehouse, 30
Bretton Woods Agreement, 5

Brock, Richard, 156
Broker, 69, 196
Brokerage firm, 69, 72, 83, 275
Brokerage statements, 276–277
    confirmation of sale, 275
    monthly activity statement, 277
    statement of open trades, 276
    statement of purchase and sale, 276
Brunthaver, Carroll, 2
Building a marketing strategy, 191–195
    decision period, 191
    futures market, not designed for farmers, 193–195
    role of futures prices, 191
Bull, bull market, 62
Business events of transactions, 27
    delivery, 27
    pricing, 27
    settlement, 27
    title transfer, 27
Butz, Earl, 8
Buying hedge, 46–48, 74, 77–78
Buying the Board, 37, 190, 304, 314

C

Calendar selling, 197
Call, 50, 217–220
Carry, carrying charges, 16, 188, 234, 240
    excess, 242–246, 253
    full, 243–246, 253
    inverted, 253
    less than, 243–246, 253
    normal, 246
Carrying charge hedging, 80, 238–242
Carryover, 138, 140, 143, 144, 184
Carter, President Jimmy, 6
Cash commodity, 27–28, 37, 41, 232, 235
Cash contract, forward sale, deferred delivery, booking, 37–40, 211, 216, 310
Cash flow, break-even strategy, 202

Cash market, 27–28, 232
Cash market speculation (*See* Speculating the cash market)
Cash sale, 27–29, 291–293
Cash sale and Buying the Board, 37
Cash settlement, 42–44, 286
Cash surplus in brokerage account, 49, 70, 131, 273
Cash versus futures over time, 240–242
Cash versus nearby futures, 243
Cattle cycle, 294–299
Change in demand, 2, 4–6, 105–107
Changing economic conditions, 2
Chart, charting, 151–155
    open interest, 152, 157–158
    price movement, 152
    volume, 152, 157–158
Chart formations, 151, 158–173
    confirmation, 171
    defining price objectives, 159
    identifying tops and bottoms, 164
    reliability, 171, 173
Charts of seasonal tendencies, 206–207
Chicago Board of Trade, 56, 63
    purposes, 57
Chicago Mercantile Exchange, 59, 60
Clearing Corporation, 57, 68
Closing out (*See* Settlement of futures)
Closing price, 65–68, 152
Cobweb theorem, 133–135
    price cycles, 133–134
    production cycles, 133
    time lag, 135
Collateral, 28, 272
Commercial storage (*See* Hired storage)
Commission, commission fees, 69, 196
Commission merchants, 88
Commodity, definition, 59
Commodity Credit Corporation, 214
Commodity Exchange Authority, 88
Commodity exchanges, 58–59
    evolution of, 56–58
    listing of, 60
Commodity futures contract (*See* Futures contract)
Commodity Futures Trading Commission (CFTC), 69, 88
Commodity loan, 7, 8, 73, 214–215
Commodity pool operator, 89
Commodity pricing problems, 183–188
Communication role of price, 17–18, 116

Components of basis, 233–234
Confirmation, chart formation, 171
Confirmation of trade, 72, 275
Conflict between individual and all farmers, 129–130
Congestion (*See* Support)
Consolidation, 161
Consumer utility, 15
Continuous marketing, 303
Convergence of prices, 234–235, 243
Corn price, 138
Corner, 89
"Correct" futures price, 62
Costs of marketing, 19
Costs of production, 131, 197, 202
Covering (*See* Offset, offsetting trades, covering)
Cow-calf production, 297–299
Crop insurance, 73
Cross-elasticity of demand, 106
Crystal ball gazing, 151, 189, 194
Cut losses, let profits run, 84
Cycles (*See* Price cycles)

D

Dealing with brokers and brokerage accounts, 198–199
Decision period for pricing, 191
Deferred payment contract, 33–34
Deficiency payments, 73, 214
Defining price objectives, 159–164
Delayed pricing, priced later, 31–34
Delivery point(s), 43, 236
Delivery settlement (*See* Cash settlement)
Delta, 218
Demand, 2, 102–107, 135
    aggregate, 104
    changes in, 105–107
    cross-elasticity, 106
    curve, 103
    elasticity, 104–105, 126–127
    farm level, 107, 125, 126
    principle of, 103
    retail level, 107, 125, 126
    schedule, 102
    understanding, 107
Demand and consumption, 107
Demand curve, 103
Demand elasticity, 104–105, 125
Demand schedule, 102
Devaluation, 5, 123
Diminishing marginal utility, 103
Dips, 155

Disposable personal income, 137
Distinction between hedging and speculation, 195
Divergence of prices, 241
Double top or bottom, 169

E

Econometrician, 135
Economics of agricultural marketing, 17–20
Effective demand, 107
Effects of demand elasticity on price and gross receipts, 125–128
Effects of price ceilings and floors, 131
El Niño, 123
Elasticity, 104, 106, 137
    cross-, 106
    demand, 104–105
    effect of time, 112
    income, 106
    supply, 111
Elasticity coefficients, 125–127
Elements of marketing, 15–17
    form utility, 17, 121
    place utility, 15, 121, 233
    time utility, 16, 121, 233
Elevator (*See* Hired storage)
Embargo, 6
Enterprise profitability, 137
Equal but opposite, 80
Equilibrium price and quantity, 114–119, 131, 133, 135
Evaluating hedge loan requests, 272
Evaluation of grain marketing, 56–59
Excess margin (*See* Account surplus)
Exchange rules and regulations, 65
Exchanges (*See* Commodity exchanges and Futures markets)
Exports, 6, 7

F

Facilitating function, 13
Farm gate price, 43, 263
Farm level demand, 107
Farmer-owned reserve, 214
Farmer participation, 83, 89
Farmer psychology and basis, 243–244
Farmers store to speculate, 188–191
Feeder purchases, 310–311
Feeder spreads, 212–213, 310
    reverse feeder spread, 213, 310
Feeding value, 138
Financing function, 14

Financing hedging, 268–277
Financing price protection, 78–79
Fixed exchange rates, 5
Flags, 160
Flexible hedging, 305–307, 310, 314
Floor broker, 83
Floor traders, 82
Food Security Act of 1985, 214
Forecasting prices, 138, 143, 145, 146, 150
Form utility, 17, 121
Formations (*See* Chart formations)
Forward contract (*See* Cash contract)
Forward sale or purchase (*See* Cash contract)
Functions of marketing, 12–14
    exchange, 12
    facilitating, 13
    physical, 12
Fundamental analysis, 135–147, 281–282
Futrell, Gene, 12, 17
Futures, 41–50, 56–93, 286–288
Futures Commission Merchant (FCM), 69
Futures contract, 40–42
Futures market, not designed for farmers, 193–195
Futures markets, 56–99
Futures price index, 2, 3

G

Gambling, 91, 92
Gaps, 161–164
    breakaway, 161
    common, 161
    exhaustion, 161
    measuring, 161
Garcia, Philip, 280
Good, Darrel, 136
Good till canceled order (GTC), 71
Government programs, 8, 73, 114, 130, 131, 140, 206, 214, 215
    acreage allotment, 130, 214
    deficiency payments, 130, 214
    non-recourse loans, 73, 214
    price support, 73, 214
    storage payments, 214
Grades, grading, 13, 28, 53, 54
Grain Futures Act of 1922, 88
Grain merchant hedging, 193 (*See also* Routine hedging)
Grain shrinkage chart (*See* Standard moisture table)

Grain Standards Act, 28

H

Habits and preferences, 106
Hand signals, 65
Harvest losses, 29, 258
Head and shoulders, 159
Hedge account, 69
Hedge to arrive, 49–50
Hedger, 61, 79, 81
Hedging, 45, 74, 78, 195, 204, 236
Hedging cattle, 283–284
Hedging hazards, 80
Heilbroner, Robert L., 121
Hieronymus, T. A., 59, 82, 136, 286
High price of day, 65, 152
Hinton, Royce, 136
Hired storage, 29, 258
Historical basis charts, 250–251
Hog/corn ratio, 197, 212, 316
Hog price, 143–145, 281–282
How trading takes place, 63–64

I

Identifying a trend, 156–157
In the money, 217
Income elasticity, 106
Income growth, 2
Inelastic demand of agricultural products, 104, 117
Initial margin, 48, 70, 269–270
Internal Revenue Service, 34, 69
Intrinsic value, 218
Inventory liquidation, 295
Inverted market, 253
Investment medium, 81
Irwin, Scott, 140, 144

K

Keeping score, 211–212, 215
Kohls, Richard L., 12

L

Law of one price (LOOP), 120–121, 236
Lender policies for hedge loans, 271–277
Leuthold, Raymond M., 280
Leverage, 85
Licensed warehouse, 29–30
Limit (up or down), 87–88
Limit order, 71

Liquid market, 313
Liquidity, 81, 82, 269–271
Live cattle price, 145–147
Livestock basis, 284–286
Livestock marketing, 280–299
    basis, 284–286
    bear spread hedging, 288–291
    cash sales, 291–293
    cattle cycle, 295–299
    fundamental price forecasting, 281–282
    hedging, 283–286
    perishable commodities, 280
Loan prices, 215
Loan program, 214
Local basis, year to year, 246
Localized future price, 237
Locals, 65
Location—where to sell, 262–263
Long, 46–48, 74–77
Long hedge (See Buying hedge)
Long the basis, 74–75
Low price of day, 65, 152

M

Maintenance margin, 48–49, 70, 269–270
Major types of farming, 16
Margin, 48–49, 70, 269–270
    initial, 48, 70
    maintenance, 48, 70
    margin call, 48, 51, 217
Margin call, 48, 51, 217, 219, 273
Marked to the market, 70
Market, 11–12, 59, 65
Market advisor, 89, 275, 313, 315
Market clearing price, 138, 150
Market forces in action, 121–124
Market if touched (MIT), 71
Market imperfections, 26
Market information, 14
Market intermediaries, 13, 14
Market liquidity, 313
Market order, 70
Market tops and bottoms, 164–168
Marketing, 11–12
Marketing, as system coordination, 17
Marketing concepts, 11
Marketing costs or charges, 20
Marketing margins, 119, 137
Marketing plan, 183, 191, 195, 196, 268, 302–316
Marketing process, 12, 17
Marketing strategy, 26, 183, 191–195,

282–284, 313–316
Marketing tools, 26, 27–51
Martin, John, 156
Martin, Larry, 280
Maturity of contracts, 58
Maximum price fluctuation, 87–88, 94–95
McCoy, John H., 284
Merchandising, 13
Messengers, 65
MidAmerica Commodity Exchange, 59, 60, 90
Mini-contracts, 59
Minimum price fluctuation, 87–88, 94–95
Moisture discounts, 28, 53
Monetary exchange rates, 2
Monitoring hedge loans, 275–277
Monthly statement, 72, 277
Moving average, 172–177, 209–211
    characteristics of, 175
    crossover, 175
    front-weighted, 172
    weighted, 174

N

Nearby delivery month, also spot month, 233, 243
New York Cotton Exchange, 60, 86
Non-recourse loans, 73, 214
Non-storable futures prices, 286–288
Notice of intention to deliver, 43

O

Offer, 65–66
Offset, offsetting trades, covering, 44
On-farm storage, 29, 258–262
OPEC, 6
Open contracts, 63
Open interest, 157–159
Open outcry, 65
Open position, 72
Opening an account, 69
Opening range, 65
Operating agreement, 273–274
Optimizing, 210
Option, 50–51, 217–220
Option buyer, 50, 217–220
Option seller, also option writer, 50, 217–220
Orders, 70–72
    good till canceled, 71
    limit order, 71
    market if touched, 71
    market order, 70
    spread, 71
    stop-loss, 71
Original margin (*See* Margin)
Out of the money, 218
Over-bought, over-sold, 151

P

Paper trading, 89, 93–95
Parallel prices, 45, 74–78, 234–236
Payment-In-Kind (PIK), 8, 215
Perfect hedge, 45, 48, 76–78
Perishable commodities, 280, 286–288
Physical risk, 72
Physicals (*See* Actuals)
Pit, 63–64
Pit trader, 82
Place and time differences, 232–233
Place utility, 15, 233
Point, 87
Population, 4, 106
Position limits (spec limits), 89
Pre-harvest hedge (*See* Pre-production pricing)
Premium, option, 50, 217
Pre-production pricing, 18, 37–40, 81, 187, 216, 237–238
Price, price level, 26–27, 183, 213, 305
Price, Robert V., 284
Price analysis, 135
Price ceiling, 131–132
Price convergence, 234–236
Price-cost squeeze, 130
Price cycles, 133–135, 294–299
    production cycle, 108, 111, 133
Price discovery, 61–63, 119–121, 137, 150
Price expectations, 185
Price floor, 51, 131–132
Price fluctuation (*See* Volatility, price)
Price forecasting, 135, 281
Price limit (*See* Limit)
Price:loan ratio, 140, 144
Price objective, 151
Price quotation, 65–68
Price ratio, critical, 111
Price risk, 72–74, 196, 236
Price strategy, 181–229
Price support, 7, 8, 214
Price taker, 18, 26, 130, 186
Price trend (*See* Trend)
Prices of substitute goods, 106
Pricing, 183–188

Pricing choices, 186–188
   price takers, 186
   routine hedgers, 186
   selective pricing/selective speculation, 187
Pricing signals, 208, 252–258
Pricing strategies, 197–217, 270, 282–284, 303–304
   analysis of, 215–217
   calendar selling, 197
   cost of production, 197
   scale-down buying, 204
   scale-up selling, 204
   seasonal pricing, 205
   trend following, 209–211
Principle of demand, 103
Principle of supply, 108
Processing, 13, 141
Processor margin, 19–20, 119, 141
Production cycle (*See* Price cycles)
Production function, 109–110
Production quotas, 130
Production risks, 38, 46, 73, 183
Profit and loss statement, 72
Profit margin hedging, 212–215, 311–312, 315
Profit maximization, 183, 194
Protein principle, 2, 5
Psychological time lag, 135, 295
Psychology of a trend, 155–156
Psychology of support and resistance, 158–159
Public elevators (*See* Hired storage)
Purcell, Wayne, 17
Purchasing power, 103
Pure and perfect competition, 129
Put option, 50, 217

R

Rally, 155, 156, 206
Range, 65–68
Ratio of cows to people, 297–298
Reading charts, 153–155
Regression analysis, 137, 145, 146
Regulation of futures trading, 88–89
Relationship of prices and cost of production, 131–133
Relative share of consumer's dollar, 19–20
Reliability, chart formations, 171–172
Resistance, chart formations, 153, 158
Restricting output, 130
Retail level demand, 107
Retracement, 164, 166

Returns to storage, 234
Reversal, 164
Reverse feeder spread, 213
Risk, 182
Risk bearing, 13–14, 72–73, 182
Risk capital, 90
Risk management, 195, 238
Risk transfer, 72–74
Role of banker (*See* Role of lender in hedging)
Role of futures prices, 191–193
Role of lender in hedging, 81, 271–277
   documentation, 274
   evaluating requests, 272
   monitoring loans, 275
   operating agreement, 273
   policies, 271
Role of price, 18, 116, 203
Role of small traders, 84
Roles of futures markets, 61–92, 203
   investment medium, 81–92
   price discovery, 61–63
   risk transfer, 72–78
Roll, rolling forward, rolling the hedge, 35, 49, 72, 252–256, 291
Round turn, 69
Routine hedging, 79, 186, 193, 194, 283
Rules for speculating, 90–91
Russia (*See* Soviet Union)

S

Saucer or bowl, 167–168
Scale-down buying, scale-up selling, 204
Scalper, 82
Seasonality, seasonal patterns, 158, 190, 205–208
Segregation of customer funds, 89
Selective hedging, 79, 187, 194, 283, 303, 310
Selective speculation, 187, 194, 283, 303
Selling hedge, 45, 48, 74–76
Settlement of futures, 42–44, 68
Settlement price, 65, 152
730-Day Marketing Plan, 26, 302–316
Shepherd, Geoffrey, 12, 17
Shift in demand, 105–107, 116
Shift in supply, 113–114, 116
   government programs, 114
   innovation, 113
   input cost, 114
   rules and regulations, 114
   technology, 113
Shifting price risk, 44, 74, 91

Short, 45, 48, 74–77
Short covering (*See* Offset, offsetting trades, covering)
"Short crop, long tail," 190, 253
Short hedge (*See* Selling hedge)
Short the basis, 74, 234, 236–240
Shortage, 131
Shortcomings of cost-of-production pricing, 203
Small traders, 81, 83, 86–87
Soviet Union, 2, 6, 121
Soybean balance sheet, 142
Soybean price, 141
Spec limits, 89
Speculate, 33, 188, 195, 269
Speculating the basis, 240, 307–310
Speculating the cash market, 81, 195, 268–269, 277, 303–305
Speculative account, 69
Speculative error, 138–139
Speculator, 61, 79, 81–87
    farmer, 83
    floor brokers, 83
    floor traders, 82
    pit traders, 82
    scalpers, 82
    small traders, 83
    spreaders, 82
Spot, or spot market, 56, 232
Spot month, 233
Spot price, 233
Spread, 234, 243, 256, 281–282
Spread order, 71, 212–213
Spread traders, 82, 245, 310
Squeezes, 89
Standard moisture table, 53
Standardization (*See* Grades, grading)
Stewart, Blair, 84
Stop order, stop-loss order, 71
Storage costs, 258–262
Storage hedge, 236, 238–240
Storage profit, 238–240, 262, 307–310
Storage return, 236, 238, 262, 307–310
Storage space merchandiser, 190, 242
Storage speculation, 188–189
Strain, J. Robert, 12, 17
Strike price, 50, 217–219
Supply, 2, 108–114, 135
    curve, 108
    definition, 108
    effect of time on elasticity, 111–112
    elasticity, 111–112
    principle, 108
    schedule, 108
Supply carryover, 137, 138–144
Supply control, 8, 130
Supply-demand balance, 116–118
Support, 155, 159
Surplus, 70, 131, 273
Switching (*See* Rolling the hedge)

T

Target price, 8
Tax treatment of futures transactions, 69
Taylor, Robert J., 172
Technical analysis, 150–179
Technical rally, 156
Technical Reliability Index, 172–173
Technical trading, 150–179
Terminal elevator (*See* Hired storage)
Texas hedge, 81, 89, 313
Theoretical basis, 240–242, 308
365-day hindsight pricing, 187
Time effect on supply elasticity, 111–112
Time utility, 15, 233
Time value, 218
Title transfer, 27, 29–51
Top the market, 164–168, 183–184, 188, 289, 303
Trade the basis, 194, 232–263
Trader, 81–84
Trading advisor (*See* Market advisor)
Trading months, 58
Trading units, 85–86, 94–95
Transportation, 13, 233, 262
Treasury Bills, 270
Trend, 153–157, 208
    down trend, 153, 155
    rules of, 156
    sideways, 153, 155
    up trend, 153, 155
Trend following, 208–211
Trend line, 153
Triangle, 161
Tri-party agreement, 275
"True" supply-demand price, 62, 280
Types of hedging, 79–80
    carrying charge, 80
    routine, 79
    selective, 79

U

Uhl, Joseph N., 12
Underlying futures contract, 50

Understanding demand, 107
Understanding grain prices, 138–142
Understanding livestock prices, 143–147
Uniform Commercial Code, 40
Unsecured creditor, 33, 34
U.S.S.R. (*See* Soviet Union)

V

Variable price limit, 88
Vertical integration, 73
Volatility, price, 2, 8, 10, 27, 113, 127, 182, 270
Volume, 152, 157–159, 313

W

Walk grain to market, 121
Wall Street Journal, The, 65–66, 247–248
Warehouse licensing and bonding, 29–30
Warehouse receipt, 29, 33, 43, 57
Weighted moving average, 172, 174
 front-weighted, 172
Whipsaw, 177, 208
World population growth, 2

Z

Zero-sum game, 78, 85, 91